Economic, Environmental, and Social Assessments of Raw Materials for a Green and Resilient Economy

Economic, Environmental, and Social Assessments of Raw Materials for a Green and Resilient Economy

Editors

Andrea Thorenz
Armin Reller

MDPI • Basel • Beijing • Wuhan • Barcelona • Belgrade • Manchester • Tokyo • Cluj • Tianjin

Editors
Andrea Thorenz Armin Reller
University of Augsburg University of Augsburg
Germany Germany

Editorial Office
MDPI
St. Alban-Anlage 66
4052 Basel, Switzerland

This is a reprint of articles from the Special Issue published online in the open access journal *Resources* (ISSN 2079-9276) (available at: https://www.mdpi.com/journal/resources/special_issues/Raw_Materials).

For citation purposes, cite each article independently as indicated on the article page online and as indicated below:

LastName, A.A.; LastName, B.B.; LastName, C.C. Article Title. *Journal Name* **Year**, *Volume Number*, Page Range.

ISBN 978-3-0365-2668-3 (Hbk)
ISBN 978-3-0365-2669-0 (PDF)

Cover image courtesy of Dr. Andrea Thorenz

© 2021 by the authors. Articles in this book are Open Access and distributed under the Creative Commons Attribution (CC BY) license, which allows users to download, copy and build upon published articles, as long as the author and publisher are properly credited, which ensures maximum dissemination and a wider impact of our publications.
The book as a whole is distributed by MDPI under the terms and conditions of the Creative Commons license CC BY-NC-ND.

Contents

About the Editors . . . . **vii**

Preface to "Economic, Environmental, and Social Assessments of Raw Materials for a Green and Resilient Economy" . . . . **ix**

Christoph Helbig, Martin Bruckler, Andrea Thorenz and Axel Tuma
An Overview of Indicator Choice and Normalization in Raw Material Supply Risk Assessments
Reprinted from: *Resources* **2021**, *10*, 79, doi:10.3390/resources10080079 . . . . **1**

Christoph Helbig, Alex M. Bradshaw, Andrea Thorenz and Axel Tuma
Supply Risk Considerations for the Elements in Nickel-Based Superalloys
Reprinted from: *Resources* **2020**, *9*, 106, doi:10.3390/resources9090106 . . . . **27**

Simon Meißner
The Impact of Metal Mining on Global Water Stress and Regional Carrying Capacities—A GIS-Based Water Impact Assessment
Reprinted from: *Resources* **2021**, *10*, 120, doi:10.3390/resources10120120 . . . . **43**

Steffen Kiemel, Simon Glöser-Chahoud, Lara Waltersmann, Maximilian Schutzbach, Alexander Sauer and Robert Miehe
Assessing the Application-Specific Substitutability of Lithium-Ion Battery Cathode Chemistries Based on Material Criticality, Performance, and Price
Reprinted from: *Resources* **2021**, *10*, 87, doi:10.3390/ resources10090087 . . . . **77**

Julia S. Nikulski, Michael Ritthoff and Nadja von Gries
The Potential and Limitations of Critical Raw Material Recycling: The Case of LED Lamps
Reprinted from: *Resources* **2021**, *10*, 37, doi:10.3390/resources10040037 . . . . **105**

Rudolf Suppes and Soraya Heuss-Aßbichler
How to Identify Potentials and Barriers of Raw Materials Recovery from Tailings? Part I: A UNFC-Compliant Screening Approach for Site Selection
Reprinted from: *Resources* **2021**, *10*, 26, doi:10.3390/resources10030026 . . . . **123**

Rudolf Suppes and Soraya Heuss-Aßbichler
How to Identify Potentials and Barriers of Raw Materials Recovery from Tailings? Part II: A Practical UNFC-Compliant Approach to Assess Project Sustainability with On-Site Exploration Data
Reprinted from: *Resources* **2021**, *10*, 110, doi:10.3390/resources10110110 . . . . **149**

Aitor Barrio, Fernando Burgoa Francisco, Andrea Leoncini, Lars Wietschel and Andrea Thorenz
Life Cycle Sustainability Assessment of a Novel Bio-Based Multilayer Panel for Construction Applications
Reprinted from: *Resources* **2021**, *10*, 98, doi:10.3390/resources10100098 . . . . **197**

About the Editors

Andrea Thorenz leads the interdisciplinary research group Resource Lab at the Institute of Materials Resource Management at the University of Augsburg. As a member of the Faculty of Mathematics, Natural Sciences, and Materials Engineering, her curriculum embodies the networking idea of the environment- and resource-oriented research projects of the Resource Lab.: After her studies in environmental science and socio-economics, she worked as a project leader and director of studies in the economy for many years.

Armin Reller is a chemist and held the chair for Resource Strategy at the University of Augsburg. For many years now, he has been involved in the implementation of interdisciplinary concepts, i.e., with the Fraunhofer Project Group Materials Recycling and Resource Strategy (IWKS). He is an explorer in the field of sustainable use of natural resources and member of the resource commission of the Federal Environment Agency as well as of the board of trustees of the development fund regarding rare metals.

Preface to "Economic, Environmental, and Social Assessments of Raw Materials for a Green and Resilient Economy"

The main future challenge our society faces is the development of a sustainable and resilient economy. This requires a decoupling of economic growth and greenhouse gas emissions. Therefore, mitigation technologies are essential to limit global warming and its related effects. Since climate change is already progressing, climate change mitigation strategies must be complemented by adaptation strategies to address the remaining risks and consequences. To obtain economic prosperity and bring future-orientated mitigation and adaptation technologies to maturity, we need primary and secondary resources as a base for functional materials. All processes related to mining, smelting, and refining must be aligned with the Sustainable Development Goals (SDGs). An essential tool in this context is the establishment of a Circular Economy, meaning a thoughtful, structured way of cascading the use of resources. To identify critical resources, we need adequate classification schemes and assessments. The assessments must consider the importance of single metals, minerals, or biogenic resources for future technologies, the potential supply risks, and the environmental and social impacts of all related products and supply chain aspects. Concerning technological and societal needs, vulnerability studies are well established. Criticality assessments quantify the supply risk and the environmental and social burdens of processes related to the mining, smelting, and refining of primary raw materials. While supply risk indicators are often based on categories such as "country and company concentration", "political risk", "future demand", and "supply reduction risk", Life Cycle Assessment (LCA) and Social LCA are state of the art quantitative tools to measure environmental and social impacts. An important field for future research is the assessment of secondary raw materials. Here, a classification scheme based on the United Nations Framework Classification for Resources (UNFC) is essential. Furthermore, in assessing secondary raw materials, a focus must be put on tailings and urban mines. Lastly, in addition to metals and minerals, biogenic materials such as straw and bark are important resources for the achievement of a carbon-neutral world. However, using natural resources such as wheat straw also addresses issues like maintaining a sustainable soil organic carbon balance, or, more generally, the environmental and social impacts of a bioeconomy. Often, ecological goals such as the "reduction of the global warming potential" and "land use" are conflicting. Therefore, to obtain a sustainable and resilient circular economy using secondary raw materials and abundant biogenic resources, we need advanced assessments, classification schemes, and optimization approaches considering economic, environmental, and social objectives and technological requirements.

Andrea Thorenz, Armin Reller
Editors

Review

An Overview of Indicator Choice and Normalization in Raw Material Supply Risk Assessments

Christoph Helbig *, Martin Bruckler, Andrea Thorenz and Axel Tuma

Resource Lab, University of Augsburg, Universitaetsstr. 16, 86519 Augsburg, Germany; martin.bruckler@student.uni-augsburg.de (M.B.); andrea.thorenz@mrm.uni-augsburg.de (A.T.); axel.tuma@wiwi.uni-augsburg.de (A.T.)
* Correspondence: christoph.helbig@wiwi.uni-augsburg.de

Abstract: Supply risk assessments are an integral part of raw material criticality assessments frequently used at the country or company level to identify raw materials of concern. However, the indicators used in supply risk assessments to estimate the likelihood of supply disruptions vary substantially. Here, we summarize and evaluate the use of supply risk indicators and their normalization to supply risk scores in 88 methods published until 2020. In total, we find 618 individual applications of supply risk criteria with 98 unique criteria belonging to one of ten indicator categories. The most often used categories of supply risk indicators are concentration, scarcity, and political instability. The most frequently used criteria are the country concentration of production, depletion time of reserves, and geopolitical risk. Indicator measurements and normalizations vary substantially between different methods for the same criterion. Our results can be used for future raw material criticality assessments to screen for suitable supply risk indicators and generally accepted indicator normalizations. We also find a further need for stronger empirical evidence of widely used indicators.

Keywords: criticality assessments; supply risk; raw material; concentration; scarcity; political instability; mineral resources

Citation: Helbig, C.; Bruckler, M.; Thorenz, A.; Tuma, A. An Overview of Indicator Choice and Normalization in Raw Material Supply Risk Assessments. *Resources* **2021**, *10*, 79. https://doi.org/10.3390/resources10080079

Academic Editor: Ben McLellan

Received: 2 July 2021
Accepted: 30 July 2021
Published: 4 August 2021

Publisher's Note: MDPI stays neutral with regard to jurisdictional claims in published maps and institutional affiliations.

Copyright: © 2021 by the authors. Licensee MDPI, Basel, Switzerland. This article is an open access article distributed under the terms and conditions of the Creative Commons Attribution (CC BY) license (https://creativecommons.org/licenses/by/4.0/).

1. Introduction

Raw material criticality assessments are carried out to identify materials of concern [1]. Their goals range from risk mitigation to hotspot analysis. The actors can be governments and companies alike. The scope of risk consideration ranges from physical accessibility to reputation damage. Even the material scope can differ from chemical elements to whole supply chains [2]. It is good practice to follow four phases for the design and communication of a criticality assessment, consisting of (i) goal and scope definition, (ii) indicator selection and evaluation, (iii) aggregation, (iv) interpretation and communication [2]. Most criticality assessments consider indicators in the two dimensions "supply risk" [3] and "vulnerability" [4]. Several different indicator categories are used for both, as identified by Schrijvers et al. [1]. However, there is little evidence for the general significance of individual risk aspects for raw material criticality [5]. Commodity prices are linked to changes in supply risk aspects, but the scale and significance level of this empirical evidence depends strongly on the specific raw material [6].

The present article is an update to an earlier review by Achzet and Helbig [3]. When that review was published, only 15 criticality assessments were available for a systematic review. In the past eight years, raw material criticality assessments have increased substantially in quantity, impact, and scope [7]. The International Round Table on Materials Criticality (IRTC) held a series of expert workshops and conducted a broad review of various criticality assessments, focusing on risk types, geographical scope, time horizons, and objectives of the methods [1]. However, their study did not cover the details of each criterion and the normalization and interpretation of each of the supply risk and vulnerability indicators [1].

Nevertheless, looking at such information is essential to guide future method developers and users in applying assessments. Such detailed information helps in the second criticality assessment phase, indicator selection, and evaluation [2]. Therefore, the present review focuses on indicator usage instead of the general goals of the methods or aggregation procedures. We provide an overview on supply risk indicator usage in all relevant criticality assessment schemes.

For this purpose, we distinguish indicator categories, criteria, measurements, and normalizations. Indicator categories are general supply risk aspects considered in assessments and may have multiple evaluation criteria. We identify frequently used indicator categories and, for each category, the most relevant supply risk criteria. The criteria need to be measured and consequentially normalized. We want to provide an overview on possible measurements and normalizations. Due to a lack of empirical evidence, we cannot provide a recommendation for best practice on each criterion. Normalization can happen with a continuous formula, stepwise normalization, or point-wise evaluation. For example, Graedel et al. [8] consider the country concentration of production in the criterion for concentration, measured with the Herfindahl–Hirschman Index (HHI), and apply a logarithmic normalization formula to evaluate this criterion on a shared supply risk score. Using such a procedure transparently and in a reproducible manner helps improve criticality assessments and follows good practice [2]. Our review fosters this transparency and reproducibility.

2. Method

Our review includes 88 supply risk assessment methods published from 1977 to 2020. The methods are published in peer-reviewed literature, research reports, working papers, books, book sections, or corporate or institutional websites. The previous reviews by Achzet and Helbig [3] and Schrijvers et al. [1] contributed to this collection. The list of studies was extended with citation chaining, considering only publications in English or German. The complete list of studies is included in Appendix A.

Most of the 88 methods are full criticality or supply risk assessments that follow the four good practice steps in criticality assessment [2]. Others are either a collection of indicators, which do not aggregate the results, or methods consisting of only a single supply risk indicator. The Supplementary Material spreadsheets additionally list publications that we did not include in our review because they were reviews, obsolete publications (which have been updated by the same authors or institutions by now), or applications of supply risk assessment methods without any methodological change. All of these exclusions avoid double-counting.

All methods included in the review were reviewed concerning their supply risk indicators. If the method additionally had a vulnerability or economic importance dimension, those indicators were not considered. For each of the 618 indicators used in the various supply risk assessments, we identify the overarching indicator category, the measurement (with minimum and maximum values) and the normalization type (normalization formula, stepwise supply risk levels, point-wise evaluation, or no normalization). The list of all indicators, including the normalization formula, supply risk levels, or evaluation points, can be found in Supplementary Material spreadsheets. Table 1 shows a glossary for the relevant terms contained in this data sheet.

The review process also included an attempt for the harmonization of terminology in supply risk assessments. For reasons of transparency, the spreadsheet in the Supplementary Material therefore also contains the original criterion name. However, in our review, harmonized criterion names are used. For example, one method may call its indicator "producer diversity", while another calls its indicator "company concentration", and both may be measured with the HHI. Therefore, company concentration is used in this case as the harmonized criterion name for both cases.

Category names are also harmonized always to indicate risk or problem, as shown in Figure 1. For example, many supply risk assessments consider some form of recycling in their method, but recycling itself is not a problem for supply risk—the contrary is the case. The lack of secondary production increases the dependence on primary production to maintain global material flows and supply chains. Therefore, all categories have received a name indicating that "more" in this indicator equals higher supply risk and criticality, even for those studies that initially assessed supply security or supply chain resilience rather than supply risks.

Figure 1. Overview of supply risk categories identified in the review. Sector width is not proportional to indicator frequency.

Table 1. Glossary of Supplementary Material spreadsheet.

Column Name	Explanation
Method	Scientific publication (peer-reviewed, technical report, book, book section, or website) with a novel approach to assess supply risk
Original criterion name	Name of indicator as it appeared in the publication
Criterion harmonized	The overarching term for indicators expressing the same risk
Category harmonized	The overarching term for indictors used to express similar risks
Year	Year of publication
Type	**Assessment**: Indicators aggregated to an overall supply risk **Collection of indicators**: Indicators assessing supply risk without aggregation **Single indicator**: Only one indicator presented
Measurement	Determination approach of an indicator
Unit	Unit of measurement
Norm. type	**Level**: Subdivision of indicator values into supply risk levels **Points**: Assignment of discrete indicator values or qualitative descriptions of indicator values to supply risk point **Normalization**: Formula to transform indicator values into a supply risk score

3. Results

The review of all 88 supply risk assessments results in a list of 618 individual indicators. These indicators can be grouped into ten indicator categories with a varying number of criterions each. Risks and criterion labeling follow a single Latin letter and a two-digit numerical code, e.g., A01 for country concentration production.

The categories are (A) concentration, (B) scarcity, (C) political instability, (D) regulations, (E) by-product dependence, (F) dependence on primary production, (G) demand growth, (H) lack of substitution options, (I) price volatility, and (J) import dependence. A total of 53 additional indicators did not fit these ten categories and therefore have been allocated to the group of other indicators (X). The review results in each of these categories are described in the following subsections one by one. Table 2 shows the indicator categories and their frequency.

Table 2. List of supply risk categories identified in the review. Categories define the leading letter (A-J, X) by order of frequency of their indicators.

Criterion Codes	Category Name	Frequency
A01–A18	Concentration	137
B01–B25	Scarcity	93
C01–C09	Political instability	75
D01–D15	Regulations	68
E01–E02	By-product dependence	44
F01–F08	Dependence on primary production	43
G01–G11	Demand growth	32
H01–H03	Lack of substitution options	26
I01	Price volatility	17
J01–J04	Import dependence	16
X01–X53	Other indicators	67

Figure 2 summarizes the use of all criteria used at least three times. It shows almost all criteria are still used nowadays, with the prominent exception of criterion B07, depletion time reserve base, which is not used anymore because most data providers discontinue reserve base data.

For each of the indicator categories, we show a graphical representation of relevant normalizations. To allow a better comparison, the original formulas are rescaled to a common "normalized supply risk score" between 0 and 100 for all categories.

3.1. Concentration (A)

The market concentration (A) is the most frequently used indicator category making up 137 of the 618 indicators (22%). The associated indicators can be grouped into a total of 18 harmonized criteria, of which the five most frequently used criteria are country concentration production (A01), company concentration (A02), country concentration reserves (A03), and country concentration import (A04). In total, these four criteria are used in 118 of 137 indicators (86%) of the concentration category.

The first appearance of concentration as a supply risk indicator dates back to 1977 when Grebe et al. [9] considered the number of countries accounting for 40%, 60%, or 80% of the global production or global reserves as a measurement of A01 and A03. These measurements were converted for both indicators into a scale from 1 to 5, indicating the extent of supply risk, whereby the exact transformation routine is not given [9]. The so-called Herfindahl–Hirschman Index (HHI) [10,11] is a much more frequent concentration measurement for criteria A01 to A04. Some publications proposed a combined indicator composed of the HHI and an indicator from another category as a weighting factor, for example, the political instability category. Another frequently used measurement is the accumulated share gathered from top countries of production or reserves, as presented by Grebe et al. [9]. Normalization approaches using the HHI measurement for A01 to A04 are presented in Figure 3. Information about the remaining harmonized criteria can be found in the Supplementary Material spreadsheets.

In the case of country concentration production (A01), the logarithmic transformation of the HHI (ranging from 0 to 10,000) into a normalized supply risk scale (ranging from 0 to 100) was applied by Graedel et al. [8] and other methods (cf. Equation (1)).

$$\text{HHI}_{normalized} = 17.5 \cdot \ln(\text{HHI}) - 61.18 \tag{1}$$

The values 17.5 and 61.18 in Equation (1) have been set by Graedel et al. [8] to fit the normalization so that an HHI value of 1800 results in a normalized score of 70 and an HHI value of 10,000 marks a normalized score of 100. Helbig et al. [12] adopted this approach with other fitting parameters, resulting in a slightly different normalization applied by three other publications.

Nassar et al. [13] and four other methods do not explicitly mention a normalization procedure for the HHI. We conclude that a simple linear transformation of the HHI into a score from 0 to 100 represents their interpretation of the country concentration best. Zhou et al. [14] also determine normalized scores by scaling the HHI values linearly, but they use the extreme values observed in their data set as thresholds.

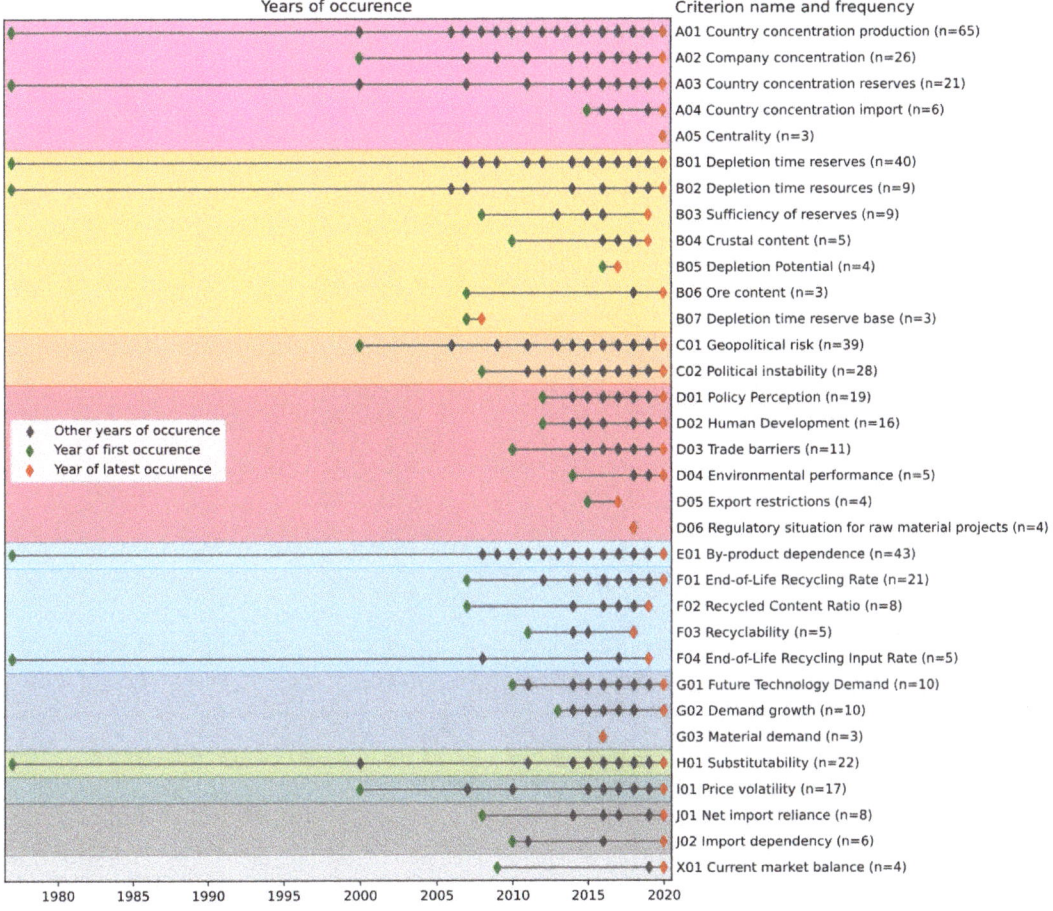

Figure 2. Occurrence of indicators by publication year. Only indicators occurring at least three times are shown. Each rhombus represents at least one occurrence.

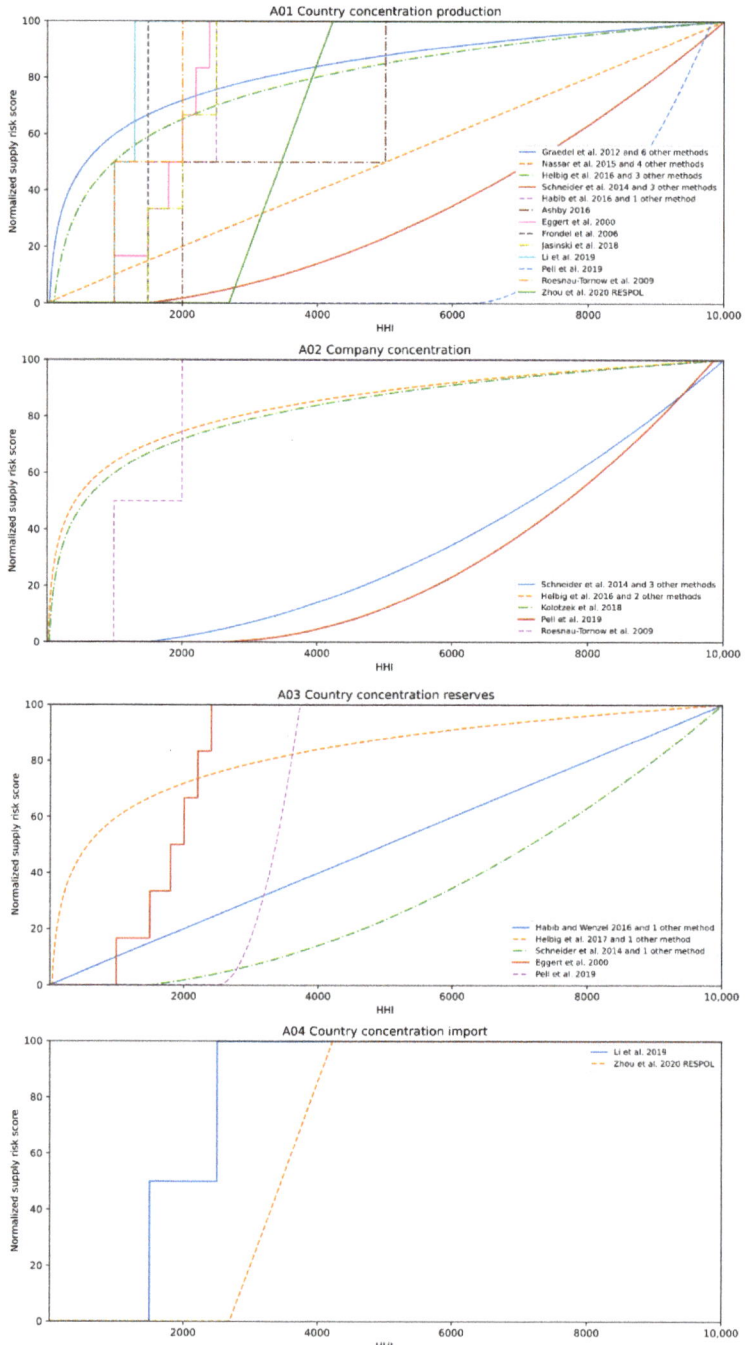

Figure 3. Selected normalization schemes of concentration criteria: (A01) country concentration production, (A02) company concentration production, (A03) country concentration reserves, (A04) country concentration import. All four criteria are measured with the Herfindahl–Hirschman Index (HHI).

Schneider et al. [15] define a threshold of 1500 for the HHI. Below this threshold, the normalized supply risk score is 0. Above 1500, the HHI is normalized by the squared ratio of the HHI value, which is also called a distance-to-target method [16]. Three other methods applied this parabolic approach. A similar approach is proposed by Pell et al. [17], but we could not fully reconstruct the normalization approach. We interpret that the HHI values were first scaled from 0 to 1 by the minimum and maximum values of the observed raw material and consequently normalized by the distance-to-target method. Based on the results, the normalization formula of Equation (2) was applied.

$$\text{HHI}_{\text{normalized}} = \left(\frac{\text{HHI} - \text{HHI}_{\min}}{\text{HHI}_{\max} - \text{HHI}_{\min}} \right)^2 \cdot 100 \qquad (2)$$

The remaining normalization schemes for A01 uses various stepwise functions with two [18] to seven levels [19]. Except for Habib et al. [20], the stepwise procedure only has single appearances.

For A02, we identified less variety in terms of measurements and normalization schemes. The most frequently used is the method of Schneider et al. [15] which was already explained for A01. The threshold of 1500 is once again used, which results in an identical normalization curve. Three other publications applied this approach. Pell et al. [17] applied the same normalization scheme with extreme values of HHI observed for A02 and the distance-to-target approach. Helbig et al. [12] adopted their normalization formula from A01 for A02 with different key points, leading to a slightly different formula (cf. Equation (3)).

$$\text{HHI}_{\text{normalized}} = 15.81 \cdot \ln(\text{HHI}) - 45.62 \qquad (3)$$

The same formula is also applied in two other publications. Kolotzek et al. [21] stuck to the key points used by Helbig et al. [12] for A01 and applied a logarithmic transformation for A02. The work from Rosenau-Tornow et al. [22] is the only study involving a level-based normalization on the HHI for A02.

A03 is also dominated by normalizations based on normalization formulas. Habib and Wentzel [23] and Nassar et al. [13] applied no transformation. Therefore, we assigned an HHI of 0 to the normalized supply risk score of 0 and an HHI of 10,000 to a score of 100. Helbig et al. [24] applied the same logarithmic transformation as Helbig et al. [12] for A01. Schneider et al. [15] also used the distance-to-target method with an HHI threshold of 1500 as for A01 and A02. Each of the three approaches is applied in one other publication. Pell et al. [17] proceeded as in A01, A02 scaling the HHI values according to the observed minimum and maximum values for A03 followed by the distance-to-target method. Eggert et al. [19] also use the same levels as in A01 to assign HHI values to supply risk levels.

For A04, only two different normalization approaches for the HHI are identified. Zhou et al. [14] applied the same curve to A04 as for A01. Li et al. [25] decreased the number of levels from four for A01 to three for A04, using thresholds of HHI 1500 and 2500. The limits of the levels applied in this approach are identical to those from Rosenau-Tornow et al. [22] for A01 and A02.

3.2. Scarcity (B)

The second-most frequently occurring indicator category is scarcity (B), for which we identified 25 different harmonized criteria. The four most common criteria are the depletion time of reserves (B01) and resources (B02), the sufficiency of reserves (B03), and the crustal content (B04). Figure 4 visualizes the normalization approaches for B01, B02, and B04. Information about the remaining harmonized criteria can be found in the Supplementary Material spreadsheets.

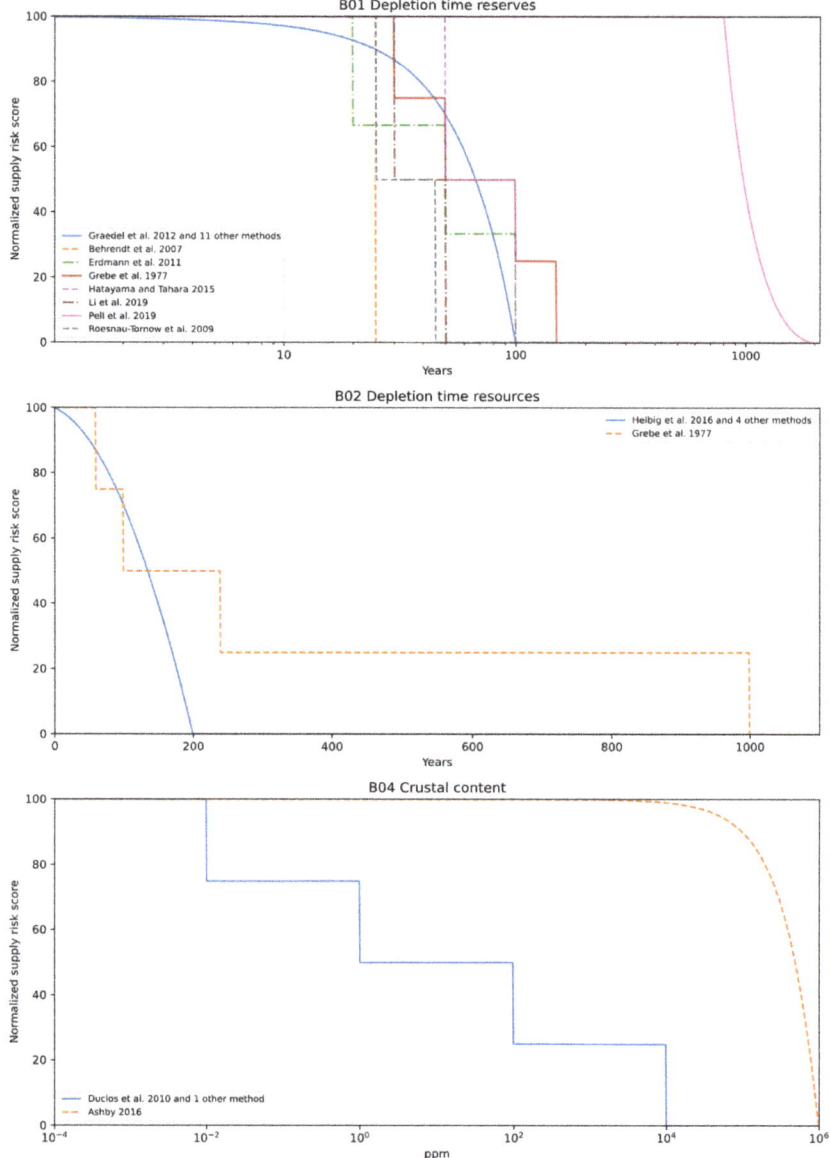

Figure 4. Selected normalization schemes of scarcity criteria: (B01) depletion time reserves, (B02) depletion time resources, and (B04) crustal content. Both depletion times are quantified in years, crustal content in parts per million (ppm). The criterion B03, the sufficiency of reserves, is not shown due to a lack of evident normalization and measurement in the respective assessments.

Both depletion time of reserves (B01) and depletion time of resources (B02) first appeared in the work of Grebe et al. [9]. The ratio between the available deposits and the current (primary) production rate determines the depletion time. Some authors also use terms such as the *static reach* for these criteria. No matter the name, the ratio is typically expressed in years. For B01, the considered deposits are available reserves, meaning the deposits are identified, and extraction is techno-economically viable. Graedel et al. [8]

presented the most frequently used normalization scheme adopted by 11 other methods (cf. Equation (4)).

$$DT_{normalized} = 100 - 0.2 \cdot DT - 0.008 \cdot DT^2 \qquad (4)$$

A parabolic function is used to assign high depletion time (DT) to low supply risk scores. Three key points are used to determine the shape of the parabola in Equation (4): A DT of 0 years leads to a normalized supply risk score of 100, whereas a value of 50 years is assigned to a score of 70 and a DT of 100 years results in a score of 0. Depletion times above 100 years are interpreted with no supply risk by Graedel et al. [8].

Pell et al. [17] applied their approach already presented for concentration criteria A01 to A03 by rescaling the inverted DT to a score from 0 to 1 with the observed minimum and maximum values and applying the distance-to-target method. The remaining publications presented for B01 in Figure 4 developed individual level-based normalization approaches. The number of levels varies from just two proposed by Behrendt et al. [26] to five in the work of Grebe et al. [9].

The depletion time of resources (B02) shows more consensus in the normalization approach. Resources, in contrast to reserves, also include inferred and sub-economic deposits; therefore, the depletion time of resources is larger than the depletion time of reserves. In most cases, the normalization of B02 is similar to that of B01. Helbig et al. [12] proposed a parabolic transformation comparable to the approach of Graedel et al. [8] for B01. For the DT of the resources, they suggest different key points by doubling the periods: A DT value of 200 years is considered as causing no supply risk at all, resulting in a supply risk score of 0, and a value of 100 years results in a score of 70. Four other publications followed this approach. The only different normalization approach found was the level-based normalization by Grebe et al. [9], consisting of five levels. Here, a DT exceeding 1000 years yields a supply risk score of 0. However, this method has never been applied by another study in our review.

For the crustal content (B04), the "abundance in earth's crust" was identified as the mainly used indicator. Two different approaches were found for normalization. Ashby [27] considers a high supply risk for materials with rare abundance in the earth's crust, but it does not propose a specific transformation into a normalized supply risk score. Nevertheless, we want to display Ashby's intention of assigning a high supply risk score to a low abundance [27]. Therefore, we conducted a simple linear transformation considering a value of 10^6 ppm as no supply risk and a value of 0 ppm as a normalized score of 100. An alternative method of normalization was applied by Duclos et al. [28] and one other method subdividing abundance values into five levels of supply risk.

3.3. Political Instability (C)

The third most used supply risk category is political instability (C), which is dominated by two harmonized criteria, namely geopolitical risk (C01) and political instability (C02). These two criteria make up 67 out of 75 cases (89%) for this category. Each of the eight remaining criteria identified is applied only once.

C01 appeared for the first time in the work of Eggert et al. [19] in 2000. To evaluate the geopolitical risk, they used the political country risk evaluation of Hermes/BMWi classification on a scale from 1 to 7. They use this classification three times for indicators in this category, each with a different weighting: the production shares, the export shares, and the reserve shares of the countries, respectively.

In contrast, C02 appeared first in the method of Morley and Eatherley [29] in 2008. They classified the percentile rank of the Worldwide Governance Indicator "Political Stability and Absence of Violence/Terrorism" (WGI_{PV}^{PR}) [30] of the largest producer into a supply risk score using a three-level normalization function. In addition, the World Bank developed five other Worldwide Governance Indicators, which are updated yearly: Government Effectiveness (WGI_{GE}), Voice and Accountability, Control of Corruption, Regulatory Quality, and Rule of Law. They are available as WGI score and WGI percentile rank [30].

The classification displays the normalization schemes used for the criteria of political instability according to their measurement: WGI score and WGI percentile rank (cf. Figure 5). In most cases, the WGI scores or ranks are weighted by production share. Other weighing factors are import shares [31] or the consideration of the largest producers [32]. The composition of a WGI dimension in combination with the HHI is described in Section 3.1. Other schemes are expressed in the Supplementary Material.

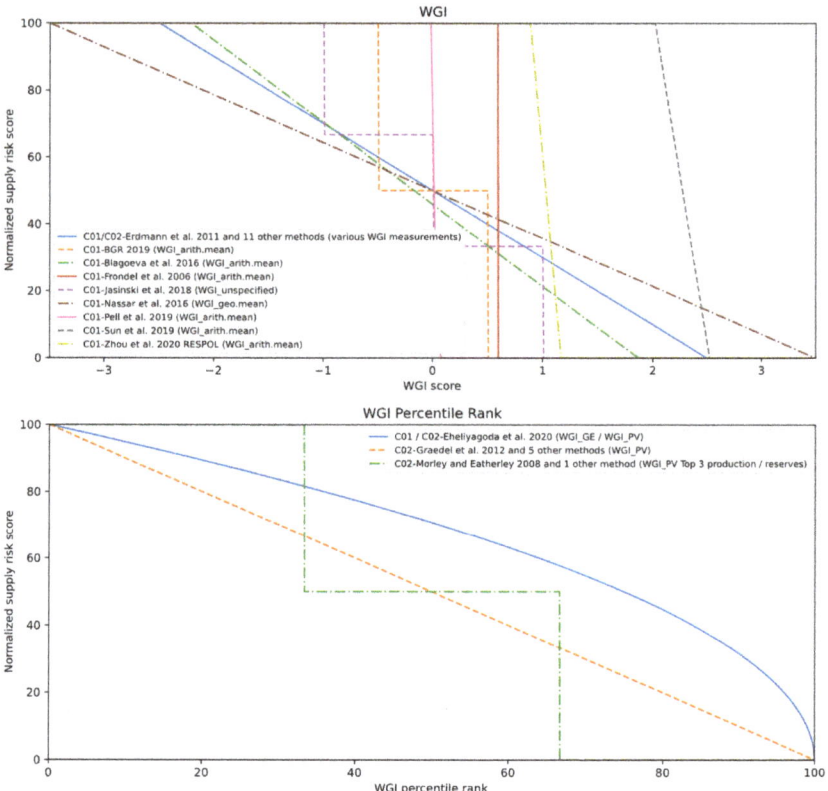

Figure 5. Selected normalization schemes of political instability criteria: (C01) Geopolitical risk and (C02) Political instability. Indicators use either the score or the percentile rank of the Worldwide Governance Indicators (WGI).

The WGI indicators in the default unit usually range from −2.5 to 2.5 [30], where a low value indicates bad governance. Consequently, the most frequently used normalization approach for both C01 and C02 is the linear transformation of WGI scores based on a hypothetical lower bound of −2.5 and upper bound of 2.5, as presented in Equation (5). After the conversion, values of −2.5 in WGI units and lower yield the highest supply risk score. Twelve other methods have adopted this transformation.

$$\text{WGI}_{\text{normalized}} = 20 \cdot (2.5 - \text{WGI}) \tag{5}$$

Diverging from this are Nassar et al. [33], who instead assume a range from −3.5 to 3.5 (cf. Figure 5). Three methods use the observed minimum and maximum WGI scores for normalizing them to supply risk scores: Blagoeva et al. [34] and Zhou et al. [14] based on the arithmetic mean of all six WGI dimensions, and Nassar et al. [13] based on the geometric mean of all six WGI dimensions. DERA [35] and Jasinski et al. [36] presented a level-based normalization with three and four levels, respectively. In addition to the approach of Erdmann et al. [32], none of the above-presented methods has been taken up so far. Sun et al. [37] used the function presented in Equation (6) to normalize the weighted arithmetic mean of the WGI-dimensions.

$$\text{WGI}_{\text{normalized}} = -1.9841 \cdot \text{WGI}_{\text{arith.mean}} + 5.7001 \quad (6)$$

For the normalization of the percentile rank of the WGI dimensions, four different methods were identified. Graedel et al. [8] simply inverted the percentile rank (on a scale from 0 to 100) by assigning the highest political stability with the lowest supply risk (cf. Equation (7)). This approach was adopted by five other methods, whereas other methods did not take up the remaining approaches for the normalization of WGI percentile ranks.

$$\text{WGI}_{\text{normalized}} = 100 - \text{WGI}^{\text{PR}} \quad (7)$$

Eheliyagoda et al. [38] developed a proceeding for both C01 and C02 using Equation (8) to invert and rescale the weighted $\text{WGI}_{\text{PV}}^{\text{PR}}$ respectively $\text{WGI}_{\text{GE}}^{\text{PR}}$ (Governance and Effectiveness).

$$\text{WGI}_{\text{normalized}} = \sqrt{(100 - \text{WGI}_{\text{PV/GE}}) \cdot 100} \quad (8)$$

3.4. Regulations (D)

The fourth most often mentioned category is regulations (D), which is used in 68 out of 618 indicators (11%). We identify policy perception (D01), human development (D02), trade barriers (D03), and environmental performance (D04) as the most prominent harmonized criteria. In contrast to most other categories, the risk from regulations has emerged more recently in the work of Thomason et al. [39] in 2010. They determined the percentage of produced goods expressed in U.S. market shares that a country intends to supply to the U.S. as a measurement of D03. In 2012, Graedel et al. [8] developed a measure to determine D01 and D02 for the first time.

It is worth mentioning that three dominant measurements of regulations have been developed within the criteria: The Policy Perception Index (PPI) in D01 is provided by the Fraser Institute and captures the influence of policies on mining activities in a country [40]. The Human Development Indicator (HDI) in D02 has been developed by the UNDP and evaluates the living conditions of a country [41]. The Environmental Performance Index (EPI) in D04 is provided by Yale University and rates the ability of a country to cope with environmental challenges [42]. All three measurements are updated annually. The associated normalization schemes are displayed in Figure 6.

Graedel et al. [8] weighted the PPI of mining regions by the respective production share. They normalized this measurement by a simple inversion subtracting the PPI from 100 according to Equation (9). The PPI ranges from 0, indicating low policy attractiveness, to 100, displaying high policy attractiveness for mining activities. Ten other methods adopted this approach in the same way.

$$\text{PPI}_{\text{normalized}} = 100 - \text{PPI} \quad (9)$$

Bach et al. [43] applied the distance-to-target method as described in Section 3.1 on the inverted and weighted PPI values using a threshold of 55. Eheliyagoda et al. [38] applied the same approach for the WGI percentile ranks of D01 and D02 (cf. Equation (8)) on the weighted PPI values. Both Zhou et al. [14] and Pell et al. [17] adopted their previously presented approaches. Zhou et al. [14] applied the normalization based on the minimum and maximum observed values of the PPI, whereas Pell et al. [17] first applied a rescaling from 0 to 1 according to the minimum and maximum observed values followed by the distance-to-target approach.

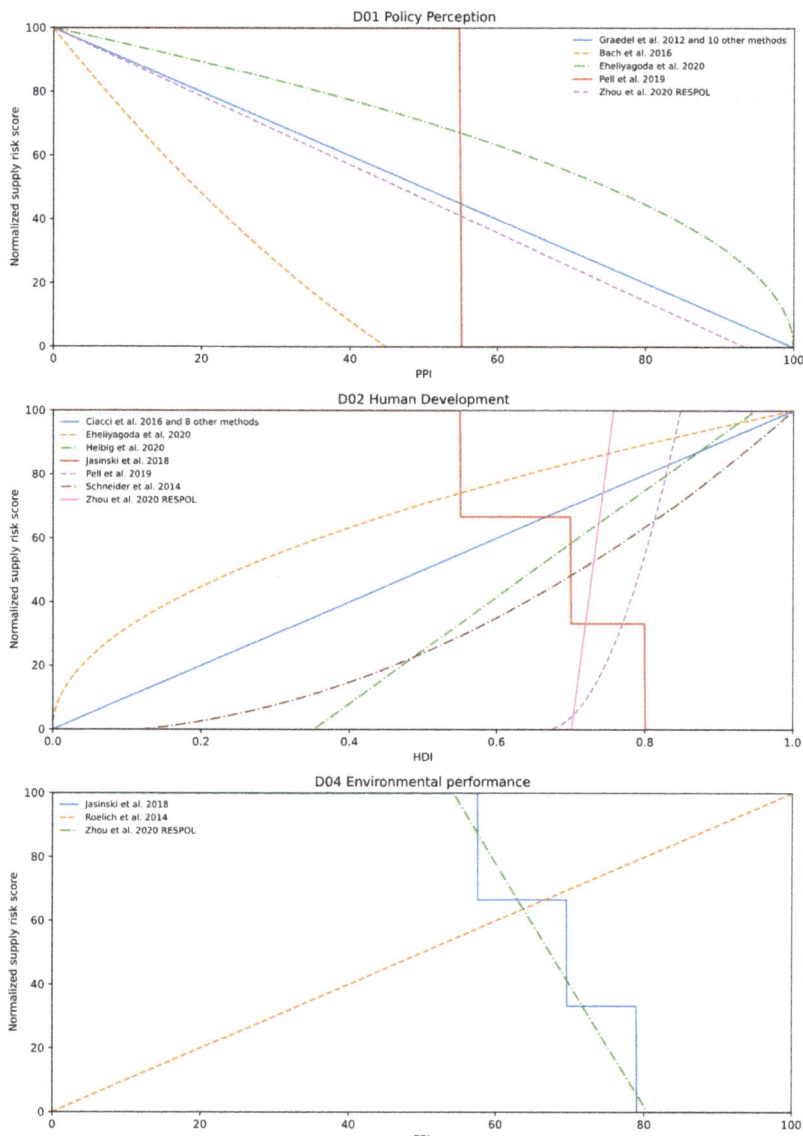

Figure 6. Selected normalization schemes of regulations criteria: (D01) policy perception, (D02) human development, and (D04) environmental performance. Policy perception is measured with the Policy Perception Index (PPI), human development with the Human Development Indicator (HDI), and environmental performance with the Environmental Performance Index (EPI).

For D02, the HDI weighted by production shares of mining countries was the most often used measurement. The HDI evaluates the three dimensions of life expectancy, educational standard, and standard of living on a scale from 0 to 1 [44]. Ciacci et al. [45] rescaled the weighted HDI values as presented in Equation (10) by a scaling factor of 100, which was followed by eight other methods.

$$\text{HDI}_{\text{normalized}} = 100 \cdot \text{HDI} \qquad (10)$$

Eheliyagoda et al. [38], Zhou et al. [14], as well as Pell et al. [17] applied their approaches on the HDI as conducted for previous criteria: Eheliyagoda et al. [38] applied the normalization formula shown in Equation (8) on HDI values, Zhou et al. [14] normalized according to the observed minimum and maximum values, Pell et al. [17] applied their combination of a normalization to a scale from 0 to 1 based on minimum and maximum values and the distance-to-target method. Schneider et al. [15] also stuck to their distance-to-target method already performed for A01–A03 with a threshold for the weighted HDI of 0.12. Helbig et al. [46] applied the formula presented in Equation (11) to normalize the weighted HDI values, resulting in 0 to 100. All of the previously mentioned studies deem high values of the HDI as high supply risk.

$$\text{HDI}_{\text{normalized}} = 100 \cdot \frac{\text{HDI} - 0.352}{0.949 - 0.352} \tag{11}$$

An opposite interpretation of the HDI was proposed by Jasinski et al. [36], resulting in an alternative normalization. They consider countries with low human development as critical because of the high probability of improving social conditions by introducing policies that disrupt mining activities. In other words, high weighted HDI values lead to a low supply risk in their study. Therefore, a stepwise normalization consisting of four levels is conducted.

The discrepancy in the interpretation of indicators continues for D04 with the EPI as a single measurement. Roelich et al. [47] used the production-weighted EPI as a measurement of D04 for the first time to describe the risk that a country has or introduces environmental policies that might restrict mining activities. Thus, an EPI value of 100 yields a high potential for restrictions in mining activities due to environmental policies. Since an EPI of 100 indicates high supply risk, no further normalization needs to be applied.

In contrast, Zhou et al. [14] and Jasinski et al. [36] oppositely interpreted the EPI. According to their normalization approaches, a higher EPI value leads to lower supply risk scores. They are less vulnerable to incidents and related supply failures because of their environmental standards [36]. While Zhou et al. [14] applied the same method as previously used for A by scaling the weighted EPI according to the minimum and maximum values observed, Jasinski et al. [36] used a four-level normalization applied for D02 with slightly different limits.

3.5. By-Product Dependence (E)

The fifth most often occurred category is by-product dependence (E), which is used in 44 out of 618 indicators (7%). We have found one dominant criterion giving the same name as the category by-product dependence (E01), which first appeared in the work of Grebe et al. [9]. A qualitative approach was used to assign the observed raw materials to a supply risk score ranging from 1 to 5. More information about the classification can be found in the Supplementary Material spreadsheets. However, the most commonly used measurement for E01 is companionality developed by Nassar et al. [13] and the companion metal fraction (CMF), which is the percentage share of a raw material produced as a by-product. Companionality (CP) evaluates the contribution of raw material to the profitability of a mine in contrast to other raw materials sourced from the same mine for all sourcing locations. The CP values are usually rescaled by a factor of 100, as shown in Equation (12) to result in a supply risk score from 0 for no risk by independent raw materials to 100 for high supply risk posed by full dependence of raw materials from other mined materials.

$$\text{CP}_i = \frac{\sum_j \left(\left(100 \cdot \left(1 - \min\left(\frac{\text{Revenue}_{ij}}{\text{Cost of sales}_j}, 1 \right) \right) \right) \cdot \text{Sales volume}_{ij} \right)}{\text{Sales volume}_i} \tag{12}$$

The normalization schemes of CMF are displayed in Figure 7. Same as for companionality, the most common approach is a multiplication by 100 to create a supply risk score ranging from 0, indicating a raw material is not produced as a by-product to 100, meaning a raw material is entirely made as a by-product. This method is used by Graedel et al. [8], followed by six other methods. Schneider et al. [15] applied the same distance-to-target approach for the categories above using a threshold of 0.2. BGS [48] and Jasinski et al. [36] proposed a stepwise three-level respectively four-level normalization. Other methods have not adopted either approach.

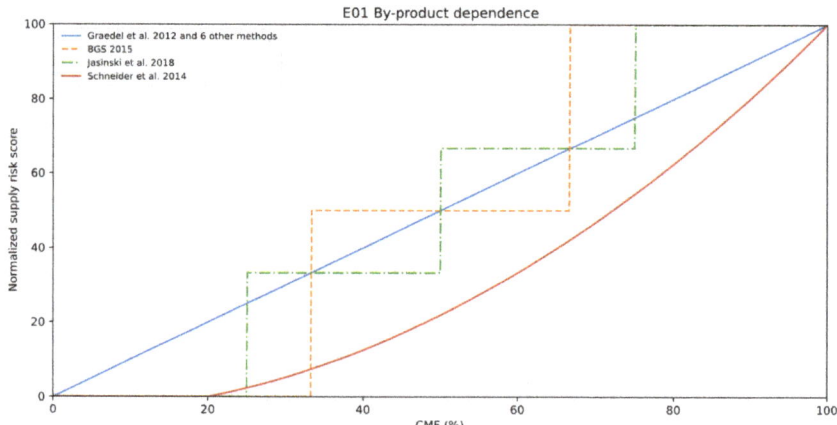

Figure 7. Normalization schemes for the by-product dependence criterion (E01) relying on the Companion Metal Fraction (CMF), given in percentages.

3.6. Dependence on Primary Production (F)

The sixth supply risk indicator category is the dependence on primary production. This terminology inverts the typically used original category name of recycling or recyclability. The inversion reflects that it is precisely the lack of recycling that increases the supply risk. The two most often used criteria in this category are the end-of-life recycling rate (F01) and the recycled content ratio (F02). The UNEP report on recycling of metals a decade ago has given a good overview on the terminology on metal cycles, including differentiation between old scrap and new scrap, and the importance of collection rates, remelting yields, and growing material demands for the measurements of recycling in global cycles [49]. The report is still often used as the data source for various supply risk assessments.

The argument for why dependence on primary production and thus a lack of secondary production, i.e., recycling, causes higher supply risk is the following: Secondary raw materials are a raw material source independent of the primary production route, in particular mining; it is available without geological exploration, with its availability depending predominantly on past material use, and it is known locally in the countries of utilization. Therefore, the availability of secondary raw materials makes shortages of primary raw materials and high market concentrations less likely.

In general, there are two schools of thought on how recycling should be measured as a criterion for dependence on primary production: Either method uses the end-of-life recycling rate (EoLRR) as the measurement or they use the recycling content ratio (RCR). Figure 8 shows the normalization schemes applied in these two measurements.

The EoLRR measures the share of end-of-life wastes collected and recycled so that the material can enter a new fabrication or manufacturing stage. Since there will always be waste flows that are not collected and thermodynamic limits to remelting yields, this EoLRR will always be smaller than 100%. The predominant normalization formula applied to the EoLRR is the naïve approach to linearly rescale the values of 0% to 100% to scores of 100 to 0 points.

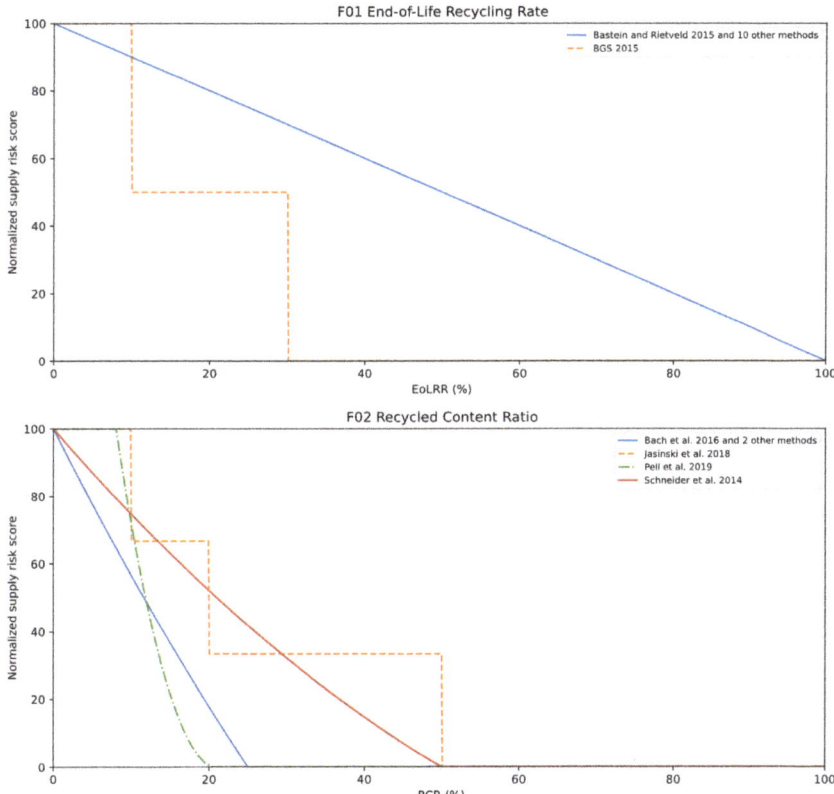

Figure 8. Selected normalization schemes for criteria for dependence on primary production: (F01) end-of-life recycling rate and (F02) recycled content ratio. Both end-of-life recycling rate (EoLRR) and recycled content ratio (RCR) are given in percentages.

In contrast, the RCR measures the share of recycled content in fabricated or manufactured goods. Since these goods can only consist of primary or secondary materials, this ratio will also be between 0% and 100%. However, because raw material markets have been growing for most materials over the past decades, the RCR will often be lower than the EoLRR. Therefore, methods already attribute no supply risk scores for any RCR over 50%.

Both EoLRR and RCR have their flaws as measurements for primary production dependence. For the EoLRR, there may be high recycling rates at end-of-life; however, these are irrelevant if the supply risks are emerging from rapidly growing future technology demand. As the EoLRR considers only recycling from old scrap, which is only formed after the use phase, there is a natural time lag between demand growth and growth of end-of-life wastes. For the RCR measurement, the ratio of recycled content may be high in a fabricated product. Still, if this all came from new scrap recycling, recycling before the use phase does not alter the primary material demand. One should be cautious that high prompt scrap formation rates with high recycling rates for prompt scrap might artificially increase the RCR without providing any risk-reducing alternative raw material source.

3.7. Demand Growth (G)

The seventh supply risk category is that of expected demand growth, in particular from future technologies. The most common approach is to relate the expected additional demand in the future and relate it to current production volumes. Angerer et al. [50] have first utilized this approach, which is a study that has been excluded from our dataset because it has been updated by Marscheider-Weidemann et al. [51].

This approach to calculate the future technology demand as a ratio between additional demand growth and current production typically needs a base year (for current production) and reference year (for future technology demand). For example, Angerer et al. originally calculated the raw material demand for various future technologies for 2030 and used 2006 as the base year. Using the ratio rather than, e.g., the quantity or value of future technology demand also allows comparing different raw materials produced in orders of varying magnitude. Not the absolute amount of material production is problematic, but rather the required relative demand growth. Since base years and reference years differ between supply risk assessments naturally, depending on their publication date and goal and scope of the evaluations, normalizations can only be compared based on the annualized additional demand growth, given in percentages (cf. Figure 9).

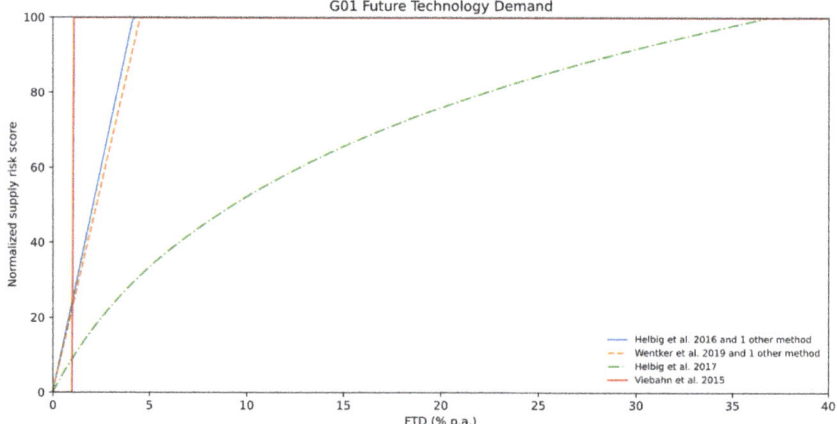

Figure 9. Normalization schemes for the future technology demand criterion (G01). Underlying measurements are annualized in future technology demand (FTD) growth per year (% p.a.).

3.8. Lack of Substitution Options (H)

The eighth indicator category for supply risk is that of lack of substitution options. A lack of viable substitutes for a material or product creates a dependency in the supply chains, reducing the system's resilience. In 22 out of 26 cases (84 %), the substitutability of raw material is used as a criterion in this category. Substitutability can happen on a material, component, assembly, or conceptual level as described by Habib and Wenzel [23] at the example of wind turbines. In particular, for high-tech applications, the substitution of materials is often limited, as developed by Nassar [52] for platinum-group metals. The most prominent evaluation of substitutability for a large set of raw materials has been published by Graedel et al. [53], who set out to identify the main applications of each element, identify possible substitute materials in each of these applications, and then evaluate the performance of that substitute. As a result of the heterogeneity of applications, these are difficult to quantify. Therefore, experts' judgment on a multi-point scale is used to conclude the substitutability score. Application shares are afterwards used to calculate a weighted average of the scores [53]. Graedel et al. were not the first and not the only ones to use such an approach for evaluating the lack of substitution options. The first to use the lack of substitution options as an indicator for supply risk was again Grebe et al. [9], however only with the straightforward classification of raw materials with "no substitution", "hardly any substitution", and "substitution" and no differentiation of application shares. The European Commission [54] and Erdmann et al. [32] also applied this concept of a weighted average of expert assessment for the main applications of the raw material. A shift from substitutability to substitution has been used for the later updates of the EU Critical Raw Materials list [55]. While this may seem to be quibbling, the difference is substantial, as substitution only considers proven and readily available substitutes. Consequentially, the supply risk scores of many raw materials in the EU criticality study increased due to this change [56,57].

The normalization scheme for lack of substitution options is trivial: typically, a linear scale is used, with no further rescaling. Therefore, no figure is shown for this indicator category.

3.9. Price Volatility (I)

The ninth indicator category for supply risk is price volatility. It was impossible to identify different criteria for this category, so all 17 cases are assigned to the same criterion I01. In detail, the measurements vary between the price volatility, the variation coefficient, and the relative price change within a specific period. All methods use a stepwise normalization of price volatility measurements to supply risk scores. Most studies use four-level to five-level normalization functions in which higher price volatility leads to higher supply risks. The only exception is the method by Eggert et al. [19], who use a seven-level decreasing normalization function. The authors were among the early supply risk assessments, and they did not explain why they evaluated low price volatility with high supply risk. Figure 10 shows the different level choices of the four methods. Other methods' schemes are shown in the Supplementary Material spreadsheets.

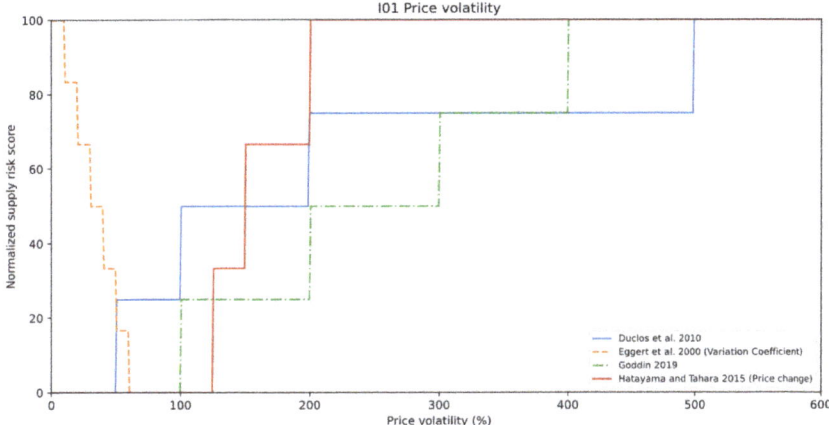

Figure 10. Normalization schemes for the price volatility criterion (I01). Price volatility is either used directly, in percentage, or calculated as variation coefficient or price change, also in percentages.

According to economic theory, the interpretation of the criterion price volatility is ambiguous, because a price increase should result from a supply–demand gap, not the reason. It is also questionable if one can anticipate future supply risks by the analysis of historical price development. Therefore, it is not surprising that this indicator category has only been used in very selected supply risk assessments and not been used consecutively by a series of methods.

3.10. Import Dependency (J)

The tenth indicator category for supply risks is import dependency. This indicator category is specifically designed for a national perspective. This category is measured with the net import reliance criterion in eight out of 16 cases (50%) (J01). The net import reliance

(NIR) is calculated as the ratio between net imports and apparent consumption, as shown in Equation (13).

$$\text{NIR} = \frac{\text{Net imports}}{\text{Apparent consumption}} = \frac{\text{Imports} - \text{Exports}}{\text{Domestic Production} + \text{Imports} - \text{Exports}} \quad (13)$$

The rationale behind using this indicator for supply risk assessments is to identify materials for which the upstream supply chain is out of the hands of domestic policy and trade. If a country has to rely on foreign exploration, extraction, or processing, it can consider their continuous operation, or the access to the materials, as less reliable. Therefore, higher net import reliance is considered with higher supply risks.

Most methods, starting with Goe and Gaustad [58], use the simple linear normalization approach where no net imports result in no supply risk, and 100% NIR results in a supply risk score of 100. Only Li et al. [59] define the steps at 40% and 70% NIR as thresholds for their three-level normalization function. Figure 11 shows the normalization functions for J01. Other criteria are shown in the Supplementary Material spreadsheets.

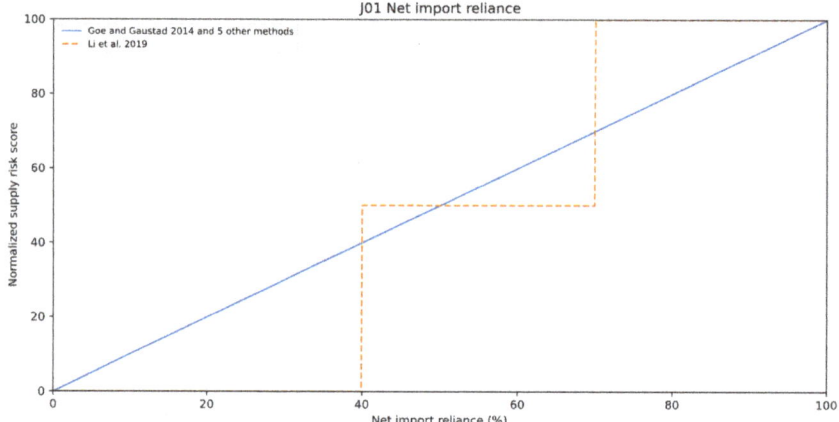

Figure 11. Normalization schemes for the net import reliance criterion (J01), which is measured in percent.

3.11. Other Indicators

Our review found an additional 67 cases of indicator uses that could not be grouped into indicator categories. Therefore, these "other" indicators are a collection of 53 widely differing criteria, none of them used more than four times in total. Those that are at least used twice are the current market balance (X01), stock keeping (X02), purchasing potential (X03), supply adequacy (X04), natural disasters (X05), economic importance (X06), the Sector Competition Index (X07), the economy of storage and transport (X08), storage complexity (X09), investment potential (X10), material cost impact (X11), and material dependency (X12).

If one wants to find patterns in this loose collection of indicators, three areas of interest may occur: stocks and storage patterns, price and cost aspects, as well as total demand and market size. However, due to the high variation in indicator application and measurements and a lack of repeated implementation in supply risk assessments, we will refrain from discussing these indicators in detail. All individual indicators are listed with their respective measurement and normalization scheme in the Supplementary Material spreadsheets.

4. Discussion and Conclusions

The variety in supply risk indicator usage is impressive. It is understandable because of the different goals and scopes of studies in our review. For example, omitting physical scarcity as a risk factor makes sense when the assessment is focused on short-term risks. Likewise, companies will be much less concerned about import dependence than nations. Therefore, even after another "five years of criticality assessments", the harmonization that Graedel and Reck asked for has not taken place [7]. However, many of Graedel and Reck's other "desirable aspects" are covered nowadays by the methods in this review.

The material scope often includes various chemical elements and biotic raw materials and minerals [60,61]. The risk factors also include geology (scarcity, by-product dependence), regulations, and geopolitics (political instability, import dependence). For example, even cultural aspects are used, the "conformity of ideological values" by Nassar et al. [62]. The substitutability or the lack thereof, the dependence on primary production, and the by-product dependence are three of the ten indicator categories. However, these categories often rely on previous assessments such as Graedel et al. [49,53] or Nassar et al. [13]. In contrast to the studies reviewed by Achzet and Helbig [3], in 2013, most of the methods in this review are now published in peer-reviewed journals, not as technical reports. Similar to the European Commission or the United States, some governmental reports undertake the split path of publishing the technical report and a peer-reviewed methodological paper in parallel [62,63].

However, the periodical update that Graedel and Reck [7] also asked for is a rare feature. Many studies are carried out by researchers at universities or other academic institutes without permanent funding for such updates. The EU and US criticality lists, updated every three to four years, are exceptions [60,64].

The transparency of some methods is hampered by the non-disclosure of data [15,65–68]. While we understand the importance of confidentiality, particularly for company reports, from a scientific perspective, this reduces the transparency and accessibility of corporate supply risk assessment reports [28,69]. Some other methods used sophisticated intermediate scores, thresholds, and renormalizations up to the point that results turned out to be irreproducible or the quantitative results simply contradict the textual explanations [17,34,62,70].

Some methods such as Zhou et al. [14], Pell et al. [17], and Bach et al. [71] use normalizations based on the specific material scope, for example setting the bounds by looking at the minimum and maximum observed measurement, leading to a distortion of the supply risk score since the score is dependent on the raw materials selected for the assessment. In other words, the supply risk score of the observed raw materials varies depending on the investigated raw materials. The integration of individual bounds depending on the values in the normalization function leads to different supply risk scores for the same indicator value. Therefore, results in between studies are not comparable, and the overall results are weakened. Adjusting indicator calculation or normalization schemes may be viable for specific purposes of custom methods. However, in these cases, full transparency and reproducibility are even more important.

The supply risk assessments in the review still often do not adequately report data uncertainties and sensitivity to methodological choices. Only very few authors undertake the effort of doing Monte Carlo simulation or other error propagation methods [8,12,31,72]. Variations of indicator choices and normalizations, which have been discussed by Erdmann and Graedel [73], are also rare.

Concluding, we want to highlight two efforts by individual researchers and one general recommendation which, in our view, would improve future supply risk assessments. Firstly, the effort from Mayer and Gleich [6] (and in many other publications from this working group) to link commonly used supply risk indicators with price variations, as the theoretical result of supply shortages through multiple regression analysis is a practical approach. One of their results was that impacts of supply risk aspects vary between chemical elements and, therefore, no universal indicator set will be found. Secondly, Hatayama and

Tahara [5] established a list of supply disruption events, which, if continued, extended to global coverage, and further evaluated could be an excellent basis for event studies. Such event studies could be used to statistically assess the likelihood of supply disruptions at various levels of supply risk indicators. For example, this would eventually allow identifying a non-linear normalization formula instead of naïve approaches or single threshold values. However, the normalization formula and thresholds are still better than the semi-quantitative approach of point-wise or step-wise normalizations used in many assessments. Supply risk indicators can be measured and should be interpreted quantitatively.

We strongly recommend updating some of the data sources commonly used in supply risk assessments. While the USGS provides annual updates to production and reserves data, and while the various political and regulatory indices are also updated annually, the data sources for by-product dependence, dependence on primary production, and lack of substitution options by now are up to a decade old. Given increasing efforts to implement a circular economy, ongoing technological development, and rapid material extraction growth, the values of these data sources are at risk of becoming outdated.

Supplementary Materials: The following are available online at https://www.mdpi.com/article/10.3390/resources10080079/s1, Table S1: List of studies, Table S2: Indicators summary, Table S3: Normalization and Thresholds, Tables S4–S7: Data for Figure 2, Tables S8–S10: Data for Figure 3, Tables S11 and S12: Data for Figure 4, Tables S13–S15: Data for Figure 5, Table S16: Data for Figure 6, Tables S17 and S18: Data for Figure 7, Table S19: Data for Figure 8, Table S20: Data for Figure 9, Table S21: Data for Figure 10.

Author Contributions: Conceptualization, C.H. and A.T. (Andrea Thorenz); formal analysis, C.H. and M.B.; data curation, M.B.; writing—original draft preparation, C.H. and M.B.; writing—review and editing, A.T. (Andrea Thorenz) and A.T. (Axel Tuma); visualization, C.H.; supervision, A.T. (Andrea Thorenz) and A.T. (Axel Tuma). All authors have read and agreed to the published version of the manuscript.

Funding: This research received no external funding.

Institutional Review Board Statement: Not applicable.

Informed Consent Statement: Not applicable.

Data Availability Statement: All underlying data is available in the Supplementary Materials.

Conflicts of Interest: The authors declare no conflict of interest.

Appendix A

Table A1. List of studies. Additional information is provided in the Supplementary Material spreadsheet.

Short Name	Year	Type	Ref.
Adibi et al. 2017	2017	assessment	[74]
Alonso et al. 2007	2007	collection of indicators	[75]
Althaf and Babbit 2020	2020	assessment	[70]
Apple 2019	2019	collection of indicators	[69]
Ashby 2016	2016	collection of indicators	[27]
Bach et al. 2016	2016	assessment	[43]
Bach et al. 2017 RESPOL	2017	assessment	[71]
Bach et al. 2017 Sustainability	2017	assessment	[76]
Bach et al. 2018	2018	assessment	[61]
Bastein and Rietveld 2015	2015	assessment	[77]

Table A1. Cont.

Short Name	Year	Type	Ref.
Bauer et al. 2011	2011	assessment	[78]
Behrendt et al. 2007	2007	assessment	[26]
Beylot and Villeneuve 2015	2015	assessment	[79]
BGS 2015	2015	assessment	[48]
Blagoeva et al. 2016	2016	assessment	[34]
Blengini et al. 2017 RESPOL	2017	assessment	[55]
Brown 2018	2018	assessment	[80]
Buchert et al. 2009	2009	assessment	[81]
Calvo et al. 2018	2018	assessment	[82]
Ciacci et al. 2016	2016	assessment	[45]
Cimprich et al. 2017	2017	assessment	[83]
Cimprich et al. 2018	2018	assessment	[84]
Coulomb et al. 2015	2015	assessment	[85]
Daw 2017	2017	assessment	[86]
DERA 2019	2019	assessment	[35]
Duclos et al. 2010	2010	assessment	[28]
European Commission 2014	2014	assessment	[56]
Eggert et al. 2000	2000	assessment	[19]
Eheliyagoda et al. 2020	2020	assessment	[38]
Erdmann et al. 2011	2011	assessment	[32]
Frenzel et al. 2017 RESPOL	2017	assessment	[87]
Frondel et al. 2006	2006	single indicator	[18]
Fu et al. 2019	2019	assessment	[88]
Gemechu et al. 2016	2016	assessment	[31]
Glöser-Chahoud et al. 2016	2016	assessment	[89]
Goddin 2019	2019	assessment	[90]
Goe and Gaustad 2014	2014	assessment	[58]
Graedel et al. 2012	2012	assessment	[8]
Graedel et al. 2015	2015	assessment	[91]
Grebe et al. 1977	1977	assessment	[9]
Habib and Wenzel 2016	2016	assessment	[23]
Habib et al. 2016	2016	single indicator	[20]
Hatayama and Tahara 2015	2015	assessment	[92]
Helbig et al. 2016	2016	assessment	[12]
Helbig et al. 2017	2017	assessment	[24]
Helbig et al. 2018	2018	assessment	[72]
Helbig et al. 2020	2020	assessment	[46]
Ioannidou et al. 2019	2019	assessment	[93]
Jasinski et al. 2018	2018	assessment	[36]
Kim et al. 2019	2019	assessment	[94]

Table A1. *Cont.*

Short Name	Year	Type	Ref.
Kolotzek et al. 2018	2018	assessment	[21]
Kosmol et al. 2018	2018	assessment	[95]
Li et al. 2019	2019	assessment	[59]
Malinauskiene et al. 2018	2018	assessment	[65]
Marscheider-Weidemann et al. 2016	2016	single indicator	[51]
Martins and Castro 2019	2019	assessment	[96]
Mayer and Gleich 2015	2015	assessment	[6]
Miyamoto et al. 2019	2019	assessment	[97]
Morley and Eatherley 2008	2008	assessment	[29]
Moss et al. 2013	2013	assessment	[98]
Nansai et al. 2015	2015	assessment	[99]
Nansai et al. 2017	2017	assessment	[100]
Nassar et al. 2015	2015	collection of indicators	[13]
Nassar et al. 2016	2016	assessment	[33]
Nassar et al. 2020	2020	assessment	[62]
NRC 2008	2008	assessment	[101]
Parthemore 2011	2011	assessment	[102]
Pell et al. 2019	2019	assessment	[17]
Pfleger et al. 2015	2015	assessment	[68]
Roelich et al. 2014	2014	assessment	[47]
Rosenau-Tornow et al. 2009	2009	assessment	[22]
Schneider et al. 2014	2014	assessment	[15]
Shammugam et al. 2019	2019	assessment	[103]
Simon et al. 2014	2014	assessment	[104]
Spörri et al. 2017	2017	assessment	[105]
Sun et al. 2019	2019	assessment	[37]
Thomason et al. 2010	2010	assessment	[39]
Tuma et al. 2014	2014	assessment	[106]
van den Brink 2020	2020	assessment	[107]
Viebahn et al. 2015	2015	assessment	[108]
Wentker et al. 2019	2019	assessment	[109]
Yan et al. 2020	2020	assessment	[110]
Yuan et al. 2019	2019	assessment	[111]
Zepf et al. 2014	2014	assessment	[112]
Zhou et al. 2019	2019	assessment	[66]
Zhou et al. 2020 JCLEPRO	2020	assessment	[67]
Zhou et al. 2020 RESPOL	2020	assessment	[14]

References

1. Schrijvers, D.; Hool, A.; Blengini, G.A.; Chen, W.-Q.; Dewulf, J.; Eggert, R.; van Ellen, L.; Gauss, R.; Goddin, J.; Habib, K.; et al. A review of methods and data to determine raw material criticality. *Resour. Conserv. Recycl.* **2020**, *155*, 104617. [CrossRef]
2. Helbig, C.; Schrijvers, D.; Hool, A. Selecting and prioritizing material resources by criticality assessments. *One Earth* **2021**, *4*, 339–345. [CrossRef]
3. Achzet, B.; Helbig, C. How to evaluate raw material supply risks—An overview. *Resour. Policy* **2013**, *38*, 435–447. [CrossRef]
4. Helbig, C.; Wietschel, L.; Thorenz, A.; Tuma, A. How to evaluate raw material vulnerability—An overview. *Resour. Policy* **2016**, *48*, 13–24. [CrossRef]
5. Hatayama, H.; Tahara, K. Adopting an objective approach to criticality assessment: Learning from the past. *Resour. Policy* **2018**, *55*, 96–102. [CrossRef]
6. Mayer, H.; Gleich, B. Measuring Criticality of Raw Materials: An Empirical Approach Assessing the Supply Risk Dimension of Commodity Criticality. *Nat. Resour.* **2015**, *6*, 56–78. [CrossRef]
7. Graedel, T.E.; Reck, B.K. Six Years of Criticality Assessments: What Have We Learned So Far? *J. Ind. Ecol.* **2016**, *20*, 692–699. [CrossRef]
8. Graedel, T.E.; Barr, R.; Chandler, C.; Chase, T.; Choi, J.; Christoffersen, L.; Friedlander, E.; Henly, C.; Jun, C.; Nassar, N.T.; et al. Methodology of Metal Criticality Determination. *Environ. Sci. Technol.* **2012**, *46*, 1063–1070. [CrossRef] [PubMed]
9. Grebe, W.H.; Krauß, U.; Kurszona, M.; Schmidt, H.; Kamphausen, D.; Liebrucks, M.; Rumberger, M.; Wettig, E.; Bäse, K.; Jägeler, F.; et al. *Ausfallrisiko bei 31 Rohstoffen: Bergwirtschaftliche und Rohstoffwirtschafliche Kriterien für das Angebot und die Nachfrage wichtiger Rohstoffe*; Bundesanstalt Für Geowissenschaften und Rohstofe, Deutsches Institut für Wirtschaftforschung, Institut zur Erforschung Technologischer Entwicklungslinien: Berlin, Germany, 1977.
10. Herfindahl, O.C. Concentration in the US Steel Industry. Ph.D. Thesis, Columbia University, New York, NY, USA, 1950.
11. Hirschman, A.O. *National Power and the Structure of Foreign Trade*; University of California Press: Berkeley, CA, USA, 1980; ISBN 9780520040823.
12. Helbig, C.; Bradshaw, A.M.; Kolotzek, C.; Thorenz, A.; Tuma, A. Supply risks associated with CdTe and CIGS thin-film photovoltaics. *Appl. Energy* **2016**, *178*, 422–433. [CrossRef]
13. Nassar, N.T.; Graedel, T.E.; Harper, E.M. By-product metals are technologically essential but have problematic supply. *Sci. Adv.* **2015**, *1*, e1400180. [CrossRef] [PubMed]
14. Zhou, N.; Wu, Q.; Hu, X.; Zhu, Y.; Su, H.; Xue, S. Synthesized indicator for evaluating security of strategic minerals in China: A case study of lithium. *Resour. Policy* **2020**, *69*, 101915. [CrossRef]
15. Schneider, L.; Berger, M.; Schüler-Hainsch, E.; Knöfel, S.; Ruhland, K.; Mosig, J.; Bach, V.; Finkbeiner, M. The economic resource scarcity potential (ESP) for evaluating resource use based on life cycle assessment. *Int. J. Life Cycle Assess.* **2014**, *19*, 601–610. [CrossRef]
16. Frischknecht, R.; Steiner, R.; Jungbluth, N. *The Ecological Scarcity Method—Eco-Factors 2006. A Method for Impact Assessment in LCA*; Federal Office for the Environment: Bern, Switzerland, 2009.
17. Pell, R.S.; Wall, F.; Yan, X.; Bailey, G. Applying and advancing the economic resource scarcity potential (ESP) method for rare earth elements. *Resour. Policy* **2019**, *62*, 472–481. [CrossRef]
18. Frondel, M.; Angerer, G.; Buchholz, P. *Trends der Angebots-und Nachfragesituation bei Mineralischen Rohstoffen*; BMWi: Berlin, Germany, 2006.
19. Eggert, P.; Haid, A.; Wettig, E.; Dahlheimer, M.; Kruszona, M.; Wagner, H. *Auswirkungen der Weltweiten Konzentration in der Bergbauproduktion auf die Rohstoffversorgung der Deutschen Wirtschaft*; Deutsches Institut für Wirtschaftsforschung: Berlin, Germany, 2000; Volume 184, ISBN 3428102738.
20. Habib, K.; Hamelin, L.; Wenzel, H. A dynamic perspective of the geopolitical supply risk of metals. *J. Clean. Prod.* **2016**, *133*, 850–858. [CrossRef]
21. Kolotzek, C.; Helbig, C.; Thorenz, A.; Reller, A.; Tuma, A. A company-oriented model for the assessment of raw material supply risks, environmental impact and social implications. *J. Clean. Prod.* **2018**, *176*, 566–580. [CrossRef]
22. Rosenau-Tornow, D.; Buchholz, P.; Riemann, A.; Wagner, M. Assessing the long-term supply risks for mineral raw materials-a combined evaluation of past and future trends. *Resour. Policy* **2009**, *34*, 161–175. [CrossRef]
23. Habib, K.; Wenzel, H. Reviewing resource criticality assessment from a dynamic and technology specific perspective—using the case of direct-drive wind turbines. *J. Clean. Prod.* **2016**, *112*, 3852–3863. [CrossRef]
24. Helbig, C.; Kolotzek, C.; Thorenz, A.; Reller, A.; Tuma, A.; Schafnitzel, M.; Krohns, S. Benefits of resource strategy for sustainable materials research and development. *Sustain. Mater. Technol.* **2017**, *12*, 1–8. [CrossRef]
25. Li, X.Y.; Ge, J.P.; Chen, W.Q.; Wang, P. Scenarios of rare earth elements demand driven by automotive electrification in China: 2018–2030. *Resour. Conserv. Recycl.* **2019**, *145*, 322–331. [CrossRef]
26. Behrendt, S.; Scharp, M.; Kahlenborn, W.; Feil, M.; Dereje, C.; Bleischwitz, R.; Delzeit, R. *Maßnahmen und Konzepte zur Lösung des Problems konfliktverschärfender Rohstoffausbeutung am Beispiel Coltan*; Umweltbundesamt: Berlin, Germany, 2007.
27. Ashby, M.F. *Materials and Sustainable Development*; Butterworth-Heinemann: Oxford, UK, 2016.
28. Duclos, S.J.; Otto, J.P.; Konitzer, D.G. Design in an Era of Constrained Resources: As Global Competition for Materials strains the Supply Chain, Companies must know where a Shortage can hurt and then plan around it. *Mech. Eng.* **2010**, *132*, 36–40. [CrossRef]

29. Morley, N.; Eatherley, D. *Material Security—Ensuring Resource Availability for the UK Economy*; C-Tech Innovations: Chester, UK, 2008.
30. Kaufmann, D.; Kraay, A.; Mastruzzi, M. The Worldwide Governance Indicators: Methodology and Analytical Issues. *World Bank Policy Res. Work. Pap.* **2010**, *5430*, 1–31. [CrossRef]
31. Gemechu, E.D.; Helbig, C.; Sonnemann, G.; Thorenz, A.; Tuma, A. Import-based Indicator for the Geopolitical Supply Risk of Raw Materials in Life Cycle Sustainability Assessments. *J. Ind. Ecol.* **2016**, *20*, 154–165. [CrossRef]
32. Erdmann, L.; Behrendt, S.; Feil, M. *Kritische Rohstoffe für Deutschland*; KfW Bankengruppe: Berlin, Germany, 2011.
33. Nassar, N.T.; Xun, S.; Fortier, S.M.; Schoeberlein, D. *Assessment of Critical Minerals: Screening Methodology and Initial Application*; Executive Office of the President of the United States: Washington, DC, USA, 2016.
34. Blagoeva, D.T.; Alves Dias, P.; Marmier, A.; Pavel, C.C. *Assessment of Potential Bottlenecks along the Materials Supply Chain for the Future Deployment of Low-Carbon Energy and Transport Technologies in the EU. Wind Power, Photovoltaic and Electric Vehicles Technologies, Time Frame: 2015–2030*; European Commission: Brussels, Belgium, 2016.
35. DERA. *DERA-Rohstoffliste 2019*; Deutsche Rohstoffagentur in der Bundesanstalt für Geowissenschaften und Rohstoffe: Berlin, Germany, 2019.
36. Jasiński, D.; Cinelli, M.; Dias, L.C.; Meredith, J.; Kirwan, K. Assessing supply risks for non-fossil mineral resources via multi-criteria decision analysis. *Resour. Policy* **2018**, *58*, 150–158. [CrossRef]
37. Sun, X.; Hao, H.; Hartmann, P.; Liu, Z.; Zhao, F. Supply risks of lithium-ion battery materials: An entire supply chain estimation. *Mater. Today Energy* **2019**, *14*, 100347. [CrossRef]
38. Eheliyagoda, D.; Zeng, X.; Li, J. A method to assess national metal criticality: The environment as a foremost measurement. *Humanit. Soc. Sci. Commun.* **2020**, *7*, 43. [CrossRef]
39. Thomason, S.J.; Atweill, R.; Bajraktari, Y.; Bell, J.; Barnett, D.; Karvonides, N.; Niles, M.; Schwartz, E. *From National Defense Stockpile (NDS) to Strategic Materials Security Program (SMSP): Evidence and Analytic Support*; Institute for Defense Analysis: Washington, DC, USA, 2010.
40. Yunis, J.; Elmira, A. Survey of Mining Companies 2020. Fraser Inst. Annu. 2018. Available online: http://www.fraserinstitute.org (accessed on 3 August 2021).
41. UNDP. *The Next Frontier: Human Development and the Anthropocene*; UNDP: New York, NY, USA, 2020; ISBN 9789211264425.
42. Wendling, Z.A.; Emerson, J.W.; de Sherbinin, A.; Esty, D.C. *2020 Environmental Performance Index*; Yalce Center for Environmental Law & Policy: New Haven, CT, USA, 2020.
43. Bach, V.; Berger, M.; Henßler, M.; Kirchner, M.; Leiser, S.; Mohr, L.; Rother, E.; Ruhland, K.; Schneider, L.; Tikana, L.; et al. Integrated method to assess resource efficiency—ESSENZ. *J. Clean. Prod.* **2016**, *137*, 118–130. [CrossRef]
44. UNDP. *2020 Human Development Report: Technical Notes*; UNEP: Nairobi, Kenya, 2020.
45. Ciacci, L.; Nuss, P.; Reck, B.K.; Werner, T.T.; Graedel, T.E. Metal Criticality Determination for Australia, the US, and the Planet—Comparing 2008 and 2012 Results. *Resources* **2016**, *5*, 29. [CrossRef]
46. Helbig, C.; Bradshaw, A.M.; Thorenz, A.; Tuma, A. Supply Risk Considerations for the Elements in Nickel-Based Superalloys. *Resources* **2020**, *9*, 106. [CrossRef]
47. Roelich, K.; Dawson, D.A.; Purnell, P.; Knoeri, C.; Revell, R.; Busch, J.; Steinberger, J.K. Assessing the dynamic material criticality of infrastructure transitions: A case of low carbon electricity. *Appl. Energy* **2014**, *123*, 378–386. [CrossRef]
48. BGS Risk. *List 2015—An Update to the Supply Risk Index for Elements or Element Groups that are of Economic Value*; British Geological Survey: Keyworth, UK, 2015.
49. Graedel, T.E.; Allwood, J.M.; Birat, J.-P.; Reck, B.K.; Sibley, S.F.; Sonnemann, G.; Buchert, M.; Hagelüken, C. *Recycling Rates of Metals—A Status Report, A Report of the Working Group on the Global Metal Flows to the International Resource Panel*; UNEP: Nairobi, Kenya, 2011.
50. Angerer, G.; Marscheider-Weidemann, F.; Lüllmann, A.; Erdmann, L.; Scharp, M.; Handke, V.; Marwede, M. *Raw Materials for Emerging Technologies*; Frauenhofer IRB Verlag: Stuttgart, Germany, 2009.
51. Marscheider-Weidemann, F.; Langkau, S.; Hummen, T.; Erdmann, L.; Tercero Espinoza, L.A.; Angerer, G.; Marwede, M.; Benecke, S. *Rohstoffe für Zukunftstechnologien 2016*; Deutsche Rohstoffagentur (DERA): Berlin, Germany, 2016.
52. Nassar, N.T. Limitations to elemental substitution as exemplified by the platinum-group metals. *Green Chem.* **2015**, *17*, 2226–2235. [CrossRef]
53. Graedel, T.E.; Harper, E.M.; Nassar, N.T.; Reck, B.K. On the materials basis of modern society. *Proc. Natl. Acad. Sci. USA* **2015**, *112*, 6295–6300. [CrossRef]
54. European Commission. *Critical Raw Materials for the EU*; European Commission: Brussels, Belgium, 2010.
55. Blengini, G.A.; Nuss, P.; Dewulf, J.; Nita, V.; Peirò, L.T.; Vidal-Legaz, B.; Latunussa, C.; Mancini, L.; Blagoeva, D.; Pennington, D.; et al. EU methodology for critical raw materials assessment: Policy needs and proposed solutions for incremental improvements. *Resour. Policy* **2017**, *53*, 12–19. [CrossRef]
56. European Commission. *Report on Critical Raw Materials for the EU: Report of the Ad hoc Working Group on Defining Critical Raw Materials*; European Commission: Brussels, Belgium, 2014.
57. European Commission. *Study on the Review of the List of Critical Raw Materials*; European Commission: Brussels, Belgium, 2017; ISBN 978-92-79-47937-3.

58. Goe, M.; Gaustad, G.G. Identifying critical materials for photovoltaics in the US: A multi-metric approach. *Appl. Energy* **2014**, *123*, 387–396. [CrossRef]
59. Li, S.; Yan, J.; Pei, Q.; Sha, J.; Mou, S.; Xiao, Y. Risk Identification and Evaluation of the Long-term Supply of Manganese Mines in China Based on the VW-BGR Method. *Sustainability* **2019**, *11*, 2683. [CrossRef]
60. Blengini, G.A. *European Commission Study on the EU's List of Critical Raw Materials-Final Report*; European Commission: Brussels, Belgium, 2020; ISBN 978-92-79-72119-9.
61. Bach, V.; Berger, M.; Forin, S.; Finkbeiner, M. Comprehensive approach for evaluating different resource types—Case study of abiotic and biotic resource use assessment methodologies. *Ecol. Indic.* **2018**, *87*, 314–322. [CrossRef]
62. Nassar, N.T.; Brainard, J.; Gulley, A.; Manley, R.; Matos, G.; Lederer, G.; Bird, L.R.; Pineault, D.; Alonso, E.; Gambogi, J.; et al. Evaluating the mineral commodity supply risk of the U.S. manufacturing sector. *Sci. Adv.* **2020**, *6*, eaay8647. [CrossRef] [PubMed]
63. Blengini, G.; Blagoeva, D.; Dewulf, J.; Torres De Matos, C.; Nita, V.; Vidal-Legaz, B.; Latunussa, C.; Kayam, Y.; Talens Peiró, L.; Baranzelli, C.; et al. *Assessment of the Methodology for Establishing the EU List of Critical Raw Materials*; European Commission: Brussels, Belgium, 2017; ISBN 9789279696114.
64. Nassar, N.T.; Fortier, S.M. *Methodology and Technical Input for the 2021 Review and Revision of the U.S.*; Unites States Geological Survey: Reston, VA, USA, 2021.
65. Malinauskienė, M.; Kliopova, I.; Hugi, C.; Staniškis, J.K. Geostrategic Supply Risk and Economic Importance as Drivers for Implementation of Industrial Ecology Measures in a Nitrogen Fertilizer Production Company. *J. Ind. Ecol.* **2018**, *22*, 422–433. [CrossRef]
66. Zhou, Y.; Li, J.; Wang, G.; Chen, S.; Xing, W.; Li, T. Assessing the short-to medium-term supply risks of clean energy minerals for China. *J. Clean. Prod.* **2019**, *215*, 217–225. [CrossRef]
67. Zhou, Y.; Li, J.; Rechberger, H.; Wang, G.; Chen, S.; Xing, W.; Li, P. Dynamic criticality of by-products used in thin-film photovoltaic technologies by 2050. *J. Clean. Prod.* **2020**, *263*, 121599. [CrossRef]
68. Pfleger, P.; Lichtblau, K.; Bardt, H.; Bertenrath, R. *Rohstoffsituation der Bayerischen Wirtschaft*; Vereinigung der Bayerischen Wirtschaft: München, Germany, 2015.
69. Apple. *Material Impact Profiles*; Apple Inc.: Los Altos, CA, USA, 2019.
70. Althaf, S.; Babbitt, C.W. Disruption risks to material supply chains in the electronics sector. *Resour. Conserv. Recycl.* **2020**, *167*, 105248. [CrossRef]
71. Bach, V.; Finogenova, N.; Berger, M.; Winter, L.; Finkbeiner, M. Enhancing the assessment of critical resource use at the country level with the SCARCE method—Case study of Germany. *Resour. Policy* **2017**, *53*, 283–299. [CrossRef]
72. Helbig, C.; Bradshaw, A.M.; Wietschel, L.; Thorenz, A.; Tuma, A. Supply risks associated with lithium-ion battery materials. *J. Clean. Prod.* **2018**, *172*, 274–286. [CrossRef]
73. Erdmann, L.; Graedel, T.E. Criticality of Non-Fuel Minerals: A Review of Major Approaches and Analyses. *Environ. Sci. Technol.* **2011**, *45*, 7620–7630. [CrossRef]
74. Adibi, N.; Lafhaj, Z.; Yehya, M.; Payet, J. Global Resource Indicator for life cycle impact assessment: Applied in wind turbine case study. *J. Clean. Prod.* **2017**, *165*, 1517–1528. [CrossRef]
75. Alonso, E.; Gregory, J.; Field, F.R.; Kirchain, R. Material availability and the supply chain: Risks, effects, and responses. *Environ. Sci. Technol.* **2007**, *41*, 6649–6656. [CrossRef]
76. Bach, V.; Berger, M.; Finogenova, N.; Finkbeiner, M. Assessing the availability of terrestrial Biotic Materials in Product Systems (BIRD). *Sustainability* **2017**, *9*, 137. [CrossRef]
77. Bastein, T.; Rietveld, E. *Materials in the Dutch Economy: A Vulnerability Assessment*; TNO: Delft, The Netherlands, 2015.
78. Bauer, D.; Diamond, D.; Li, J.; McKittrick, M.; Sandalow, D.; Telleen, P.U.S. *Department of Energy Critical Materials Strategy*; Diane Publishing: Collingdale, PA, USA, 2011.
79. Beylot, A.; Villeneuve, J. Assessing the national economic importance of metals: An Input-Output approach to the case of copper in France. *Resour. Policy* **2015**, *44*, 161–165. [CrossRef]
80. Brown, T. Measurement of mineral supply diversity and its importance in assessing risk and criticality. *Resour. Policy* **2018**, *58*, 202–218. [CrossRef]
81. Buchert, M.; Schüler, D.; Bleher, D. *Critical Metals for Future Sustainable Technologies and Their Recycling Potential*; United Nations Environment Programme: Berlin, Germany, 2009.
82. Calvo, G.; Valero, A.; Valero, A. Thermodynamic Approach to Evaluate the Criticality of Raw Materials and Its Application through a Material Flow Analysis in Europe. *J. Ind. Ecol.* **2018**, *22*, 839–852. [CrossRef]
83. Cimprich, A.; Young, S.B.; Helbig, C.; Gemechu, E.D.; Thorenz, A.; Tuma, A.; Sonnemann, G. Extension of geopolitical supply risk methodology: Characterization model applied to conventional and electric vehicles. *J. Clean. Prod.* **2017**, *162*, 754–763. [CrossRef]
84. Cimprich, A.; Karim, K.S.; Young, S.B. Extending the geopolitical supply risk method: Material "substitutability" indicators applied to electric vehicles and dental X-ray equipment. *Int. J. Life Cycle Assess.* **2018**, *23*, 2024–2042. [CrossRef]
85. Coulomb, R.; Dietz, S.; Godunova, M.; Nielsen, T.B. Critical Minerals Today and in 2030: An Analysis for OECD Countries. *OECD Environ. Work. Pap.* **2015**, *91*, 49. [CrossRef]
86. Daw, G. Security of mineral resources: A new framework for quantitative assessment of criticality. *Resour. Policy* **2017**, *53*, 173–189. [CrossRef]

87. Frenzel, M.; Mikolajczak, C.; Reuter, M.A.; Gutzmer, J. Quantifying the relative availability of high-tech by-product metals—The cases of gallium, germanium and indium. *Resour. Policy* **2017**, *52*, 327–335. [CrossRef]
88. Fu, X.; Polli, A.; Olivetti, E. High-Resolution Insight into Materials Criticality: Quantifying Risk for By-Product Metals from Primary Production. *J. Ind. Ecol.* **2019**, *23*, 452–465. [CrossRef]
89. Glöser-Chahoud, S.; Tercero Espinoza, L.A.; Walz, R.; Faulstich, M. Taking the Step towards a More Dynamic View on Raw Material Criticality: An Indicator Based Analysis for Germany and Japan. *Resources* **2016**, *5*, 45. [CrossRef]
90. Goddin, J.R.J. Identifying Supply Chain Risks for Critical and Strategic Materials. In *Critical Materials-Underlying Causes and Sustainable Mitigation Strategies*; World Scientific Publishing Co. Pte. Ltd.: Singapore, 2019; pp. 117–150.
91. Graedel, T.E.; Harper, E.M.; Nassar, N.T.; Nuss, P.; Reck, B.K. Criticality of metals and metalloids. *Proc. Natl. Acad. Sci. USA* **2015**, *112*, 4257–4262. [CrossRef]
92. Hatayama, H.; Tahara, K. Criticality Assessment of Metals for Japan's Resource Strategy. *Mater. Trans.* **2015**, *56*, 229–235. [CrossRef]
93. Ioannidou, D.; Pommier, R.; Habert, G.; Sonnemann, G. Evaluating the risks in the construction wood product system through a criticality assessment framework. *Resour. Conserv. Recycl.* **2019**, *146*, 68–76. [CrossRef]
94. Kim, J.; Lee, J.; Kim, B.; Kim, J. Raw material criticality assessment with weighted indicators: An application of fuzzy analytic hierarchy process. *Resour. Policy* **2019**, *60*, 225–233. [CrossRef]
95. Kosmol, J.; Müller, F.; Keßler, H. The Critical Raw Materials Concept: Subjective, Multifactorial and Ever-Developing. In *Factor X*; Lehmann, H., Ed.; Springer: Cham, Germany, 2018; pp. 71–92. ISBN 9783319500799.
96. Martins, F.; Castro, H. Significance ranking method applied to some EU critical raw materials in a circular economy—Priorities for achieving sustainability. *Procedia CIRP* **2019**, *84*, 1059–1062. [CrossRef]
97. Miyamoto, W.; Kosai, S.; Hashimoto, S. Evaluating Metal Criticality for Low-Carbon Power Generation Technologies in Japan. *Minerals* **2019**, *9*, 95. [CrossRef]
98. Moss, R.L.; Tzimas, E.; Willis, P.; Arendorf, J.; Tercero Espinoza, L.A. *Critical Metals in the Path towards the Decarbonisation of the EU Energy Sector*; Publications Office of the European Union: Luxembourg, 2013.
99. Nansai, K.; Nakajima, K.; Kagawa, S.; Kondo, Y.; Shigetomi, Y.; Suh, S. Global Mining Risk Footprint of Critical Metals Necessary for Low-Carbon Technologies: The Case of Neodymium, Cobalt, and Platinum in Japan. *Environ. Sci. Technol.* **2015**, *49*, 2022–2031. [CrossRef]
100. Nansai, K.; Nakajima, K.; Suh, S.; Kagawa, S.; Kondo, Y.; Takayanagi, W.; Shigetomi, Y. The role of primary processing in the supply risks of critical metals. *Econ. Syst. Res.* **2017**, *29*, 1–22. [CrossRef]
101. U.S. National Research Council. *Minerals, Critical Minerals, and the U.S. Economy*; The National Academies Press: Washington, DC, USA, 2008.
102. Parthemore, C. *Elements of Security: Mitigating the Risks of U.S. Dependence on Critical Minerals*; Center for a New American Security: Washington, DC, USA, 2011.
103. Shammugam, S.; Rathgeber, A.; Schlegl, T. Causality between metal prices: Is joint consumption a more important determinant than joint production of main and by-product metals? *Resour. Policy* **2019**, *61*, 49–66. [CrossRef]
104. Simon, B.; Ziemann, S.; Weil, M. Criticality of metals for electrochemical energy storage systems—Development towards a technology specific indicator. *Metall. Res. Technol.* **2014**, *111*, 191–200. [CrossRef]
105. Spörri, A.; Wäger, P. Metal Risk Check. Available online: https://www.metal-risk-check.ch/ (accessed on 2 March 2021).
106. Tuma, A.; Reller, A.; Thorenz, A.; Kolotzek, C.; Helbig, C. *Nachhaltige Ressourcenstrategien in Unternehmen: Identifikation kritischer Rohstoffe und Erarbeitung von Handlungsempfehlungen zur Umsetzung einer Ressourceneffizienten Produktion*; Deutsche Bundesstiftung Umwelt: Augsburg, Germany, 2014.
107. van den Brink, S.; Kleijn, R.; Sprecher, B.; Tukker, A. Identifying supply risks by mapping the cobalt supply chain. *Resour. Conserv. Recycl.* **2020**, *156*, 104743. [CrossRef]
108. Viebahn, P.; Soukup, O.; Samadi, S.; Teubler, J.; Wiesen, K.; Ritthoff, M. Assessing the need for critical minerals to shift the German energy system towards a high proportion of renewables. *Renew. Sustain. Energy Rev.* **2015**, *49*, 655–671. [CrossRef]
109. Wentker, M.; Greenwood, M.; Asaba, M.C.; Leker, J. A raw material criticality and environmental impact assessment of state-of-the-art and post-lithium-ion cathode technologies. *J. Energy Storage* **2019**, *26*, 101022. [CrossRef]
110. Yan, W.; Cao, H.; Zhang, Y.; Ning, P.; Song, Q.; Yang, J.; Sun, Z. Rethinking Chinese supply resilience of critical metals in lithium-ion batteries. *J. Clean. Prod.* **2020**, *256*, 120719. [CrossRef]
111. Yuan, Y.; Yellishetty, M.; Muñoz, M.A.; Northey, S.A. Toward a dynamic evaluation of mineral criticality: Introducing the framework of criticality systems. *J. Ind. Ecol.* **2019**, *23*, 1264–1277. [CrossRef]
112. Zepf, V.; Simmons, J.; Reller, A.; Ashfield, M.; Rennie, C. *Materials Critical to the Energy Industry. An Introduction*, 2nd ed.; BP p.l.c.: London, UK, 2014.

Article

Supply Risk Considerations for the Elements in Nickel-Based Superalloys

Christoph Helbig [1,*], Alex M. Bradshaw [2,3], Andrea Thorenz [1] and Axel Tuma [1]

1. Resource Lab, University of Augsburg, Universitaetsstr. 16, 86159 Augsburg, Germany; andrea.thorenz@mrm.uni-augsburg.de (A.T.); axel.tuma@wiwi.uni-augsburg.de (A.T.)
2. Max Planck Institute for Plasma Physics, Boltzmannstraße 2, 85748 Garching, Germany; alex.bradshaw@ipp.mpg.de
3. Fritz Haber Institute of the Max Planck Society, Faradayweg 4-6, 14195 Berlin, Germany
* Correspondence: christoph.helbig@wiwi.uni-augsburg.de

Received: 20 July 2020; Accepted: 26 August 2020; Published: 31 August 2020

Abstract: Nickel-based superalloys contain various elements which are added in order to make the alloys more resistant to thermal and mechanical stress and to the adverse operating environments in jet engines. In particular, higher combustion temperatures in the gas turbine are important, since they result in higher fuel efficiency and thus in lower CO_2 emissions. In this paper, a semi-quantitative assessment scheme is used to evaluate the relative supply risks associated with elements contained in various Ni-based superalloys: aluminium, titanium, chromium, iron, cobalt, niobium, molybdenum, ruthenium, tantalum, tungsten, and rhenium. Twelve indicators on the elemental level and four aggregation methods are applied in order to obtain the supply risk at the alloy level. The supply risks for the elements rhenium, molybdenum and cobalt are found to be the highest. For three of the aggregation schemes, the spread in supply risk values for the different alloy types (as characterized by chemical composition and the endurance temperature) is generally narrow. The fourth, namely the cost-share' aggregation scheme, gives rise to a broader distribution of supply risk values. This is mainly due to the introduction of rhenium as a component starting with second-generation single crystal alloys. The resulting higher supply risk appears, however, to be acceptable for jet engine applications due to the higher temperatures these alloys can endure.

Keywords: superalloy; rhenium; turbine; supply risk; metal; single-crystal

1. Introduction

Single crystal nickel-based superalloys are state of the art materials for the hot sections of high-pressure turbines that contain the blades, vanes, shrouds and nozzles. They not only withstand the high temperatures generated by fuel combustion in a jet engine, but also endure the extreme mechanical stress. They are also resistant to corrosion [1]. To achieve this result, Ni-based superalloys can contain up to 15 alloying elements, including Al, Ti, Cr, Fe, Co, Nb, Mo, Ru, Ta, W, and Re, often in small quantities. The role of each element depends on the overall composition. As described in detail by Darolia [2], the elements can be added in order to (i) reinforce the solid solution-strengthened gamma (γ) matrix, (ii) form and strengthen the cuboid-shaped gamma prime (γ') precipitates, (iii) form a protective scale and provide for its adhesion, (iv) avoid topologically close-packed phases, (v) minimise the density increase or (vi) increase oxidation resistance and hot-corrosion resistance. The book by Reed [3] provides an overview of the history and properties of superalloys. In general, Ni-based superalloys can be classified into wrought, cast, power-processed, directionally solidified, and single-crystal superalloys; the latter can be further divided into six consecutively numbered "generations" [4].

The casting of aircraft turbine blades consisting of alloy single crystals may be seen as an outstanding achievement of materials technology [5].

The global demand for superalloys is dominated by the aviation industry. Further applications are in gas turbines for power generation and ship turbines [6], but it is the growth of the aviation industry that determines the overall demand. The manufacturer Airbus announced in 2019 that it expects a demand for 39,000 new aircraft over the next two decades, thus doubling the global fleet size from 23,000 to 48,000 aircraft for passenger and freight transport [7]. Although passenger travel activity has dropped sharply in 2020 due to the COVID-19 pandemic, a long-term recovery and a return to rapid growth in the aviation industry are expected.

An aircraft usually has two to four engines, for which the main requirements are thrust, reliability, low noise generation and high fuel efficiency. The turbine is driven by the energy transfer from the hot compressed gases to the rotating blades, after re-direction through static nozzle guide vanes [6]. In particular, the high-pressure turbine blades are mostly single-crystal superalloys, with complicated geometries allowing for continuous cooling of the blades during operation. There is currently no suitable substitute for superalloys in this function, although they may be replaced at some time in the future by ceramic matrix composites, which are expected to be able to endure even higher temperatures [8,9].

A significant property of superalloys is their ability to withstand "creep" which is an irreversible deformation of the alloy occurring after prolonged exposure to heat and mechanical strain. The key material performance parameter for this property is the so-called "endurance", or "creep life" temperature. The latter is the highest temperature at which the alloy can endure creep testing under specified conditions of temperature and pressure. The creep life temperature has increased by about 25–30 °C in each single-crystal generation [2,10]. The majority of single-crystal superalloys at present in use belong to the second and third generations, which are capable of enduring around 1000 °C [2]. Turbine entry temperatures may well be even higher than the endurance temperatures of the blade materials, as a result of special coatings and continuous cooling of the blades. Thanks to the decades-long development of superalloys, in particular at companies like General Electric, Pratt & Whitney, and Rolls-Royce, turbines operate today at substantially higher turbine entry temperatures of about 1500 °C and therefore higher thermodynamic efficiencies and reduced fuel consumption than a few decades ago [5,11,12]. Roughly speaking, a 30 °C higher engine temperature can increase the efficiency of a jet turbine by up to 0.5%, with the potential to reduce fuel costs by about 20,000 USD per year per engine [6]. Higher engine temperatures played an important role in reducing average fuel burn in new aircraft by 45% from 1968 to 2014 [13]. New commercial jet aircraft in the 1970s used to have an average fuel burn of more than 40 g per passenger-km. In the 2010s, the fuel burn, which is directly linked to greenhouse gas emissions, has been reduced to about 26 grams fuel per passenger-km. Despite these achievements, the aviation industry is still at risk of falling behind its own fuel efficiency goals [13].

Despite their even higher endurance temperature, the steps to the fourth, fifth and sixth generations have not been taken, or perhaps, have not yet been taken. According to Schafrik [14] and Pollock [15], this reluctance on the part of turbine design engineers is due to the perception that metals of very low abundance in the Earth's crust such as rhenium, will soon become more difficult to extract and, as a result, noticeably depleted, with concomitant steep price rises. Rhenium, for example, is mostly recovered as a by-product from molybdenum concentrates obtained in turn from copper porphyry deposits. Its crustal abundance is estimated to lie between 0.2 and 2 ppb [16]. Apart from superalloys the other major use of Re is as a component of bimetallic petroleum-reforming platinum catalysts. Following actual price increases for rhenium of up to a factor five in the first decade of this century (see Millensiffer et al. [16] for a figure showing this curve), turbine manufacturers began to take notice. One of the measures taken by General Electric, for example, has been the development of a new low-Re superalloy René N515 [15] with considerably less rhenium (1.2%wt Re) and with similar mechanical properties to the second generation René N5 [2]. The response of GE to perceived shortages of rhenium by minimizing the amount of critical metals in superalloys was described by Griffin and colleagues [17] as an example of successful company-level management strategies for combating raw material criticality. General Electric's researchers have reported several times on the strategy of the corporation concerning critical raw materials, in particular rhenium, in reports and scientific articles [18–20]. Superalloy producer Cannon-Muskegon has also introduced new alloys containing low Re or even no Re [21], whereas Pratt & Whitney appear to have a secure Re supply with long-term delivery contracts [22]. Darolia [2] also stresses the poor environmental properties and higher densities of the fourth to sixth-generation alloys, which in addition to the higher costs, are additional reasons for their rejection by turbine designers. For aerospace applications, weight is critical to fuel consumption and therefore dense alloys are also a disadvantage, in particular for the rapidly rotating blades.

The present paper deals with semi-quantitative estimates of the comparative supply risks associated with superalloys, whereby one aspect, namely the rhenium component, is of particular interest. Despite a possibly increased future use of recycled material, an increase in demand for rhenium would have a strong effect on the market price. Considerations of supply risks and, in a broader sense, raw material criticality on a technology-level have previously been assessed, for example, for thin-film photovoltaic cells [23,24], lithium-ion battery materials [25], steel [23], the Ni-based second generation single-crystal superalloy CMSX-4 [26], or bulk metallic glasses [27]. The supply risks associated with superalloys are compared based on the average chemical composition of various Ni-based superalloy types. These are the, mostly older, polycrystalline alloy types, "wrought", "powder-processed", "conventionally cast" and "directionally solidified", as well as six generations of single-crystal alloys and a group of newly developed low Re-containing single-crystal alloys. The next section describes the characteristics of the superalloy types in terms of the constituent elements, the contribution of these metals to the raw material costs of the alloy, and the endurance temperatures of superalloys. The method section briefly summarizes the supply risk approach used for technology-level assessments. The results and discussions section shows the supply risk scores on the elemental level (compared with raw material prices) and their aggregation to give the final scores at the alloy level (compared with endurance temperatures). The article ends with some brief conclusions.

2. Characteristics of Ni-Based Superalloy Types

Before commencing with the supply risk analysis, it is instructive to look briefly at the list of alloying elements and to note their function and properties. The selection of the superalloy types for the assessment in the present paper results in a list of eleven alloying elements in addition to Ni itself: Al, Ti, Cr, Fe, Co, Nb, Mo, Ru, Ta, W, and Re. Elements with lower concentrations, normally less than 0.5%wt, are not considered. All 11 of the above alloying elements are added either to strengthen the γ matrix or to promote the formation of, and strengthen, the γ' precipitates [2]. Al and Cr provide resistance to corrosion by forming a protective oxide layer. Re and Ru improve creep properties. Ru also has a positive effect on the high temperature rupture strength. The list of these observations is long [2], but in some instances the addition of certain elements can also have a concentration-dependent adversary effect.

Figure 1 shows the average density of superalloy types and their average chemical composition, as a compilation of the literature data. The values are averages for each of the eleven superalloy types, based on up to seven representative alloys already discussed in reviews on superalloy materials [2–4,10,28,29]. The sixth generation of single-crystal alloys is an exception, because the alloy called TMS-238 is the only of this type. The chemical compositions as well as the densities for the individual alloys can be found in the Supplementary Material (Table S1).

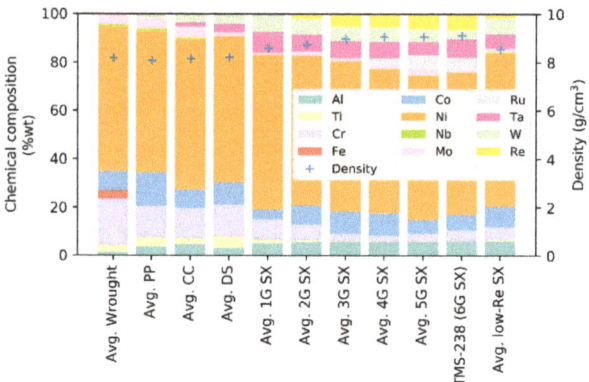

Figure 1. Average chemical composition of each superalloy type (left-hand scale) and its density (right-hand scale). PP: Powder-Processed. CC: Conventionally Cast. DS: Directionally Solidified. SX: Single-Crystal. Data are in the Supplementary Material (Tables S1 and S4) [2–4,10,28].

In all superalloy types considered here, nickel makes up more than half of the weight. The wrought alloy Inconel 718 is the only nickel-iron-based superalloy in the selection. Consequently, iron and niobium only appear in the average for wrought superalloys. Rhenium and ruthenium were introduced in the second and fourth generations, respectively, of single-crystal superalloys. Typical second-generation single-crystal alloys contain about 3%wt rhenium. This concentration was subsequently increased to about 6%wt in the third generation. The fourth generation is characterized by small additions of Ru which were increased in the fifth generation to about 5%wt. For the sixth generation a further optimization of ingredients took place, in particular to provide increased oxidation resistance [30]. The material content for Al, Co, Ta, and W is rather stable over time. Ti, and Mo are only used in small quantities. The chromium content of average superalloy types decreased throughout the single-crystal generations, but, more recently, has increased again in the sixth-generation alloy TMS-238 and in the new, low Re-containing superalloys.

Densities of superalloys range from 8.2 to 9.2 grams per cubic centimeter. There has been a progressive shift to denser materials in each single-crystal generation. On the other hand, the development of the new, low Re-containing alloys has had the effect of reducing the density [2]. The Supplementary Material (Table S4) gives an overview of the superalloys used to calculate the average composition for each type of superalloy and of the data sources for mass-share and density.

Figure 2 shows as a histogram (scale on the left) the specific raw material costs of the superalloy types per unit volume of superalloy. For this diagram, the mass content of each (average) alloy from Figure 1 is multiplied by the specific material costs of each element. Raw material prices are averaged for the year 2015 from trading-day specific market data [31,32] and are listed in the Supplementary Material (Table S5) as well as later in Figure 3. The raw material costs of the wrought, powder-processed, conventionally cast and directionally solidified alloy types are largely determined by the nickel values. The total raw material price, for example for wrought alloys, is therefore comparatively low at about 100 USD per liter of volume. While nickel is still the main component in terms of mass for the single-crystal superalloys, it is responsible for only a small share of the material costs. Starting with the second generation single-crystal alloys, rhenium raw material prices are the main factor in the alloy material costs. Considering single-crystal superalloys of the second and third generations, rhenium gives rise to the third highest material costs (after nickel and ruthenium) in the whole jet engine, including the fans, compressors, combustors and low-pressure turbines. The addition of ruthenium, starting with the fourth generation, has further increased the specific material costs. TMS-238 in the sixth generation has 60% of its raw material costs determined by rhenium and 30% by ruthenium, with all elements in total costing about 2400 USD per liter. All the other elements contained within make up less than 10% of the total raw material costs of the superalloy. However, single crystal superalloys of the fourth generation and beyond have so far not been used in commercial aircraft. Instead, there has been a considerable research effort in newly developed low-Re superalloys [33], which also do not contain ruthenium. Figure 2 shows that material costs are lower than for the second or third single-crystal generation, but the alloys cannot compete with the thermal endurance of the fourth to sixth generations.

Raw material costs (USD/cm³)
1% creep life at 1000h/137MPa (°C)
Al
Ti
Cr
Fe
Co
Ni
Temp.
Nb
Mo
Ru
Ta
W
Re
Avg. Wrought
Avg. PP
Avg. CC
Avg. DS
Avg. 1G SX
Avg. 2G SX
Avg. 3G SX
Avg. 4G SX
Avg. 5G SX
TMS-238 (6G SX)
Avg. low-Re SX

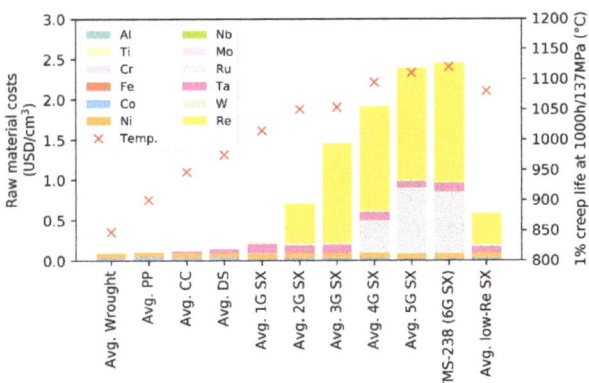

Figure 2. Average raw material costs (left-hand scale) and 1% creep life temperature at 1000 h and 137 MPa of each superalloy type (right-hand scale). PP: Powder-Processed. CC: Conventionally Cast. DS: Directionally Solidified. SX: Single-Crystal. Data are in the Supplementary Material (Tables S1 and S5) [2–4,10,28].

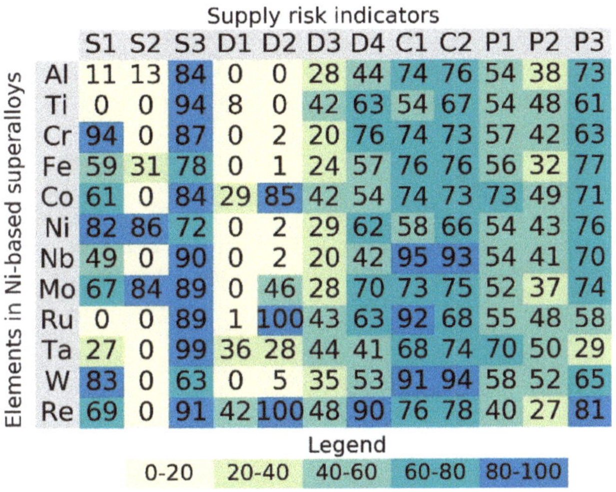

Supply risk indicators
Elements in Ni-based superalloys
S1 S2 S3 D1 D2 D3 D4 C1 C2 P1 P2 P3
Al 11 13 84 0 0 28 44 74 76 54 38 73
Ti 0 0 94 8 0 42 63 54 67 54 48 61
Cr 94 0 87 0 2 20 76 74 73 57 42 63
Fe 59 31 78 0 1 24 57 76 76 56 32 77
Co 61 0 84 29 85 42 54 74 73 73 49 71
Ni 82 86 72 0 2 29 62 58 66 54 43 76
Nb 49 0 90 0 2 20 42 95 93 54 41 70
Mo 67 84 89 0 46 28 70 73 75 52 37 74
Ru 0 0 89 1 100 43 63 92 68 55 48 58
Ta 27 0 99 36 28 44 41 68 74 70 50 29
W 83 0 63 0 5 35 53 91 94 58 52 65
Re 69 0 91 42 100 48 90 76 78 40 27 81
Legend
0-20 20-40 40-60 60-80 80-100

Figure 3. Supply risk values for all twelve indicators and all twelve elements after normalization. S1: Static Reach Reserves. S2: Static Reach Resources. S3: End-of-Life Recycling Input Rate D1: Future Technology Demand. D2: By-Product Dependence. D3: Sector Competition. D4: Substitutability. C1: Country Concentration. C2: Company Concentration. P1: Political Stability (WGI-PV). P2: Policy Perception (PPI). P3: Regulation (HDI).

A typical high-pressure single-crystal turbine blade of the second or third generation contains about 15 g of rhenium and weighs in total about 300 g. In a Rolls-Royce Trent XWB jet engine, there are 68 such single-crystal turbine blades. A single-crystal blade is operational in a jet engine for approximately 25,000 h before overhaul [1]. Therefore, about 20 kg of Ni-based single-crystal components are contained in such an engine and need to be replaced at least once throughout the lifetime of the aircraft. Given the raw material prices of 2015, this adds up to up to 3000 USD raw material value for the Ni-based single crystal superalloys in each jet engine if, e.g., a third-generation alloy is used for the turbine blades. There are likely to be more superalloys used in the engine for vanes and nozzles; these are exposed to the same thermal and environmental stress as the blades, but less mechanical stress. For example, Pratt & Whitney's new engine family, the PW1000G, is estimated to contain in total over 5 kg of rhenium, which sums up to raw material costs of 12,000 USD per engine [34].

Figure 2 additionally shows the gradually increasing average creep life temperatures which the various alloy types can endure (data points, scale on the right) [2]. "Creep life" tests can be carried out for different mechanical stress (higher stress leads to lower creep life temperature), for different durations (longer time leads to lower creep life temperature) or at different creep tolerance levels (higher creep tolerance leads to higher creep life temperature). The figure shows the average of the estimated maximum creep life temperature for which the superalloy shows a maximum of 1% deformation ("creep") after a 1000 h test duration and 137 MPa mechanical stress as a compilation of the literature data. The focus on these latter conditions enables an easier comparison of alloy types; in the literature there appears to be no set of standard conditions for performing such tests.

3. Supply Risk Assessment Method

The evaluation method used in this article to assess the supply risks associated with Ni-based single-crystal superalloys follows the approach presented in detail in previous articles by the authors [24,25]. The present description of the methodology thus focusses on the essential features of the evaluation method and the decisions to be made that are specific to the case of the assessment of superalloys. The method is based on the Augsburg method of criticality assessment [35], but with a small modification concerning the sector competition index [36], which has recently been introduced. Firstly, it evaluates the relative supply risk of twelve elements contained in various Ni-based superalloys (Al, Ti, Cr, Fe, Co, Ni, Nb, Mo, Ru, Ta, W, and Re). Secondly, it goes on to sum the supply risk "scores" at the alloy level for the various types of superalloy developed over the past few decades (wrought, powder-processed, conventionally cast, and directionally-solidified alloys; first to fifth generation single-crystal, TMS-238 as the only sixth-generation single-crystal, and low-Re single-crystal alloys). Excluded from the evaluation are alloying elements usually present in the superalloys with a mass-share of less than 0.5%wt, such as B, C, Y, Zr, or Hf. Even these minor constituents may be important for the material properties [2], but their influence on the supply risk assessment would be negligible (see also Section 4.5).

The relative supply risk for the elements is divided into four categories: (i) risk of supply reduction, (ii) risk of demand increase, (iii) market concentration risk and (iv) political risk. The risk categories each contain two to four indicators. The Supplementary Material (Table S2) contains more details of each indicator as well as its application and normalization onto a common scale of 0 (lowest supply risk) to 100 (highest supply risk). These "final" numbers are to be interpreted as relative supply risk scores, i.e., they are only to be compared with other supply risk scores derived in the context of this article. They are not estimates of the absolute likelihood of supply being unable to meet demand within a specific risk scenario.

The weighting of the twelve indicators differs from previous articles [24,25] insofar as each category is weighted with 25% of the total score, and all indicators within one category are weighted equally. Therefore, each of the four indicators in the category "demand increase risk" determines 6.3% of the final supply risk score of each element, and each of the two indicators in the category "market concentration risk" is weighted with 12.5%. We refrained from carrying out a sector-specific analytic hierarchy process similar to that carried out in previous studies, because those results showed that essentially the same conclusions would have been drawn, if equal weighting of the indicators had been applied [24,25]. Table 1 gives an overview of the twelve supply risk indicators used in the evaluation and their respective weightings.

Table 1. The supply risk indicators considered in this article and their weightings. Additional information on each indicator is available in the Supplementary Material (Tables S1 and S2) [37]. τ_{R1}: Static reach of reserves in years; τ_{R2}: Static reach of resources in years; *EoLRIR*: End-of-life recycling input rate in percent; $\delta_{t,t'}$: Annual growth factor from future technology demand; *SCI*: Sector Competition Index in points; *HHI*: Herfindahl–Hirschman-Index; *WGI*: Worldwide Governance Indicators Political Stability and Absence of Violence; *PPI*: Policy Perception Index; *HDI*: Human Development Index.

Category	ID	Indicator	Normalization	Weightings
Risk of Supply Reduction	S1	Static Reach Reserves	$S1 = 100 - 0.2\tau_{R1} - 0.008\tau_{R1}^2$	1/12 = 8.3%
	S2	Static Reach Resources	$S2 = 100 - 0.1\tau_{R2} - 0.002\tau_{R2}^2$	1/12 = 8.3%
	S3	End-of-Life Recycling Input Rate	$S3 = 100 - EoLRIR$	1/12 = 8.3%
Risk of Demand Increase	D1	Future Technology Demand	$D1 = 1000 \cdot \delta_{t,t'}$	1/16 = 6.3%
	D2	By-Product Dependence	$D2 = 100 \cdot Companionality$	1/16 = 6.3%
	D3	Sector Competition	$D3 = SCI$	1/16 = 6.3%
	D4	Substitutability	$D4 = Substitutability$	1/16 = 6.3%
Market Concentration Risk	C1	Country Concentration	$C1 = 21.64 \ln HHI_{country} - 99.31$	1/8 = 12.5%
	C2	Company Concentration	$C2 = 15.81 \ln HHI_{company} - 45.62$	1/8 = 12.5%
Political Risk	P1	Political Stability (WGI-PV)	$P1 = 20 \cdot (2.5 - WGI)$	1/12 = 8.3%
	P2	Policy Perception (PPI)	$P2 = 100 - PPI$	1/12 = 8.3%
	P3	Regulation (HDI)	$P3 = 100 \cdot \frac{HDI - 0.352}{0.949 - 0.352}$	1/12 = 8.3%

The second step of the assessment determines the supply risk score on the alloy level, i.e., for each of the superalloy types, by aggregating the results for the individual elements. The results are displayed using the four different possibilities for aggregation: the simple arithmetic mean (Equation (1)), the arithmetic mean with mass-share weighting (Equation (2)), the arithmetic mean with cost-share weighting (Equation (3)) and the "maximum" approach (Equation (4)). For the simple arithmetic mean, each element has the same weighting in the calculation. Mass-share weighting considers each element according to the contribution of its mass; cost-share weighting considers both its mass and the raw material costs. The maximum method considers only the element with the highest supply risk score.

$$SR_{mean} = \frac{\sum_{i \in Alloy} SR_i}{\sum_{i \in Alloy} 1} \quad (1)$$

$$SR_{mass} = \sum_i m_i SR_i \quad (2)$$

$$SR_{cost} = \sum_i p_i m_i SR_i \quad (3)$$

$$SR_{max} = \max_{i \in Alloy} SR_i \quad (4)$$

This supply risk assessment scheme is applied with the ultimate aim of comparing the results on the alloy level with the key technical performance parameter for superalloys, namely, the endurance temperature. Given comparable density, similar environmental properties and roughly the same prices, alloy types have a competitive advantage if they can endure higher temperatures at similar supply risks, or if they show similar endurance temperatures at substantially lower levels of supply risk. If, however, higher endurance temperatures come at the cost of higher levels of supply risk, a trade-off situation pertains, and further discussion is required.

4. Results and Discussions

4.1. Supply Risk Data

As explained in the previous section, the supply risk assessment starts with the determination of the values for all twelve indicators for each of the twelve metals under consideration. With the exception of Co, Ru, and Re (see below), the metals are mined in their own right, not as by-products. Mining production is often reported in terms of the tonnage of the corresponding mineral or ore: Al is mined as bauxite, Ti as ilmenite or rutile, and Fe as iron ore. Table 2 gives a summary of the indicator values, or supply risk scores, in the units as calculated. More details can be found in the Supplementary Material (Table S2).

The static reach of the reserves of the twelve elements ranges from values of 35 years for tungsten and 36 years for nickel to about a thousand years for ruthenium. Nickel also has the lowest value for the static reach of the resources with 60 years. Cobalt has values of over 1000 years. Nb, Ru, and Ta have values of at least 200 years without a specific figure being given for the quantity of the resources [38]. Static reaches are interpreted as a measure of the market pressure for further mineral prospecting and subsequent mining activity [25]. End-of-life recycling input rates are highest for tungsten with 37%, and lowest for tantalum with only 1% [39].

Among the twelve elements evaluated, future technology demand is expected to be particularly important for Re, Ta, and Co. It is expected that there will be 150%, 120%, and 90% additional demand, respectively, for these three metals from future technologies in 2035, compared to production in 2013 [40]. For Al, Cr, Fe, Ni, Mo, and W, there is no additional demand expected from rapidly expanding future technologies. Re and Ru are only produced as by-products [41]. Rhenium is derived mainly as a by-product in molybdenum mining, with the company MolyMet in Chile being the main producer. Ruthenium is a platinum group metal and can only be separated in refiners for platinum or palladium. Al, Ti, Cr, Fe, Ni, and W are almost entirely main mining products. Sector competition is less of an issue for the metals contained in the superalloys. Rhenium has the highest value with 48 points, because it is also a component of a reforming catalyst used in the petrochemical industry. Lowest sector competition values are observed for Cr, Nb, and Fe with 20–24 points [36]. These are metals which are used mainly in the steel and steel alloying industry with a comparatively lower added value for each use. So-called substitutability is mostly an issue for Re, Cr, and Mo, for which there are hardly any other possible materials. For Ta, Nb, and Al, in contrast, substitutes are available; they are characterised by values of less than 50 points [42].

Market concentration is measured at the company level as well as at the national level. On the Herfindahl–Hirschman-Index (HHI) scale ranging from 0 to 10,000 [43,44], the country-based concentration of production is low for Ti and Ni with values below HHI 1500. The highest country concentrations are obtained for Nb, Ru and W with values above HHI 6000. Niobium is mainly produced in Brazil, ruthenium in South Africa and tungsten in China [38]. W and Nb also have the highest company concentrations with HHI values above 6000 [45]. Low company concentrations are observed for Ni, Ti, and Ru [45].

Table 2. Compilation of supply risk indicators on the elemental level before normalization. For an explanation of the indicators and further information on assumptions concerning the data, see Supplementary Material (Table S2). Data sources: [36,38–42,45–48]. ⊕: Higher figures indicate higher risk. ⊖: Lower figures indicate higher risk.

Indicator	Dimension	Risk	Al	Ti	Cr	Fe	Co	Ni	Nb	Mo	Ru	Ta	W	Re
S1	years	⊖	94	107	16	60	58	36	68	52	1029	84	35	50
S2	years	⊖	184	258	384	161	1201	60	>200	68	>200	>200	306	221
S3	%	⊖	16	6	13	22	16	27	10	11	11	1	37	9
D1	%	⊕	0	20	0	0	90	0	2	0	3	120	0	150
D2	%	⊕	0	0	2	1	85	2	2	46	100	28	5	100
D3	qualitative	⊖	28	42	20	24	42	29	20	28	43	44	35	48
D4	qualitative	⊖	44	63	76	57	54	62	42	70	63	41	53	90
C1	HHI	⊕	3057	1221	3033	3321	3141	1450	8266	2889	6958	2346	6679	3374
C2	HHI	⊕	2221	1317	1854	2269	1902	1191	6441	2183	1373	2002	6920	2533
P1	qualitative	⊖	−0.24	−0.21	−0.36	−0.33	−1.20	−0.21	−0.20	−0.15	−0.29	−1.04	−0.44	−0.44
P2	qualitative	⊖	61	51	58	68	50	56	59	62	51	49	47	73
P3	qualitative	⊕	0.79	0.72	0.73	0.82	0.78	0.81	0.79	0.79	0.70	0.53	0.75	0.84

Political risk is determined by an evaluation of political stability in producing countries according to three distinct categories: stability, the perception of policy towards mining and the possibility of stronger regulation. The producing countries for all twelve metals are, on average, estimated as rather unstable with negative values of the "Political Stability and Absence of Violence/Terrorism" (WGI-PV) indicator of the Worldwide Governance Indicators [46]. Particularly, the high share of production of Co and Ta in the Democratic Republic of Congo is of concern. Cobalt and tantalum are also the metals with the lowest values on the Policy Perception Index. In contrast, the high share of production in Chile results in a high Policy Perception Index of 73 points for rhenium [47]. The producing countries have the highest Human Development Index for rhenium with 0.84 and nickel with 0.81. Tantalum-producing countries can be considered least "developed" with an average value of only 0.53 [48].

4.2. Normalization and Weighting

As described above, the next step is to normalize the values of each indicator to a common scale and then to apply the weighting of the indicators. The supply risk scores for the twelve indicators for each of the twelve elements following normalization are shown in Figure 3 (values are also given in Table S2 in the Supplementary Material). High values (up to 100) indicate a high supply risk. The highest average supply risk scores of the 12 elements are observed for the end-of-life recycling input rate (on average 85 points), the company concentration (77) and the country concentration (76). In contrast, future technology demand (10 points) and static reach of resources (18) have the lowest average supply risk scores. The spread of the supply risk scores is lowest for the risk emerging from policy perception (standard deviation of 7.5 points with a range of 26 points). The by-product dependence with 39 points standard deviation and scores ranging from 0 to 100 has the highest spread. The average supply risk score for all categories and all twelve elements is 54 points.

4.3. Supply Risk on the Elemental Level

Following normalization and weighting, the aggregation of the indicator values gives the relative supply risk scores for each of the twelve elements (Al, Ti, Cr, Fe, Co, Ni, Nb, Mo, Ru, Ta, W, and Re), as shown in Figure 4, where they are plotted against the raw material price. This semi-log plot allows us to check if the supply risks are already sufficiently taken into account by the commodity prices. This would be the case, if there was a high coefficient of determination (the R^2 value) close to 1 in the statistical analysis of the linear trend between the logarithm of the price and the supply risk. From Figure 4 we note that rhenium (63 points), molybdenum (61), and cobalt (60) show the highest aggregated supply risks. In contrast, titanium (44) and aluminium (46) show the lowest supply risks. However, the spread in the aggregated supply risk values is only 20 points on a 0–100 scale for this set of twelve metals. This already tells us that the spread of the aggregated results on the alloy level cannot be larger than 20 points and will, most likely, be considerably narrower.

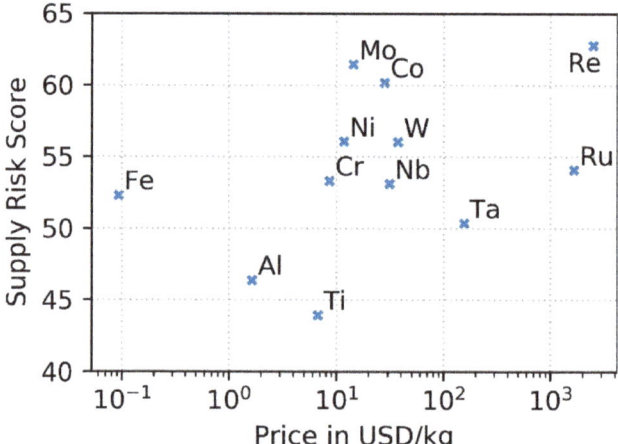

Figure 4. Supply risk score on the elemental level and the raw material price in a semi-logarithmic plot. Values are given in Tables S2 and S3 in the Supplementary Material.

In addition, rhenium happens to be the metal both with the highest price and the highest supply risk. However, the R^2 value of the linear trend calculated from these twelve elements is only 0.18 and, therefore, high supply risks do not necessarily result in high prices. This observation is important, however, because supply risk assessments are intended to be test cases for the likelihood of future supply disruption events, expressed in physical shortages or price increases (not current prices) [49,50].

4.4. Supply Risk on the Alloy Level

In order to compare the results for different superalloys, supply risk scores on the elemental level need to be aggregated to give comparative supply risks on the alloy level (which we have previously also referred to as the "technology" level [24,25]). The results for the different superalloys and the four different aggregation methods in the present work are shown in Figure 5; the exact values of the supply risk scores can be found in Table S6 in the Supplementary Material. All four data sets are plotted against the average approximate 1% creep life temperature for 1000 h and 137 MPa (already introduced in Figure 2) to display the potential trade-off between the thermal properties of the alloy type and the supply risk.

In the case of the simple arithmetic mean (Figure 5A), for which case each element contained has the same weighting, the fifth and sixth single-crystal generations give the highest supply risks with 56 points. Wrought, cast and directionally solidified alloys as well as the Re-free first generation of single-crystal superalloys show a somewhat smaller supply risk of 54 points. The mass-share approach (Figure 5B) results in a strong contribution from the nickel supply risk, so that the differences between the different generations are even smaller with values of 55 or 56 points. Applying the "maximum" approach (Figure 5D) is not very helpful and, at the most, allows us only to differentiate between the Re-containing (63 points) and the Re-free superalloy types (61 points).

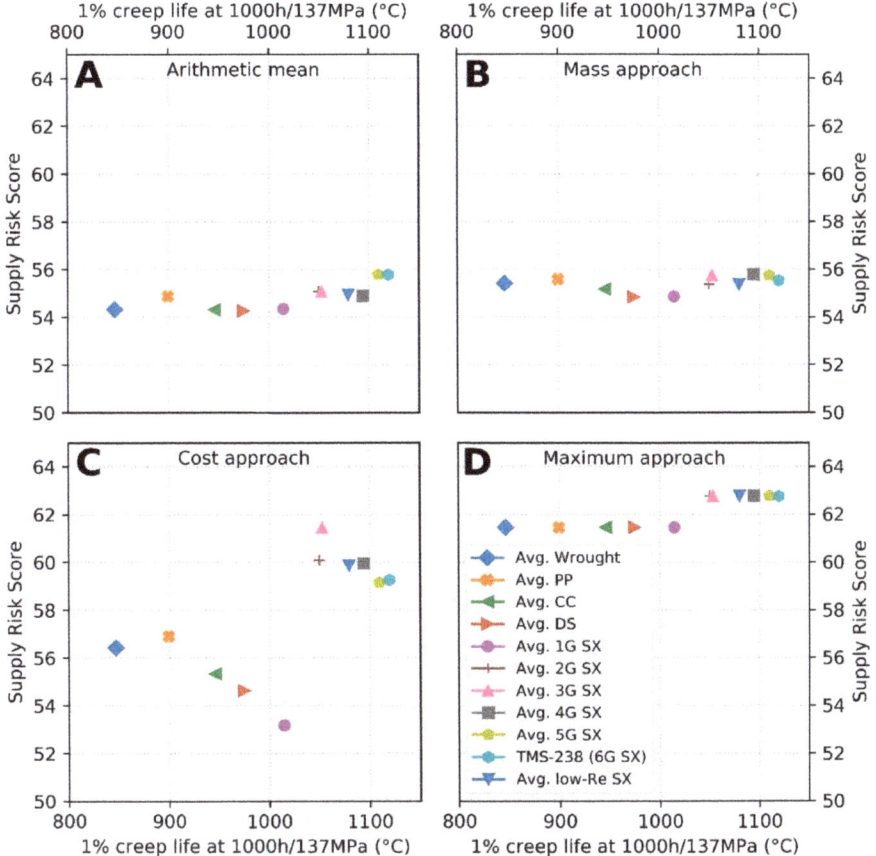

Figure 5. (**A**) Supply risk score and endurance temperature on the technology level, using the arithmetic mean as the aggregation scheme. (**B**) Results using the mass approach. (**C**) Results using the cost approach. (**D**) Results using the maximum approach. Raw data are given in Tables S1 and S6 in the Supplementary Material.

Compared to the thin-film photovoltaic and Li-ion battery materials [24,25,51], the spread in the supply risk scores for superalloys on the technology level is thus small, in particular for the aggregation schemes arithmetic mean, mass-share aggregation and maximum approach. The supply risk values for arithmetic mean and mass share schemes remain close to 55 points with little or no correlation with creep life. This results from the averaging over a large number of alloying elements with similar supply risk values (see Figure 4). Moreover, the list of alloying elements employed in each case (the chemical composition) does actually vary from alloy to alloy, but not strongly. Rhenium, for example, the element with the highest supply risk score, is contained in all alloy types from the second single-crystal generation onwards. As far as the arithmetic mean, mass-share and maximum approaches are concerned, there is no trade-off between creep life temperature and supply security for the alloy types.

On the other hand, the supply risk scores in the cost-share aggregation scheme (Figure 5C) have a substantially larger spread than in the other three schemes, namely 53 to 61 points, largely because of the difference in raw material costs for the alloy types (see Figure 2). (Note that the cost-based approach considers both the mass and the raw material price.) The contribution to raw material costs differs between the alloy types to a much larger extent and therefore the supply risk scores in the cost-share aggregation scheme at the technology level also have a substantially larger spread. The alloys of the single crystal third generation with supply risk scores of 61 points, have the highest supply risk. The lower supply risk for the similarly expensive ruthenium content in the fourth generation leads to a slightly reduced supply risks score of 60 points in this approach, despite a high rhenium content. The Re-free first generation single-crystal superalloys and the directionally solidified superalloys have only 53 and 55 points, respectively. Figure 5C shows that there are two groups of alloy type: The group up to the first single-crystal generation (which contained no rhenium) with lower supply risk in the cost-share aggregation, but also lower creep life temperatures, and all other Re-bearing single-crystal superalloys. Consequently, there is a trade-off between creep life and supply security in this specific perspective.

The results clearly show that the numerical supply risks at the technology level are indeed very similar for the different alloys on the basis of three of the four aggregation procedures. However, there may be factors, in this case raw material costs, which, because of their importance and the time period over which they are relevant, may deserve special attention within the aggregation scheme. From the methodological point of view, the result tells us that we should probably look more closely at the concept of "cost" and how it fits into a more general description of supply risk, as it is applied, for example, in the present paper. From the perspective of the airline or a jet engine manufacturer, the total costs of operation need to be considered. For aircraft and jet engines, the costs of operation are heavily determined by in-flight costs, i.e., fuel consumption. If, as demonstrated for Ni-based superalloys, the risk of cost increases for raw material supply is the main concern in terms of supply risks, then potential extra costs for supply need to be compared with savings potential during the operations phase. We have described above the potential for obtaining such savings from higher creep life temperatures. The higher supply risk score of Re-containing single-crystal superalloys therefore seems acceptable.

The fact that the cost-share perspective is a key factor for the evaluation of supply risks of Ni-based superalloy elements, stresses the importance of rhenium supply for the aviation industry. This includes management of rhenium supply risks throughout the supply chain from molybdenum mining to turbine producers. The industry is apparently aware of these Re supply risks [17], and mitigation strategies range from the development of low-Re superalloys, to recycling efforts, new separation technologies and long-term supply contracts. It should be noted that rhenium is traded mainly over-the-counter instead of on the free market [16]. Almost half of the annual production of about 50 metric tons comes from Chile where one supplier dominates. As we have seen from the indicators used above, there are several factors potentially contributing to supply risk, including the political situation in producer countries, international conflicts, the existence of monopolies or oligopolies, other high-tech applications of the element concerned and, also, "geochemical scarcity". The latter concept covers the possible decline in ore grades, more difficult mining conditions and the increasing demand for energy and/or water. It is also sometimes referred to as "mineral depletion". Generally speaking, mineral depletion is not (yet) a significant factor in the mining industry [52], although attention often focusses on the so-called static reach of the resources, i.e., the ratio of identified global resources to annual production rate.

4.5. Limitations

The use of the adjectives "relative" and "semi-quantitative" for the supply risk assessment scheme applied both here and in previous work [24,25] deserves comment. The numbers obtained on elemental and alloy level are relative supply risks scores and, therefore, should only be compared to scores obtained in this article. Consideration of the list of indicators in Table S2 of the Supplementary Material reveals that some lend themselves quickly and simply to a quantitative treatment. End-of-life recycling rate, substitutability and by-product dependence are cases in point. For other indicators, for example, those assessing market concentration and political risks, there is often no other alternative but to use risk assessments which have, at least in part, a strong subjective component. Hence, our emphasis on the word "semi-quantitative".

When considering further "limitations" of the work, it is necessary not just to consider overarching, "global" problems, but also to look at those cases where the application of the model gives rise to specific difficulties. Firstly, we note that supply risk considerations in the present assessment are based on the elements actually contained in superalloy types, not those just used in the production of the alloys. The indicator data cover the three raw material production stages: mining, smelting and refining (if applicable). Further processing of intermediates or semi-finished products before the manufacturing of Ni-based superalloys is not considered. This is justified by the present focus on the material supply risks, rather than on general supply chain risk assessments [53]. Secondly, the assessment as shown does not cover potential supply risks for elements contained with less than 0.5%wt, such as boron, carbon, yttrium, zirconium or hafnium. These elements would only have small effect on the overall results, in particular when using the mass-share and cost-share aggregation schemes. Moreover, these elements may not always be contained in all of the individual alloys of one superalloy type. Thirdly, semi-quantitative indicator-based supply risk assessments depend on the selection and weighting of the indicators and on whether there is a dynamic assessment [54]. The indicator choice of supply risk has been discussed by Achzet and Helbig [55] and, more recently, for criticality assessments in general by Schrijvers and colleagues [50]. The indicators used here do not constitute a dynamic assessment, but rather a snapshot in time, or "static" assessment [56]. For example, while the static reaches extend at least a few decades into the future, they are calculated from the recent production rate and current estimations of reserves or resources. The data used by the indicator calculations used here are based on the year 2015, whenever available. Unfortunately, the data for the recycling rate and all four indicators in the category "Risk of demand Increase" are not available on an annual basis.

5. Conclusions

Using a previously developed semi-quantitative assessment scheme [24], we have evaluated the supply risks associated with elements contained in average Ni-based superalloy types. Based on the twelve indicators in four supply risk categories, rhenium, molybdenum and cobalt are found to have the highest supply risk scores, titanium and aluminium the lowest. In the aggregations for arithmetic mean, mass-share aggregation and maximum approach, the supply risk scores of the superalloy generations are very similar. Only in the cost-share approach do the single-crystal superalloys from the second generation onwards show a substantially higher supply risk than other alloy types, because of the increased share of rhenium (up to 6%wt). Despite having a reduced Re content, the new low-Re generation is still within the group of higher supply risk alloy types, showing a substantially higher supply risk than first-generation single-crystal or non-single-crystal alloy types.

We conclude, however, that the increased costs and the relatively small increases in the supply risk scores for fourth to sixth generation single crystal superalloys are not so high that these higher generations would not be used at all. Admittedly, the supply risks are higher in the cost-share approach, but alloy composition and the fuel consumption also need to be considered. The higher generations are particularly suited to reduce costs for airlines from fuel consumption. Therefore, in the case of Ni-based superalloys, managing the supply risks of rhenium is more important than avoiding those supply risks. On the company level, these management options include hedging, stockpiling, alternative suppliers, material substitution, material and technology development, and ongoing assessment of material supply risks [17].

Following thin-film photovoltaics and Li-ion battery materials [24,25] this has been the third application of the present supply risk assessment scheme to a potential supply problem in the "hightech" sector. Future applications should also review the choice of indicators, based on recent reviews of state-of-the art in criticality assessments [50] and on upcoming reviews of evidence-based supply risk indicators.

Supplementary Materials: The following are available online at http://www.mdpi.com/2079-9276/9/9/106/s1, Table S1: Alloys info, Table S2: Elements info, Table S3: Elements prices, Table S4: Chemical composition, Table S5: Cost contribution, Table S6: Supply risk alloys.

Author Contributions: Conceptualization, C.H. and A.M.B.; methodology, C.H. and A.T. (Andrea Thorenz); software, C.H.; formal analysis, C.H.; data curation, C.H.; writing—original draft preparation, C.H. and A.M.B.; writing—review and editing, A.T. (Andrea Thorenz) and A.T. (Axel Tuma); visualization, C.H.; supervision, A.T. (Andrea Thorenz) and A.T. (Axel Tuma). All authors have read and agreed to the published version of the manuscript.

Funding: This research received no external funding.

Acknowledgments: We thank Ralph Gilles, Surendra Kumar Makineni, Winfried Petry, Dierk Raabe und Joachim Rösler for very useful discussions on superalloys.

Conflicts of Interest: The authors declare no conflict of interest.

References

1. Langston, L. Each blade a single crystal. *Am. Sci.* **2015**, *103*, 30. [CrossRef]
2. Darolia, R. Development of strong, oxidation and corrosion resistant nickel-based superalloys: Critical review of challenges, progress and prospects. *Int. Mater. Rev.* **2019**, *64*, 355–380. [CrossRef]
3. Reed, R.C. *The Superalloys: Fundamentals and Applications*; Cambridge University Press: Cambridge, UK, 2006; ISBN 9780521859042.
4. Pollock, T.M.; Tin, S. Nickel-based superalloys for advanced turbine engines: Chemistry, microstructure, and properties. *J. Propuls. Power* **2006**, *22*, 361–374. [CrossRef]
5. Langston, L.S. Single-crystal turbine blades earn ASME milestone status. *Mach. Des.* **2018**, *90*, 46–52.
6. *The Jet Engine*; Rolls-Royce Plc: Manchester, UK, 2005.
7. Airbus Airbus Forecasts Need for over 39,000 New Aircraft in the Next 20 Years. Available online: https://www.airbus.com/newsroom/press-releases/en/2019/09/airbus-forecasts-need-for-over-39000-new-aircraft-in-the-next-20-years.html (accessed on 12 February 2020).
8. Schafrik, R.; Sprague, R. Superalloy technology—A perspective on critical innovations for turbine engines. *Key Eng. Mater.* **2008**, *380*, 113–134. [CrossRef]
9. Steibel, J. Ceramic matrix composites taking flight at GE Aviation. *Am. Ceram. Soc. Bull.* **2019**, *98*, 30–33.
10. Long, H.; Mao, S.; Liu, Y.; Zhang, Z.; Han, X. Microstructural and compositional design of Ni-based single crystalline superalloys—A review. *J. Alloy. Compd.* **2018**, *743*, 203–220. [CrossRef]
11. Xu, L.; Bo, S.; Hongde, Y.; Lei, W. Evolution of rolls-royce air-cooled turbine blades and feature analysis. *Procedia Eng.* **2015**, *99*, 1482–1491. [CrossRef]
12. Perepezko, J.H. The hotter the engine, the better. *Science* **2009**, *326*, 1068–1069. [CrossRef]
13. Kharina, S.; Rutherford, D. *Fuel Efficiency Trends for New Commercial Jet Aircraft: 1960 to 2014*; International Council on Clean Transportation: Washington, DC, USA, 2015.
14. Schafrik, R.E. Materials for a non-steady-state world. *Metall. Mater. Trans. B* **2016**, *47*, 1505–1515. [CrossRef]
15. Pollock, T.M. Alloy design for aircraft engines. *Nat. Mater.* **2016**, *15*, 809–815. [CrossRef] [PubMed]
16. Millensiffer, T.A.; Sinclair, D.; Jonasson, I.; Lipmann, A. Rhenium. In *Critical Metals Handbook*; Gunn, G., Ed.; John Wiley & Sons Ltd.: Hoboken, NJ, USA, 2014; pp. 340–360.
17. Griffin, G.; Gaustad, G.; Badami, K. A framework for firm-level critical material supply management and mitigation. *Resour. Policy* **2019**, *60*, 262–276. [CrossRef]
18. Ku, A.Y.; Loudis, J.; Duclos, S.J. The impact of technological innovation on critical materials risk dynamics. *Sustain. Mater. Technol.* **2018**, *15*, 19–26. [CrossRef]
19. Ku, A.Y.; Hung, S. Manage Raw Material Supply Risks. *Chem. Eng. Prog.* **2014**, *110*, 28–35.
20. Duclos, S.J.; Otto, J.P.; Konitzer, D.G. Design in an era of constrained resources: As global competition for materials strains the supply chain, companies must know where a shortage can hurt and then plan around it. *Mech. Eng.* **2010**, *132*, 36–40. [CrossRef]
21. Wahl, J.B.; Harris, K. New single crystal superalloys, CMSX®-8 and CMSX®-7. *Proc. ASME Turbo Expo* **2014**, *6*, 179–188. [CrossRef]
22. Lipmann, A. Pratt & Whitney/Molymet Write Largest Deal in Rhenium History. Available online: https://www.lipmann.co.uk/articles/metal-matters/pratt-whitney-molymet-write-largest-deal-in-rhenium-history/ (accessed on 12 February 2020).

23. Graedel, T.E.; Nuss, P. Employing considerations of criticality in product design. *JOM* **2014**, *66*, 2360–2366. [CrossRef]
24. Helbig, C.; Bradshaw, A.M.; Kolotzek, C.; Thorenz, A.; Tuma, A. Supply risks associated with CdTe and CIGS thin-film photovoltaics. *Appl. Energy* **2016**, *178*, 422–433. [CrossRef]
25. Helbig, C.; Bradshaw, A.M.; Wietschel, L.; Thorenz, A.; Tuma, A. Supply risks associated with lithium-ion battery materials. *J. Clean. Prod.* **2018**, *172*, 274–286. [CrossRef]
26. Goddin, J.R.J. Identifying supply chain risks for critical and strategic materials. In *Critical Materials-Underlying Causes and Sustainable Mitigation Strategies*; World Scientific Publishing Co. Pte. Ltd.: Singapore, 2019; pp. 117–150.
27. Mota, R.M.O.; Graedel, T.E.; Pekarskaya, E.; Schroers, J. Criticality in bulk metallic glass constituent elements. *JOM* **2017**, *69*, 2156–2163. [CrossRef]
28. Sato, A.; Harada, H.; Yeh, A.-C.; Kawagishi, K.; Kobayashi, T.; Koizumi, Y.; Yokokawa, T.; Zhang, J.-X. A 5th generation sc superalloy with balanced high temperature properties and processability. In Proceedings of the Superalloys 2008 (Eleventh International Symposium), Seven Springs, PA, USA, 14–18 September 2008; pp. 131–138.
29. Caron, P.; Khan, T. Evolution of Ni-based superalloys for single crystal gas turbine blade applications. *Aerosp. Sci. Technol.* **1999**, *3*, 513–523. [CrossRef]
30. Kawagishi, K.; Yeh, A.; Yokokawa, T.; Kobayashi, T.; Koizumi, Y.; Harada, H. Development of an oxidation-resistant high-strength sixth-generation single-crystal superalloy TMS-238. In Proceedings of the Superalloys 2012, Seven Springs, PA, USA, 9–13 September 2012; pp. 189–195.
31. Bloomberg. Available online: www.bloomberg.com (accessed on 14 December 2019).
32. Fastmarkets Metal Bulletin. Available online: www.metalbulletin.com (accessed on 14 December 2019).
33. Fink, P.J.; Miller, J.L.; Konitzer, D.G. Rhenium reduction—Alloy design using an economically strategic element. *JOM* **2010**, *62*, 55–57. [CrossRef]
34. Roskill Rhenium Outlook to 2029, 11th Edition. Available online: https://roskill.com/market-report/rhenium/ (accessed on 12 February 2020).
35. Kolotzek, C.; Helbig, C.; Thorenz, A.; Reller, A.; Tuma, A. A company-oriented model for the assessment of raw material supply risks, environmental impact and social implications. *J. Clean. Prod.* **2018**, *176*, 566–580. [CrossRef]
36. Helbig, C.; Kolotzek, C.; Thorenz, A.; Reller, A.; Tuma, A.; Schafnitzel, M.; Krohns, S. Benefits of resource strategy for sustainable materials research and development. *Sustain. Mater. Technol.* **2017**, *12*, 1–8. [CrossRef]
37. Helbig, C. *Metalle im Spannungsfeld technoökonomischen Handelns: Eine Bewertung der Versorgungsrisiken und der dissipativen Verluste mit Methoden der Industrial Ecology*; Universität Augsburg: Augsburg, Germany, 2019.
38. USGS. *Mineral Commodity Summaries 2018*; U.S. Geological Survey: Reston, VA, USA, 2018.
39. Graedel, T.E.; Allwood, J.M.; Birat, J.-P.; Reck, B.K.; Sibley, S.F.; Sonnemann, G.; Buchert, M.; Hagelüken, C. *Recycling Rates of Metals—A Status Report, A Report of the Working Group on the Global Metal Flows to the International Resource Panel*; UNEP: Nairobi, Kenya, 2011.
40. Marscheider-Weidemann, F.; Langkau, S.; Hummen, T.; Erdmann, L.; Tercero Espinoza, L.A.; Angerer, G.; Marwede, M.; Benecke, S. *Rohstoffe für Zukunftstechnologien 2016*; Deutsche Rohstoffagentur (DERA): Berlin, Germany, 2016.
41. Nassar, N.T.; Graedel, T.E.; Harper, E.M. By-product metals are technologically essential but have problematic supply. *Sci. Adv.* **2015**, *1*, e1400180. [CrossRef] [PubMed]
42. Graedel, T.E.; Harper, E.M.; Nassar, N.T.; Reck, B.K. On the materials basis of modern society. *Proc. Natl. Acad. Sci. USA* **2015**, *112*, 6295–6300. [CrossRef] [PubMed]
43. Herfindahl, O.C. *Concentration in the US Steel Industry*; Columbia University: New York, NY, USA, 1950.
44. Hirschman, A.O. *National Power and the Structure of Foreign Trade*; University of California Press: Berkeley, CA, USA, 1980; ISBN 9780520040823.
45. Buchholz, P.; Huy, D.; Liedtke, M.; Schmidt, M. *DERA-Rohstoffliste 2014*; Deutsche Rohstoffagentur (DERA): Berlin, Germany, 2015.
46. Kaufmann, D.; Kraay, A. Worldwide Governance Indicators. Available online: http://info.worldbank.org/governance/wgi/index.aspx (accessed on 1 December 2015).
47. Jackson, T.; Green, K.P. *Annual Survey of Mining Companies 2016*; Fraser Institute: Vancouver, BC, Canada, 2017.
48. UNDP. *Human Development Report 2015*; UNDP: New York, NY, USA, 2015.

49. Frenzel, M.; Kullik, J.; Reuter, M.A.; Gutzmer, J. Raw material 'criticality'—Sense or nonsense? *J. Phys. D Appl. Phys.* **2017**, *50*, 123002. [CrossRef]
50. Schrijvers, D.; Hool, A.; Blengini, G.A.; Chen, W.-Q.; Dewulf, J.; Eggert, R.; van Ellen, L.; Gauss, R.; Goddin, J.; Habib, K.; et al. A review of methods and data to determine raw material criticality. *Resour. Conserv. Recycl.* **2020**, *155*, 104617. [CrossRef]
51. Helbig, C.; Bradshaw, A.M.; Wietschel, L.; Thorenz, A.; Tuma, A. Corrigendum to "Supply risks associated with lithium-ion battery materials". *J. Clean. Prod.* **2019**, *221*, 899–903. [CrossRef]
52. Tilton, J.E. *On Borrowed Time: Assessing the Threat of Mineral Depletion*; Resources for the Future: Washington, DC, USA, 2002.
53. Chopra, S.; Sodhi, M.S. Managing risk to avoid supply-chain breakdown. *MIT Sloan Management Review*, 15 October 2004.
54. Erdmann, L.; Graedel, T.E. Criticality of non-fuel minerals: A review of major approaches and analyses. *Environ. Sci. Technol.* **2011**, *45*, 7620–7630. [CrossRef]
55. Achzet, B.; Helbig, C. How to evaluate raw material supply risks—An overview. *Resour. Policy* **2013**, *38*, 435–447. [CrossRef]
56. Yuan, Y.; Yellishetty, M.; Mudd, G.M.; Muñoz, M.A.; Northey, S.A.; Werner, T.T. Toward dynamic evaluations of materials criticality: A systems framework applied to platinum. *Resour. Conserv. Recycl.* **2020**, *152*, 104532. [CrossRef]

© 2020 by the authors. Licensee MDPI, Basel, Switzerland. This article is an open access article distributed under the terms and conditions of the Creative Commons Attribution (CC BY) license (http://creativecommons.org/licenses/by/4.0/).

Article

The Impact of Metal Mining on Global Water Stress and Regional Carrying Capacities—A GIS-Based Water Impact Assessment

Simon Meißner

Environmental Science Center, University of Augsburg, Universitaetsstr. 1a, 86159 Augsburg, Germany; simon.meissner@wzu.uni-augsburg.de

Citation: Meißner, S. The Impact of Metal Mining on Global Water Stress and Regional Carrying Capacities—A GIS-Based Water Impact Assessment. *Resources* 2021, 10, 120. https://doi.org/10.3390/resources10120120

Academic Editor: Ben McLellan

Received: 23 September 2021
Accepted: 15 November 2021
Published: 26 November 2021

Publisher's Note: MDPI stays neutral with regard to jurisdictional claims in published maps and institutional affiliations.

Copyright: © 2021 by the author. Licensee MDPI, Basel, Switzerland. This article is an open access article distributed under the terms and conditions of the Creative Commons Attribution (CC BY) license (https:// creativecommons.org/licenses/by/ 4.0/).

Abstract: The consumption of freshwater in mining accounts for only a small proportion of the total water use at global and even national scales. However, at regional and local scales, mining may result in significant impacts on freshwater resources, particularly when water consumption surpasses the carrying capacities defined by the amount of available water and also considering environmental water requirements. By applying a geographic information system (GIS), a comprehensive water footprint accounting and water scarcity assessment of bauxite, cobalt, copper, iron, lead, manganese, molybdenum, nickel, uranium and zinc as well as gold, palladium, platinum and silver was conducted to quantify the influence of mining and refining of metal production on regional water availability and water stress. The observation includes the water consumption and impacts on water stress of almost 2800 mining operations at different production stages, e.g., preprocessed ore, concentrate and refined metal. Based on a brief study of mining activities in 147 major river basins, it can be indicated that mining's contribution to regional water stress varies significantly in each basin. While in most regions mining predominantly results in very low water stress, not surpassing 0.1% of the basins' available water, there are also exceptional cases where the natural water availability is completely exceeded by the freshwater consumption of the mining sector during the entire year. Thus, this GIS-based approach provides precise information to deepen the understanding of the global mining industry's influence on regional carrying capacities and water stress.

Keywords: metals; mining; water stress; water scarcity; water footprint accounting; life cycle assessment (LCA); geographic information system (GIS); raw materials criticality assessment

1. Introduction

Driven by growing demand and technological development, the consumption of natural resources has been increasing significantly within recent decades and is still expected to grow in the future. In particular, high-technology applications require a large variety of minerals and metals, which in some cases are referred to as critical raw materials due to increasing concerns about their limited availability and potential supply shortages. Hence, primarily during the last two decades, criticality assessment methods have been developed and constantly evolved to screen mineral commodity markets in order to identify raw materials of concern [1,2]. However, the global expansion of resource extraction, particularly mining and refining of metals, is also characterized by environmental concerns as the mining and refining of technology-relevant metals have significant impacts on ecosystems. For instance, the construction and operation of mining facilities may lead to long-term impacts such as loss of vegetation and faunal habitats, modification of landforms, changes in soil profiles or modifications to surface and subsurface drainage [3]. As a consequence, the latest criticality assessment methods of raw materials have been extended by environmental criteria to determine the ecological impacts of mining activities as well [4]. Besides energy consumption and the release of greenhouse gas emissions, particularly addressing climate change effects, recent studies emphasized the impacts resulting from the water consumption of the mining industry. Water-related impacts of individual mining operations affecting hydrological systems adjacent to the excavation sites are especially of increasing concern as local freshwater sources are essential for providing considerable amounts of freshwater for both human purposes and environmental needs.

Ore mining and ore processing basically consume large quantities of water. For instance, mining operations require large pumping, treating, heating and cooling water systems, which are intense energy consumers as well [5]. On a global and even national scale, however, the consumption of freshwater for mining and refining activities accounts for a small portion of the overall water use. Even in relatively dry, mining-intensive countries such as Australia, Chile and South Africa, mine water consumption accounts for only 2–4.5% of national water demand [6,7]. However, on a local level, significant impacts on freshwater resources can be observed, notably affecting both the quantity and quality of freshwater availability within the entire mining area. For example, acid rock drainage, leaks from tailings, waste rock dumps or direct disposal of tailings into waterways may contaminate surface and groundwater bodies [8–12]. Due to the fact that the global mining industry is increasingly confronted with declining ore grades, the industry is forced to access deposits and ore bodies of lower quality, hence requiring larger volumes of water to be utilized per ton of metal produced [13,14]. As a consequence, mining-related water consumption can put severe strains on local water supplies by competing with other water consumers, especially in areas characterized by significant water scarcity. Furthermore, particularly due to climate change effects, many water-scarce areas as well as many mining operations in particular are expected to be confronted with increasing water stress conditions within the next two decades [15,16]. Consequently, the mining industry will have to address these challenges by intensifying water management activities, taking into account both mining-related water supply risks as well as water shortages affecting the entire mining area that may likely result in future water conflicts [17–20].

Owing to the fact that water is a substantial resource for mining operations [21], the mining sector is a large industrial water user that is growing rapidly all over the world. Although the usage of water in the mining industry shares many of the characteristics of other industrial water consumers, it has some distinctive features. For example, mining projects cannot freely choose where to operate since mining is limited to locations abundant in geologically concentrated minerals and ore bodies which are economically and technically feasible. Consequently, mining companies often operate in sensitive or challenging environments facing the full spectrum of ecological and hydrological contexts, e.g., in arid regions of central Australia or the Chilean Andes, the tropical and sub-tropical areas in Indonesia and the sub-arctic areas of Finland, Canada or Russia [15,22].

Therefore, in recent years intense research has been conducted to improve the detailed understanding of the complex interactions between the mining industry and its wide range of impacts on freshwater resources. Primarily within the last decade, great efforts have been made to quantify mining-related water use, mainly based on life cycle assessment (LCA) studies focusing on selected metals and minerals including different settings of usually applied mining and refining methods (cf. [23–37]). Data on mining-related inventories of water use provided by these studies are primarily based on case studies but are increasingly sourced from sustainability reports provided by mining companies as well. However, despite detailed case studies and growing data availability, robust information on specific water consumption in the extractive industry as well as a global overview of the intensity of the impacts of mining on regional environments and water resources in particular are still lacking.

With regard to this, the article addresses the following question: To what extent is the global mining industry exposed to water stress and what impact does industrial mining have on water resources at global and regional scales? It also takes into account mining's influence on the carrying capacity of regional hydrological systems, which affects sufficient supply of freshwater at a local scale. However, limited water resources are also of rising concern to the mining industry as well, particularly as water shortages may potentially affect the production and supply of global raw material markets. This in turn leads to economic risks for global markets, whose demand for raw materials and mining commodities is constantly increasing. In conclusion, a deeper understanding of the complex interactions between natural resource extraction and potential water conflicts is of rising importance to both achieve secure supplies of raw materials to global and national markets as well as to establish sustainable management and development strategies in the mining industry.

With the example of 14 selected mineral commodities—namely bauxite, cobalt, copper, iron, lead, manganese, molybdenum, nickel, uranium and zinc and the precious metals gold, palladium, platinum and silver—a comprehensive water footprint assessment has been conducted in this survey by applying established methods of LCA and geographic information systems (GIS) to identify the water consumption of the global mining industry as well as the intensity and relevance of mining-related water stress at a regionally explicit level. Based on this approach (Figure 1 in Section 2.3), more precise information for decision makers in business and politics can be provided, thereby helping to understand the 'water–resource nexus' and the complex interactions between global raw materials markets and their impacts at the local level, especially affecting regional carrying capacities.

2. Materials and Methods

To gain deeper knowledge of how the global mining industry interacts with local water resources, a brief overview of the established water impact assessment methods being applied is given in this section. Particularly methods and indicators used for water stress and water scarcity determination, also referred to as 'water footprint assessment', are introduced. Furthermore, the GIS model provided for conducting a comprehensive water footprint assessment is introduced as well.

2.1. Water Footprint Assessment

In recent years, several assessment tools and water scarcity metrics have been developed and introduced to assess water consumption and the resultant impacts on water availability [38], whereof the most prominent will be introduced briefly in the following sections. The most widely used water assessment tool in research as well as in the industry is the 'water footprint methodology' applying the water scarcity metric 'water stress index'. Even though different assessment approaches have been established under this terminology, the most important versions are those aligning with international standards of life cycle analysis, especially regarding ISO 14040 [39] and ISO 14046 [40]. ISO 14046 in particular addresses water impact assessment, providing information about principles, requirements and guidelines for conducting and reporting water footprint assessments of products, processes and organizations. These standards have been significantly influenced and improved by years of scientific debate. Methods for quantifying water consumption or water use as well as measuring water-specific impacts associated with agricultural and industrial production systems have especially been advanced notably over the past two decades, particularly by the introduction and development of the water footprint methodology.

For instance, John Anthony Allan initially introduced the concept of 'embodied water' and 'virtual water' in 1993 to describe the water consumption needed to produce goods and services (cf. [41,42]). These concepts were consequently advanced when Arjen Hoekstra in 2003 established the term 'water footprint', additionally taking into account the geographical and temporal characteristics of the virtual water use [34,43]. As a result, the methods and data sources available to perform water footprint assessments have been notably developed, particularly through standardization efforts by the Water Footprint Network [44,45]. All these continuous efforts finally resulted in standardized assessment methods such as ISO 14046 mentioned above.

Overall, profound improvements have been made in the development of various environmental as well as water impact assessment methods, even though the application of these methods has primarily been focused on understanding the environmental impacts resulting from water consumption in the production of water-intensive commodities from agriculture [34]. These efforts have provided a wide range of robust data concerning the quantification of water use in agricultural production at local and global scales as well as along agricultural supply chains. However, while studies on freshwater use and their impacts have primarily focused on agricultural production as the major water-consuming sector, water use in industrial production has gained momentum in recent years and has been discussed with more emphasis [46]. Thus, the assessment of industrial water consumption and its consequences is highly needed, particularly in the field of resource extraction, which is often conducted in regions characterized by water stress or even water scarcity. Unfortunately, only few studies have considered the interactions between the mining industry and water resources on a regional scale and upscaled these findings to a global level to provide further information on water use in the mining sector operating worldwide. However, as these interactions are often complex and site-specific, a combined understanding of the local water contexts of individual mining sites as well as the global perspective is required before associated risks can be adequately addressed [16].

2.2. Water Stress and Water Scarcity Determination

In the context of their socioeconomic development, many countries and regions are increasingly facing challenges related to water. Physical water shortages and water quality deterioration in particular are among the problems of growing concern and thus requiring further action to be addressed [47]. In general, probably the most widely used indicator measuring water stress and water scarcity is the 'Falkenmark indicator', introduced by Falkenmark et al. [48]. It is basically defined as 'the fraction of the total annual runoff within a given area potentially available for human use, resulting in a certain volume of water available per person calculated in m^3 per capita and year' [49]. According to Falkenmark et al., a value of 1700 m^3 per capita and year of renewable freshwater was originally proposed as the threshold for water scarcity—i.e., when approaching the threshold of 1700 m^3 per capita and year, increasing water conflicts are to be expected, also defined as 'water stress'. Consequently, increasing water stress usually leads to intensified competition for water amongst the users within in a particular region, also referred to as social or economically induced water stress [49–51].

Since the introduction of the Falkenmark indicator, a variety of additional indicators have been developed and proposed for characterizing water use impacts, particularly within the LCA framework, which is the most widely used approach to assess the environmental impacts of production systems. However, the water stress index (WSI) and the water scarcity index or water scarcity footprint (WSFP) are amongst the most prominent and well-established indicators. For example, water stress, which is introduced and applied in LCA as a water stress indicator, is commonly defined as the ratio of total annual water use (WU) in relation to hydrological water availability (WA). According to Vanham et al. [52], WA is usually measured as freshwater renewal rate or runoff, whereof a specific volume of water representing environmental water needs can be deducted occasionally. Particularly in the latter case, this proportion of water availability is commonly referred to as the environmental flow requirement (EFR) or environmental water requirement (EWR). WU is typically measured as either gross or net water abstraction from water sources. If water withdrawal is used as an indicator of WU (=gross water abstraction), the resultant WSI is termed the 'withdrawal-to-availability ratio' (WTA), whereas in the case that water consumption (also termed blue water footprint according to Hoekstra et al. [44,45]) is used as an indicator of WU (=net water abstraction), the resultant WSI is termed the 'consumption-to-availability ratio' (CTA) [47]. Depending on the calculation model and data availability, WSI, WU and WA are generally calculated on an annual or intra-annual basis, usually considering monthly periods. In conclusion, commonly used WSI indicators are calculated as WTA or CTA by the following equation:

$$\text{WTA}_i / \text{CTA}_i = \frac{\sum_j \text{WU}_{ij}}{\text{WA}_i} \quad (1)$$

and when taking into account EWR, as follows:

$$\text{WTA}_{i\ (EWR)} / \text{CTA}_{i\ (EWR)} = \frac{\sum_j \text{WU}_{ij}}{\text{WA}_i - \text{EWR}_i} \quad (2)$$

WTA and CTA basically consist of a hydrological (WA) and a socioeconomic (WU) component, quantifying annual water availability (WA_i) within a particular area or watershed, i, and water use for different users, j (WU_{ij}), from industry, energy supply, mining, agriculture and households for a particular watershed, i. Moreover, WTA and CTA are often demarcated by defined threshold levels. According to Falkenmark and Gunnar [53], Raskin et al. [54] and Rockström et al. [55], a defined spatial area, e.g., a country or watershed, is termed 'severely water scarce' and thus highly water-stressed if the ratio of annual withdrawal or consumption to annually available water in the given area exceeds 40%. If this ratio is within 20–40% the area is considered as 'water scarce', and with a ratio of 10–20% as 'moderate water scarce'; if the ratio lies below 10% the area is described as 'low water scarce'. A comparative review of Vanham et al. [52] showed that these threshold levels were also adopted by the UN report 'Comprehensive assessment of the freshwater resources of the world' [56] and are widely used in the literature (cf. [57–62]). In addition, the European Commission (EC) and the European Environmental Agency (EEA) also applied these threshold levels in their Water Exploitation Index (WEI) [63,64].

However, Pfister et al. [65] introduced and established the water stress index for water impact assessment in LCA by advancing the calculation of WSI. Ranging from 0 to 1, this modified WSI serves as a characterization factor to calculate the water scarcity index as a midpoint category entitled 'water deprivation'. This also includes an impact factor termed 'water scarcity footprint' (WSFP), which is defined as WU multiplied by the WSI of a particular area, thus weighting water consumption with a region-specific water scarcity index [66].

Despite the fact that there has been criticism and debate about the strengths and weaknesses of the conceptualization of this type of impact category, e.g., by Hoekstra [67] and Pfister et al. [68], it is still the most widely established approach to assess water impacts within LCA [34]. Slightly modified conceptualizations and categories of water impact assessment, including WSI, have recently been applied by Gassert et al. [69], providing calculation data on WA, WU and WSI consisting of various indicators to describe water risks. However, contrary to Pfister et al., Gassert et al. termed WSI as Baseline Water Stress (BWS), measuring the ratio of total annual water withdrawal ($WU_{withdrawal}$) in relation to the average annual available water ($WA_{mean(1950,2010)}$) using a long-term data series (1950–2010) to reduce the effect of multi-year climate cycles. Additionally, while Pfister et al. score WSI from 0 to 1, Gassert et al. score WSI between 0 and 5. For example, raw WSI values, r, were normalized using the following equation [69,70]:

$$\text{WSI}_{BWS} = \max\left(0, \min\left(5, \frac{\ln(r) - \ln(0,1)}{\ln(2)} + 1\right)\right), \quad (3)$$

Gassert et al. defined five WSI_{BWS} categories between 0 and 5 (with ≤ 1 = lowest category with less than or equal to 10% of WTA or CTA and r > 4 = highest category with higher than 80% of WTA or CTA) including the following threshold levels to determine WSI_{BWS}:

- 0–1: 'low' water stress (<10%). The overall water consumption within a given area is lower than 10% of natural runoff. If taking into account EWR, runoff is defined as WA minus EWR, which is not or is slightly affected by water consumption.

- 1–2: 'low–medium' water stress (10–20%). Total water consumption is rated between 10 and 20% of natural runoff minus EWR, which is affected moderately.
- 2–3: 'medium–high' water stress (20–40%). Total water consumption is rated between 20 and 40% of natural runoff minus EWR, which is expected to be modified significantly.
- 3–4: 'high' water stress (40–80%). Total water consumption is rated between 40 and 80% of natural runoff minus EWR, which is seriously affected and modified.
- 4–5: 'extremely high' water stress (>80%). The basin's overall water consumption exceeds 80% of natural runoff minus EWR, violating the environmental water needs in case of exceeding water availability by 100% (= EWR-related threshold). As many mining operations are located in remote areas which are arid but simultaneously characterized by low water use, thus having less competition amongst water users, these areas are not comparable to the regular definition of WSI. Nevertheless, mining operations have to be aware of localized impacts, particularly with respect to environmental water needs [71]. As a consequence, this category includes 'arid and low water use', differing from the established water stress definition but assuming that environmental water needs are violated regardless of the amount of water used.

In summary, Pfister et al. and Gassert et al. developed widely used and helpful approaches to address water-related challenges [65,69], but applied in different ways. While the concept of water stress by Pfister et al. is mainly used as a categorization factor in LCA to derive a water scarcity index, the concept by Gassert et al. is mainly developed as a standalone indicator used to meet the growing concerns from private and public sectors in addressing issues of water scarcity.

Since most water scarcity metrics that were initially developed mainly consider water quantities for socioeconomic purposes, only little attention has been given to the water needs of nature itself in the past, even though Sullivan [72] pointed out that depleted freshwater resources are directly linked to impacts on ecosystems and ecosystem degradation in particular. Any index used in water scarcity or water impact assessment should therefore include the condition of ecosystems and the thresholds to be taken into account to maintain sustainable levels of natural water availability [49]. Meanwhile, it is generally recognized that environmental water needs have to be included in water impact assessments in order to take account of sustainability requirements in water use, as it is recommended in SDG 6 and particularly in SDG 6.4.2 (level of water stress: freshwater withdrawal as a proportion of available freshwater resources) [73]. However, besides its relevance for sustaining a wide range of ecosystem services, EWR also has direct and indirect links to other specific SDGs, such as SDG 14, 'life below water', or SDG 15, 'life on land' [52]. Consequently, ERW has been increasingly included in water impact assessment.

A well-established approach to integrating environmental water needs in water impact assessment is the calculation of 'the quality, quantity, and timing of water flows that are required to maintain the components, functions, processes, and resilience of aquatic ecosystems which provide goods and services to people' [74] (p. 80), [75], usually defined as environmental water requirement (EWR) and environmental flow requirement (EFR). For instance, Smakhtin et al. initially developed a water stress indicator recognizing EWR (EFR) as an important parameter of available freshwater [76]. In this case, mean annual runoff (MAR) was used as a proxy for total water availability in a given area, and EWR (EFR) was expressed as a percentage of the long-term mean annual river runoff that should be reserved for environmental needs in this particular area or watershed [49]. Due to the variety and complexity of occurring ecosystems, even within small-scale watersheds, quantification of EWR is not uniform, as protectable aquatic ecosystems or ecosystem services depend on different amounts and qualities of freshwater. For advanced determination of EWR thresholds, particularly with regard to water impact assessment, several methods have been introduced and discussed (cf. [49,76–80]). Thus, there are different recommendations for EWS thresholds in the existing literature, varying considerably between authors and across river regimes [52]. For instance, Richter et al. [9] suggested an EWR of 80% of the monthly mean runoff as an obligatory standard, whereas Pastor et al. [78] proposed an EWR of between 25 and 46% of the mean annual runoff. Furthermore, many studies recommended river regimes as the appropriate scale of choice when it comes to quantification and determination of EWR levels [78], but globally uniform EWR recommendations at a spatially explicit level are still lacking.

However, Sood et al. calculated global EWR estimates for SDG target indicators, presenting continent-wide cumulative annual flow and groundwater abstraction to mean water runoff ratios (Table 1) [73].

Table 1. EWR and sustainable groundwater abstraction on continent-level (data sourced from [73]).

Continental Region	Annual Flow ($km^3 a^{-1}$) and Percentage of Natural Flow (%) to Be Preserved as EWR	Sustainable Groundwater Abstraction ($km^3 a^{-1}$) and Percentage of Natural Recharge (%)
Asia	10,178.2 (57.0)	110.3 (3.4)
North America	3656.3 (55.2)	30.3 (1.9)
Europe	1489.7 (52.8)	20.0 (1.7)
Africa	5032.1 (70.2)	14.3 (0.7)
South America	11,242.9 (73.4)	24.0 (0.6)
Oceania	240.4 (35.1)	2.6 (1.0)
Australia	251.0 (48.4)	1.9 (1.3)
Global	**32,090.6 (63.0)**	**203.3 (1.6)**

According to the work of Sood et al. [73], globally, 63% of the natural flow needs to be maintained to protect ecosystems and eco-services. While South America and Africa are required to maintain more than 70% of the natural flow, in Australia and Oceania, where rivers are more degraded, 48.4 and 35.1% need to be maintained, respectively. Additionally, the percentage of annual groundwater abstraction on a global scale was estimated to be 1.6%. However, even if global estimates at the continental level are provided, EWR still has to be estimated for individual watersheds or watershed groups in order to consider comparable characteristics in terms of river regimes and environmental attributes. Especially with regard to mining, it is highly relevant to consider environmental water requirements at the watershed level, and particularly at the local level, as the lack of sustainable water management in mining operations can significantly alter local and even regional groundwater characteristics as well as the base flow characteristics of surface watercourses, especially in the case of large-scale mining projects.

For example, due to relatively constant mine-site water discharges, such as from mill operations, dewatering or the diversion of water from one watershed to another, river base flows within the mining area can be extended or elevated, possibly leading to disruption of the relationships between surface water and groundwater systems even though converting temporary water systems to perennial waters or vice versa. In particular, through the conversion of river systems from an ephemeral stream to a perennial stream, mainly caused by dewatering, mining operations can significantly affect the natural ecological systems that depend on seasonal flow variations. This might consequently result in reduced biodiversity of local as well as downstream aquatic systems [71]. As a consequence, considering EWR in water impact assessments in the mining industry is highly recommended. However, there are only very few harmonized spatial data on regional or grid-based EWRs available. As the emphasis of this article is on conducting a spatio-temporal analysis of the global water impact of mining on a regional scale, the EWR recommendations of Sood et al. [73] at the continental level are used as approximation values due to the fact that these recommendations directly align with the SDG framework.

2.3. Mining Data and System Boundaries Applied for a Water Impact Assessment

To conduct a regionalized water impact assessment of the global mining industry, a geographic information system (GIS) allowing data processing, statistical calculation and evaluation of mining activities as well as correlation of spatially explicit water characterization factors such as WSI at different spatial resolution scales was utilized. Figure 1 shows the layout and structure of the GIS model, including the data categories and impact factors applied.

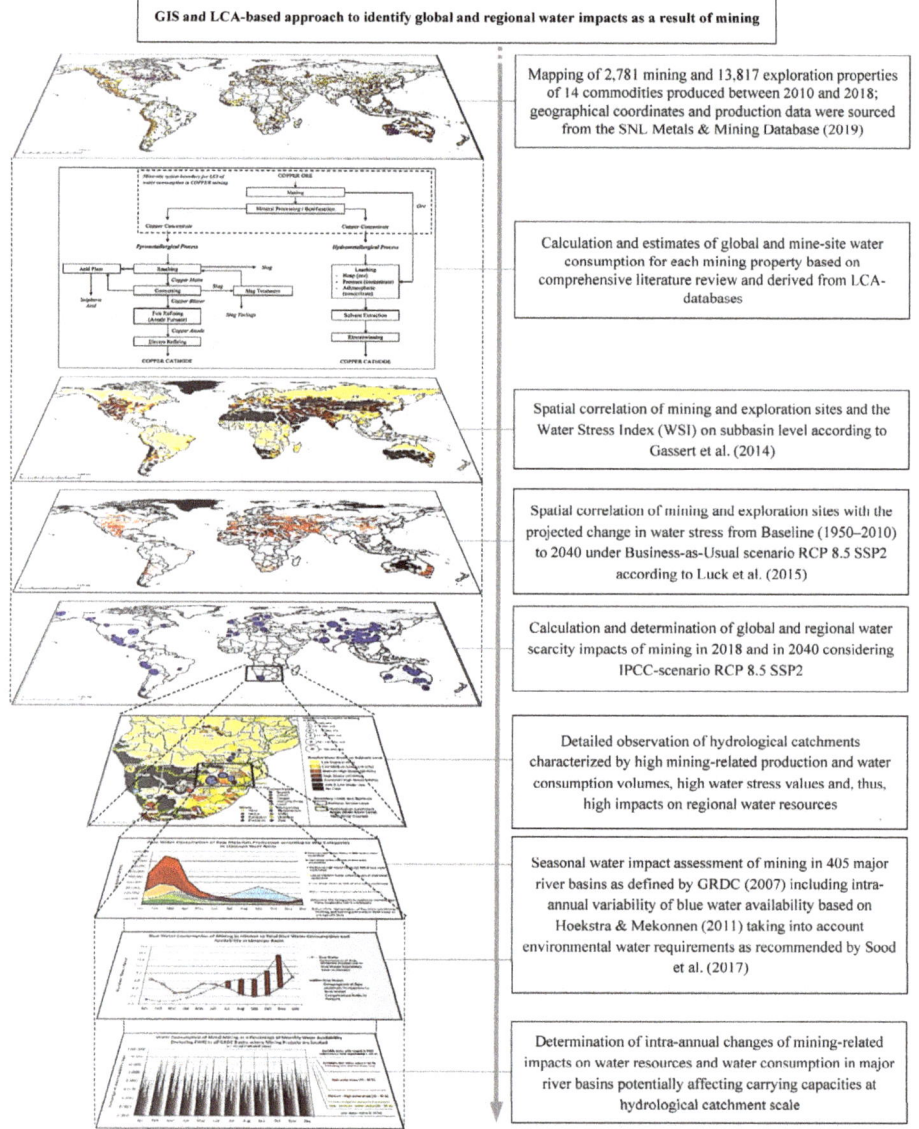

Figure 1. Structure of the GIS model applied to conduct a regionalized water impact assessment of the global mining industry. (Geographical coordinates and production data were sourced from the SNL Metals & Mining Database (2019) [81]; Water Stress Index was calculated according to Gassert et al. (2014) [69]; projected water stress in IPCC-scenario RCP 8.5 SSP2 was obtained from Luck et al. (2015) [82]; seasonal water impact was conducted based on major river basin scale as defined by GRDC (2007) [83]; data on basin-related water availability was taken from Hoekstra and Mekonnen (2011) [84] and modified by environmental water requirements as proposed by Sood et al. (2017) [73].

The following datasets were implemented and correlated in the GIS model:
- Geographic coordinates and individual mine-site production data of a time period between 2010 and 2018 for 2783 mining operations producing preprocessed ores or concentrates of bauxite, cobalt, copper, iron ore, lead, manganese, molybdenum, silver, U_3O_8 (uranium concentrate or yellow cake) and zinc as well as the refined metals gold, nickel, palladium and platinum. In addition, 13,817 exploration projects and 11,500 development projects of all 14 commodities were also considered.
- Specific water consumption volumes per t mining commodity based on a comprehensive review of LCA databases as well as recent studies on water footprints in the mining sector (shown in Tables 2 and 3) and annual water consumption volumes of each individual mining operation according to LCA calculations and mine-site production volumes.
- Water stress index (WSI) at the sub-basin level according to Gassert et al. [69], obtained from the Aqueduct Project, and WSI at the major river basin level as defined by the Global Runoff Data Centre (GRDC) [83]. In total, water stress of approximately 25,000 basin units, representing ~15,000 sub-basins and 405 major river basins according to the GRDC, was implemented.
- Mean annual and monthly water availability as well as total water consumption volumes of all water-consuming sectors at the sub-basin and major river basin levels (sourced from Gassert et al. [69] and Hoekstra and Mekonnen [84]).
- Calculated water scarcity footprints (WSFPs) of each mining operation according to the definition of Pfister et al. [65] and Ridoutt and Pfister [66], aligning with the LCA framework.
- Based on calculations by Luck et al. [82], estimated changes in water stress by 2030 and 2040 and projected water stress in 2030 and 2040 at the sub-basin level were implemented in the GIS considering three IPCC climate change scenarios (RCP4.5/SSP2, RCP8.5/SSP2 and RCP8.5/SSP3) to derive estimations of the water stress that mining operations may be confronted with in the next two decades.

In order to conduct a water impact assessment on different regional scales, spatial correlation of all datasets in the GIS is required. However, it has to be pointed out that, particularly in the case of using LCA data, the calculation of water use for water impact determination related to a particular mining location or region also has to distinguish between the different production techniques applied. Furthermore, depending on the geological setting of a deposit as well as the given ore grades jointly occurring in the metal-bearing rocks, mining operations usually produce several metal commodities simultaneously. Thus, the water consumption of a particular mining operation has to be determined according to all minerals involved at a particular mining place and along the entire production pathways and processing steps.

Despite these data-related challenges, the use of GIS allows the observation of mining-related water use and its impacts from different perspectives. For instance, from the perspective of a region in which mining activities are performed, it is rather of interest to assess water-related impacts resulting from the cumulative overall water consumption of all mining operations and mining commodities produced than assessing the water consumption resulting from the production of a single mining commodity. Hence, depending on the interest of choice, it might be useful to distinguish between the observation of environmental impacts caused by a single raw material—as usually conducted by raw material criticality assessments which primarily consider demand-side risks of raw materials markets—and impacts resulting from the mining operations within a particular region as a whole. The latter case in particular is mostly relevant for implementing environmental management and protection strategies, representing region-specific interests (=supply-side perspective) which are often neglected in traditional raw material criticality assessments.

In conclusion, when it comes to data quality and availability for water footprint accounting in the mining sector, most of the available LCA data are related to single raw material production, which has to be taken into account when conducting a water impact assessment of all mining activities located within a particular mining area and thus affecting the hydrological systems of an entire region. Therefore, in the water impact assessment conducted in this study, representing the perspective of a particular region or basin, all mining locations within the given region including all water consumptions of the mining commodities produced are summed and considered as region-specific cumulative water consumption.

2.4. Water Consumption in Mining and Refining of Metal Raw Materials

The first step towards calculating water demand in the mining sector is to develop a comprehensive and detailed understanding of a mine's production system and also its key water flows. The extraction and treatment of mineral ores to manufacture metal concentrates is the primary step in the production chain of minerals and metals and is usually carried out at the mining location. Concentrates are saleable products of ore dressing, whereby valuable metals which are recovered through mining operations are separated from waste rock, enriched prior to transportation off a mine site and shipped to the markets. In many mining operations, ore is crushed and milled to recover valuable mineral components from the ore. Depending on the mineralogy of ores and the physical and chemical properties of the concerned minerals or metals, this processing step comprises many treatment methods, which all aim to extract a wide range of valuable materials from the ore. Enrichment techniques usually applied for this purpose are gravity concentration, magnetic concentration and, most commonly, froth flotation, which is usually the most water-intense processing step of a mining project in terms of water consumption [5].

Due to the fact that mining operations are essential for supplying raw materials to high-technology industries and, furthermore, are frequently located in areas characterized by limited water supply as well as increased risks of climate change effects [15], a growing number of studies on water consumption in the mining sector have been performed in recent years. However, the outcomes of the studies in terms of water consumption per commodity unit vary slightly and, in some cases, even significantly. This is mainly due to the individual definition of the mine-site system boundaries used but also due to the different mining technologies and water efficiencies applied at the mining sites which have been observed. This could also include different mine types (open cut or underground), ore mineralogy, mill configurations and designs (e.g., flotation or hydrometallurgical-based concentration methods), water qualities used, project ages, climate settings (arid, temperate, tropical), types of energy resources used and, finally, whether a smelter and refinery were also included in the operation observed [13,85]. These variations also have a significant influence on the amount of freshwater required for the production of mineral concentrates or refined metals. Thus, there is a wide range of water consumption values of the mineral commodities, both between and within commodity types [13]. However, these inventories at least provide increasingly precise estimations of the intensity of water consumption in particular production processes considering different production steps, such as mining, smelting and refining, as well as different production pathways, e.g., pyrometallurgical and hydrometallurgical processing types.

To conduct a water impact assessment of the mining industry, in this study, data on water consumption in ore mining and production of metal concentrates as well as refined metals were collected from numerous sources—primarily scientific publications and LCA databases providing several datasets for the selected metals and production stages. In addition, company websites and environmental reports were consulted to cross-check the water consumption values from the literature. However, the number of publications observing the water consumption of metal mining and refining is still very limited. A summary of the outcomes of the studies conducted by different authors is given in Table 2, showing the water consumption values for all selected commodities, including the main literature and data sources used. The dataset includes minimum and maximum ranges as well as the calculation of an average specific water consumption value per t metal-eq. contained in the correspondent concentrate or per refined metal. Based on the global production volumes (Table 3) and the specific mean water consumption values, the total global water consumption was derived for each metal and its concentrates.

Table 2. Summary of water consumption values in the mining and refining of selected metals based on a comparative literature review including LCA databases.

Processing Stage of the Mining Commodity	Minimum Range (in m³/t)	Maximum Range (in m³/t)	Averaged Water Consumption (in m³/t Metal Commodity)	Global Water Consumption (in Mm³)	Reviewed Data Sources Providing LCA-Based Inventory Data on Specific Water Consumption Values per t Metal Commodity
Ore and Metal Concentrate					
Bauxite					
Preprocessed ore	0.320–0.395 [27]	0.447–0.578 [23,86,87]	0.447	88.9 [1] (123.1 [2])	International Aluminium Institute [23]; Gunson [27]; Frischknecht et al. (Ecoinvent, bauxite, at mine, GLO #1063) [86]; Buxmann et al. [87]
Cobalt					
Concentrate	40.72–170.84 [88]–258.00 [89]	364.00–(802.00) [27]	208.40 (327.12)	24.5 [1] (28.9 [2])	Gunson [27]; Dai et al. [88]; Shahjadi et al. [89];
Copper					
Concentrate	9.673–10.446 [86,90]; 28.000 [89]; 36.100–37.594 [27,33,86,89,90]; 40.000 [91]–42.403 [86]	67.081 [86,90]; 70.400–99.550 [27,91]	43.235	859.4 [1] (885.9 [2])	Gunson [27]; Northey et al. [33]; Frischknecht et al. (copper concentrate at beneficiation) [86]; Shahjadi et al. [89]; Fritsche [90]; Pena and Huijbregts (incl. SX-EW) [91]
Iron Ore					
Fines	0.210–0.874 [26,27,92,93]	1.519 [86]–1.529 [90]; 3.000 [93]	1.371	1382.3 [1] (1878.1 [2])	Ferreira et al. [26]; Gunson [27]; Frischknecht et al. (Fe at beneficiation, GLO #1100) [86]; Fritsche [90]; Haque and Norgate [92]; Tost et al. [93]
Lead					
Concentrate	(0.528) [90]–3.995 [27]	8.222 [89]–8.485 [27]–11.754 [86]	6.597	19.9 [1] (33.3 [2])	Gunson [27]; Frischknecht et al. (lead concentrate at beneficiation, GLO #1104) [86]; Shahjadi et al. [89]; Fritsche [90]
Manganese					
Concentrate	1.390 [86]	1.418 [90]	1.404	62.7 [1] (85.9 [2])	Frischknecht et al. (manganese concentrate at beneficiation, GLO #1110) [86]; Fritsche (2005) (manganese concentrate, GLO 2003–2004) [90]

Table 2. Cont.

Processing Stage of the Mining Commodity	Minimum Range (in m³/t)	Maximum Range (in m³/t)	Averaged Water Consumption (in m³/t Metal Commodity)	Global Water Consumption (in Mm³)	Reviewed Data Sources Providing LCA-Based Inventory Data on Specific Water Consumption Values per t Metal Commodity
Molybdenum					
Concentrate	52.2–209.6 [86]	382.0 [27]–490.5 [86]–797.0 [27]	240.9	55.0 [1] (78.5 [2])	Gunson [27]; Frischknecht et al. (molybdenum concentrate, GLO #1117, RER #5858, RAS #5859) [86]
Ore and Metal Concentrate					
Silver					
Concentrate	1621–1805 [27,89,90]	3128 [27]	1713	41.1 [1] (44.6 [2])	Gunson [27]; Shahjadi et al. [89]; Fritsche (Xtra-silver concentrate) [90]
Uranium					
Concentrate (U$_3$O$_8$)	46.20–100.00 [11,86,94]; 505.00–2478 [11,27,90,94]	6000–8207 [11,86,94]	2746	17.1 [1] (17.1 [2])	Mudd [11]; Gunson [27]; Frischknecht et al. (uranium oxide RNA #5988, RNA #5989) [86]; Fritsche (uranium oxide) [90]; Mudd et al. [94]
Zinc					
Concentrate	11.07 [89]–13.10 [27]–13.36 [86]	24.65 [27]	11.93	114.2 [1] (154.5 [2])	Gunson [27]; Frischknecht et al. (zinc concentrate at beneficiation, GLO #1157, SE #10099) [86]; Shahjadi et al. [89]
Refined Metal					
Gold					
Metal	79,949–152.630–174.780–190,558 [86]	259,290–288,140 [95]; 309,110 [27]–347,910 [86]; 392,686 [93]–427,696 [90]; 453,305 [27]–477,000 [96]	265,861	712.79 [1] (814.8 [2])	Gunson [27]; Frischknecht et al. (gold at refinery #10110-14) [86]; Fritsche [90]; Tost et al. [93]; Norgate and Haque [95]; Mudd [96]
Nickel					
Metal	80.6 [86]–107 [13]–138.0 [27]	187.36–193.0 [86]–240.0 [27]–258.2 [86]	193.8	355.3 [1] (441.4 [2])	Mudd [13]; Gunson [27]; Frischknecht et al. (primary nickel, GLO #35, GLO #1121, ZA #1124, RU #1125) [86]
Palladium					
Metal	56,779–127,172 [27,86,90]	273,523–327,874 [86,90]	210,713	45.8 [1] (46.3 [2])	Gunson [27]; Frischknecht et al. (primary at refinery, ZA #1128, RU #1129) [86]; Fritsche (primary at refinery, ZA, RU) [90]

Table 2. Cont.

Processing Stage of the Mining Commodity	Minimum Range (in m³/t)	Maximum Range (in m³/t)	Averaged Water Consumption (in m³/t Metal Commodity)	Global Water Consumption (in Mm³)	Reviewed Data Sources Providing LCA-Based Inventory Data on Specific Water Consumption Values per t Metal Commodity
Platinum					
Metal	169,968–200,000 [27,86,90]	406,998–487,876 [27,86,90]	313,496	67.3 [1] (67.3 [2])	Gunson [27]; Frischknecht et al. (primary at refinery, ZA #1134, RU #1135) [86]; Fritsche (primary at refinery, ZA, RU) [90]

[1] The water consumption refers to the global production share considered in this study (see Table 3). [2] Estimated global water consumption assuming 100% of global production is taken into account.

Concerning the data used from the literature, Gunson [27] performed the most comprehensive survey on water withdrawal and consumption in the mining industry, representing 23 mineral and metal commodities in total. The author collected data from 65 mining companies, which reported water-related data for up to 155 mining sites in 2009. Further comprehensive studies on the water consumption of mining operations have been carried out by Northey et al. (cf. [14,16,33]), predominantly focusing on copper, gold, lead and zinc. Additionally, a recent study on water consumption in copper mining was conducted by Lutter and Giljum [28], who carried out a comprehensive data survey on Chilean copper production, collecting data on water consumption from 31 Chilean copper mines and also distinguishing between different types of water used. In addition, LCA databases, e.g., Ecoinvent [86] and PROBAS [90], also provided water consumption values for different mining and refining processes.

In this study, data on water consumption in bauxite and iron production were related to preprocessed ore, whereas water consumption in cobalt, copper, lead, manganese, molybdenum, silver, uranium and zinc production is related to metal concentrates which are commonly produced at the mining site. As mine-site metal production is usually reported as the metal contained in the concentrate produced, production and water consumption values for each metal are described as metal equivalents (metal-eq.) and water consumption per t metal-eq., respectively. As the metal contents in concentrates vary between metals depending on the ore grades and the processing techniques applied as well as the literature reviewed, a wide range of specific water consumption values and, therefore, a wide range of global water consumption volumes are reported in the present study. Nickel, however, is considered an outlier because in comparison to the preceding concentrates, there is lack of reliable LCA data on water consumption for producing nickel concentrates. Thus, water footprint accounting of nickel refers to refined ferronickel. By contrast to the concentrates mentioned above, precious metals such as gold, palladium and platinum are very frequently processed in smelters and refineries located adjacent to the mining site due to their comparatively high economic value, even in the case of low production volumes. Thus, in this study, water footprint accounting for gold, palladium and platinum includes the mining, smelting and refining of pure metals, assuming that the entire production process is located at one particular production site or is adjacent, located within a region characterized by similar water availability and water stress. However, silver was excluded from this rule and calculated as concentrate, as it is often mined in combination with lead and zinc, and thus, considerable amounts of silver are commonly refined from lead and zinc concentrates, which usually takes place at specific refining plants off-site from regular mining locations. Therefore, due to the unevenly defined system boundaries referring to different on-site production steps, comparison of water consumption between all metals considered is not reliable, except within ore categories, concentrates and refined metals.

In conclusion, it can be shown that there is significant variation in the results of almost every commodity between the different publications. In some cases, the variation of water consumption values reported can be within a factor of 5 to 10, e.g., for copper concentrate with a range of 9.7 to 99.6 m^3 per t Cu-eq. in concentrates, or even within a factor of roughly 180, e.g., for uranium mining with a range from 46.2 to 8207.0 m^3 per t U$_3$O$_8$. This variation is due to different mine types, ore types and given ore grades and thus the different processing pathways applied. Furthermore, different system boundaries were determined by the consulted studies, conducted for different mine sites and time periods. Some calculations are based on different definitions regarding the terms water withdrawal and water consumption. Additionally, some studies are based on individual mine-site surveys, while others cover a wide range of mine-site specific data derived from mining companies' sustainability reports. Moreover, the challenge of the lacking coherence of mine water use in the literature and LCA databases is described in almost every study conducted, such the studies of Gunson (2013) [27], Northey et al. (2016) [14] and Tost et al. (2018) [93]. As a consequence, the comparison of data within the literature is very limited [97], particularly between all commodities observed.

Despite these given uncertainties, Table 2 shows that refined platinum (313,496 m^3 per t), gold (265,861 m^3 per t) and palladium (210,713 m^3 per t) are characterized by the largest specific water consumption values per t on average and that there are significant variations in the calculation depending on the production pathways considered. In contrast to refined metal production, water consumption in the production of concentrates is comparatively low. For example, the production of cobalt concentrates accounts for 208.4 m^3 per t, copper concentrates for 43.2 m^3 per t and zinc concentrates for 11.9 m^3 per t, whereas concentrates of manganese (1.40 m^3 per t), iron ore fines (1.37 m^3 per t) and bauxite (0.45 m^3 per t) are characterized by the lowest water consumption per t. Compared to this, silver and uranium have very high water consumption per t concentrate, accounting for 1713 and 2746 m^3, respectively. However, considering global production rates in 2018 (Table 3), iron ore fines (1382.3 Mm^3), copper concentrate (859.4 Mm^3) and gold (712.8 Mm^3) represented the largest overall volumes of water consumption of the observed commodities. Additionally, based on the global production volumes, the overall water consumption estimated for all metals considered in this study was approximately 4000.14 Mm^3 in 2018. To put these numbers into a global context, the Food and Agriculture Organization (FAO) and the World Water Assessment Programme (WWAP) estimated the global water withdrawal at 4000 Gm^3 in 2014 and global water consumption at 1400 Gm^3 [98]. These comparative numbers show that in absolute terms, and at the global level, the overall dimension of pressure put on water resources by mining is comparatively low. However, the local impacts on water resources will be increasing in the future, as the global demand for metals increases and geological accessibility as well as ore grades decline. Furthermore, it is expected that climate change will put additional pressure on local water resources [15]. Thereby, in the following section, the global and regional impacts of the water consumption in mining will be observed in detail.

3. Results

3.1. Mining-Related Water Stress and Global Water Scarcity Impact

To be able to conduct a precise water impact assessment of the global mining industry, particularly taking into account individual mining as well as hydrological specifications at a spatially explicit level, the exact geographical coordinates of all mining projects had to be identified first. In this study, coordinate data of individual deposits and mining sites for bauxite, cobalt, copper, gold, iron ore, manganese, molybdenum, nickel, palladium, platinum, silver, uranium and zinc were obtained from the SNL Metals & Mining Database [81]. Additionally, coordinates in terms of longitude and latitude data were reviewed and cross-checked from a range of further sources, including the mindat.org database of the Hudson Institute of Mineralogy [99], governmental geological authorities such as the United States Geological Survey (USGS) [100], the British Geological Survey (BGS) [101] and the geoscience portal of the Australian Geoscience Information Network (AUSGIN) of the Australian government [102] as well as the scientific literature and company-based technical, environmental and sustainability reports. Based on this review, Figure 2 shows the spatial location of all mining sites considered in this study which produced at least one of the selected commodities in 2018. Furthermore, the locations of all mining sites are shown in relation to the sub-basin water stress as calculated and published by Gassert et al. (2014) [69].

Depending on the particular mining commodity, the total production volume of all mining operations considered represented at least 60 to 100% of the global production volume (Table 3). Overall, in addition to 2783 mining operations, 13,817 exploration sites and 11,500 mining development projects were identified and considered in this GIS-based survey. However, depending on the type of ore, which may contain several elements mined simultaneously, in certain cases, there are partial overlaps of production coordinates due to co-production of by-products. For example, molybdenum, zinc, gold and silver are often mined as by-products in copper production; lead and silver are typically produced jointly within the zinc mining process and gold, cobalt and platinum group metals (PGMs) are frequently produced as co-elements of nickel mining, etc.

Every production site usually represents several processing stages, including ore mining, beneficiation and enrichment, finally providing metal concentrates with 30–40% of metal content that are transported or shipped to adjacent smelters or refineries. In exceptional cases, such as for precious metal mining, there are usually on-site smelters and refineries included, producing refined metals nearby the mining operation with high purity levels. Based on the combination of mining coordinates and LCA studies providing water consumption calculations of commonly applied mining and processing technologies, impacts on water resources can be assessed in GIS for each commodity as ore, concentrate and refined metal at different spatial scales: mine sites at the local level as well as at regional, national and even global levels.

Thus, GIS-based water impact assessment allows the precise location of particular production processes along the mining and refining pathway and therefore the determination of individual water stress profiles for each individual commodity, showing the water-specific conditions under which mineral extraction and processing take place. The overall results of this assessment are comparatively illustrated in Figure 3, showing the commodity-wise WSI determination of each mining operation regarding commodity production in t metal-eq., water consumption and the resultant water scarcity impact in m^3. The commodity profiles clearly show that in the case of each commodity, mining locations are operating under a wide range of different water stress conditions, ranging from 'low' (0–1) to 'extremely high' water stress, including 'arid areas characterized by low water use' (4–5). Furthermore, it is indicated that in some cases, the WSI profile varies significantly between small-, medium- and large-scale production mining properties.

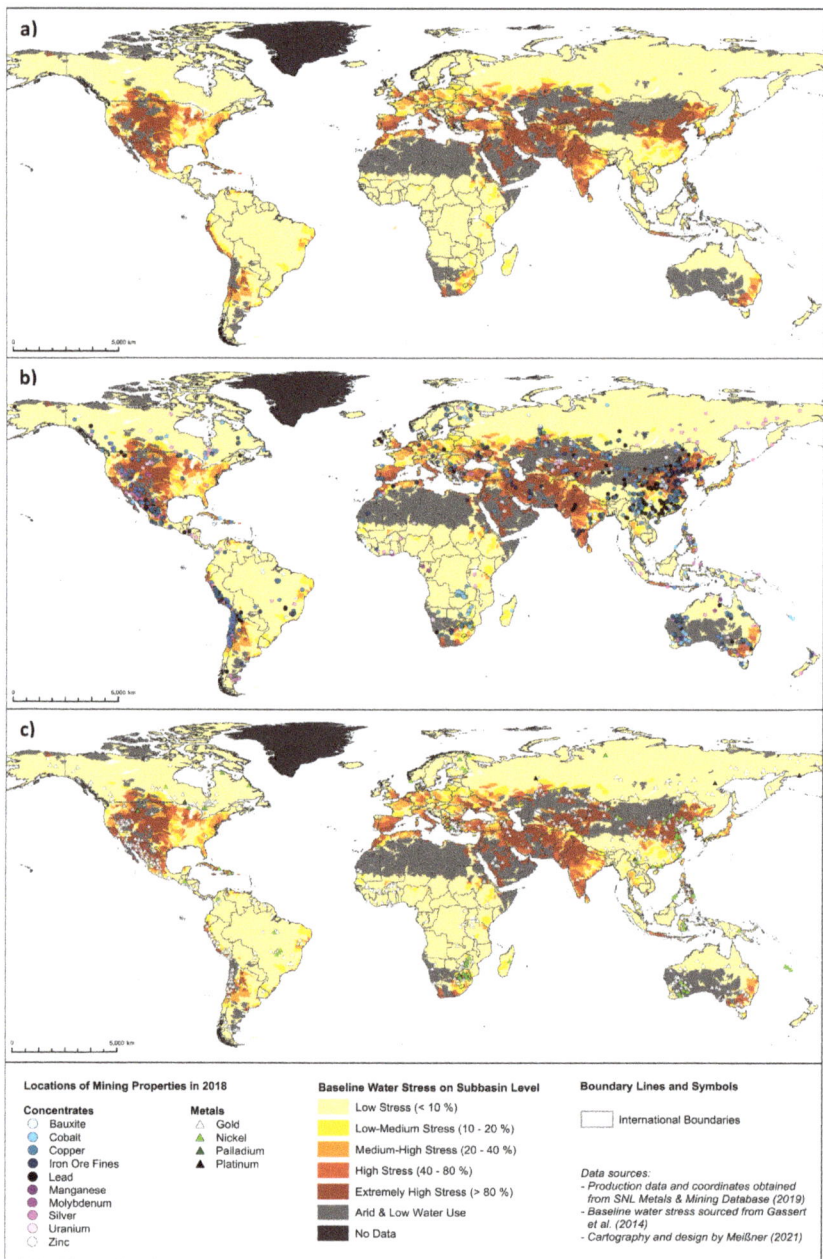

Figure 2. Locations of mining sites in 2018 in relation to water stress: (**a**) water stress at sub-basin level according to data sourced from Gassert et al. (2014) [69]; (**b**) mining locations of metal concentrates; (**c**) mining and production locations of refined metals, both related to water stress; mining and production coordinates were obtained from SNL Metals & Mining Database [81]. (Detailed maps of each individual commodity related to water stress are provided in the Supplementary Materials, see Figure S2a–n; GIS and mining data of Figure 2 and Figure S2a–n are also provided in the Supplementary Materials spreadsheet, see Table S1).

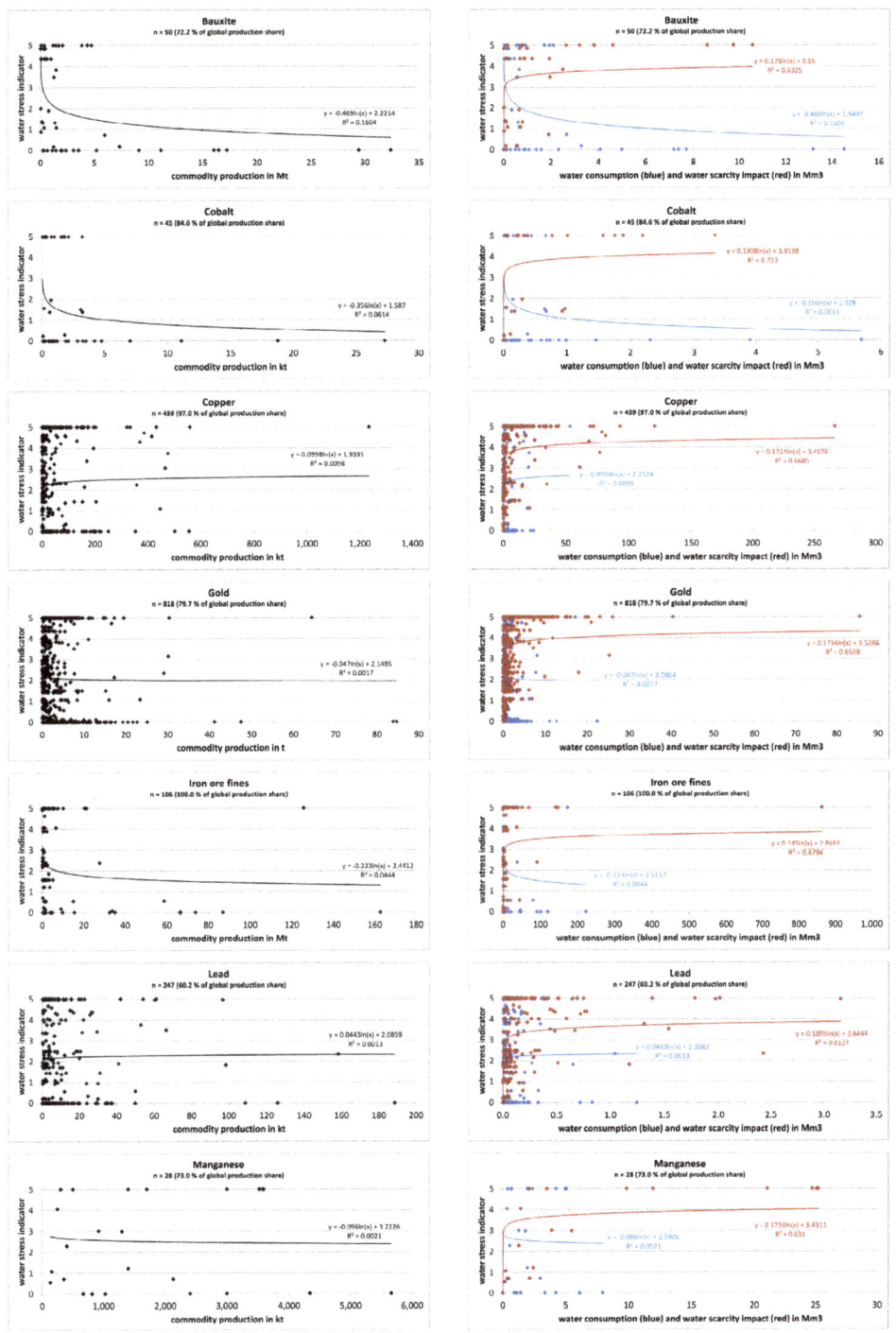

Figure 3. *Cont.*

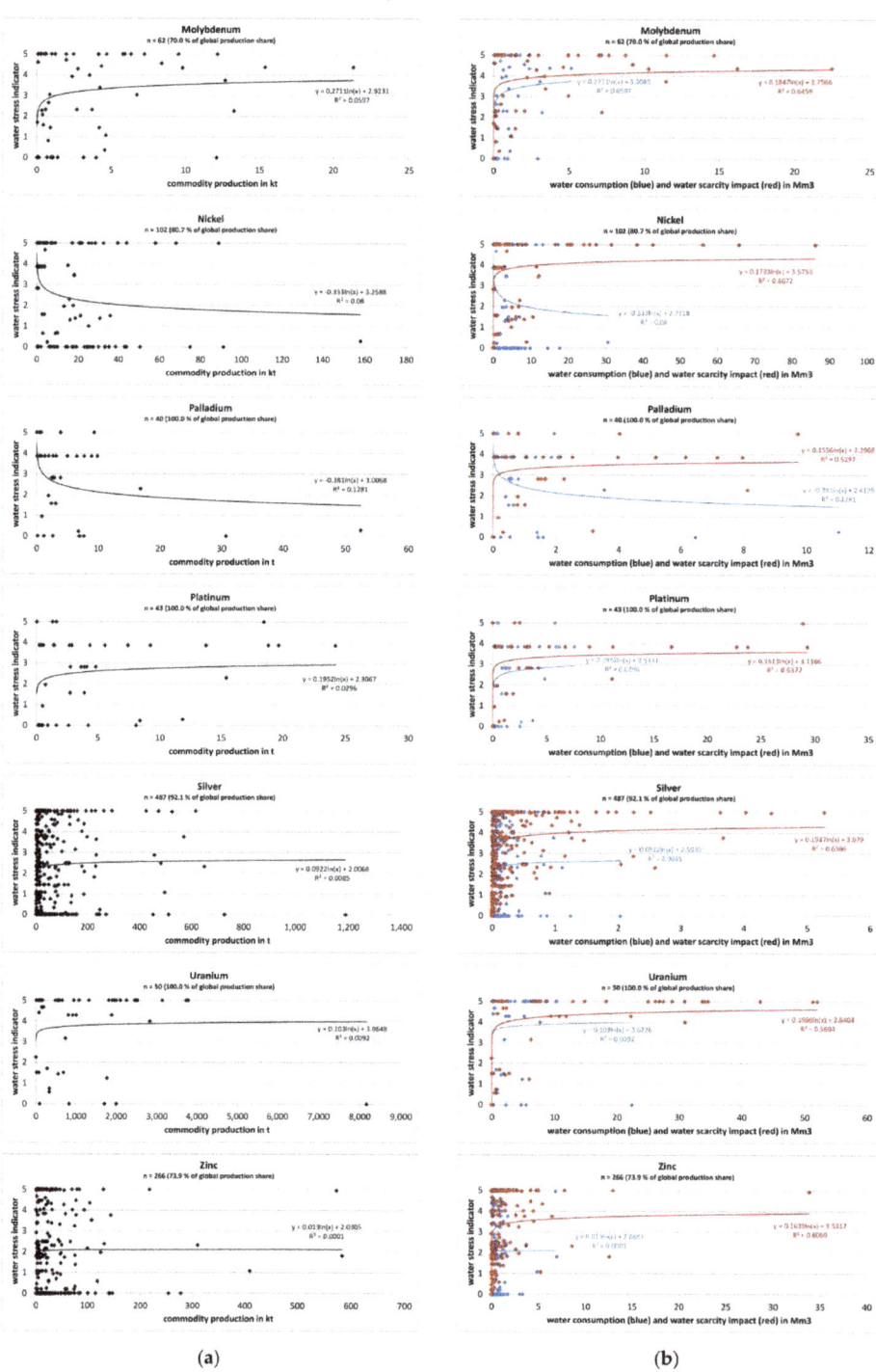

Figure 3. Commodity-wise comparison of mine-site water stress index (WSI) (**a**) in relation to the mine-site production volumes in 2018 and (**b**) in relation to water consumption and mine-site water scarcity footprint (WSFP). Production data

was sourced from the SNL Metals & Mining Database [81]. Each mining site is classified according to water stress, the categorization of which is defined as follows: 0–1 = water stress < 10%; 1–2 = water stress 10–20%; 2–3 = water stress 20–40%; 3–4 = water stress 40–80%; 4–5 = water stress > 80%, including arid areas with low water use. If the production volumes and water consumption of a commodity are independent from water stress and thus more evenly scattered over all water stress categories, this is indicated by low r-squared values, as particularly exemplified by copper, gold, silver and zinc. As large water scarcity effects of mining usually occur in areas characterized by high water stress levels, the r-squared value of the water scarcity impact is typically high. Hence, the higher the production and respective water consumption are in high water stress areas, the higher the water scarcity impact will be, as particularly shown by cobalt, copper, gold and nickel. (Additional information and calculation data of Figure 3 are provided in the Supplementary Material spreadsheet, see Tables S1 and S2).

For example, bauxite is predominantly mined under 'low' water stress from mines with an annual production output of more than 5 Mt, whereas the locations of mining operations with an annual production of lower than 5 Mt are spread over all WSI categories and therefore represent a higher water scarcity footprint (WSFP) on average than larger production facilities. Thus, the overall water scarcity impact of bauxite mining is clearly driven by mining operations producing less than 5 Mt under higher water stress conditions, which is very similar to cobalt and iron ore mining. In the case of palladium, mining sites with an annual production of lower than 10 t are mostly operating under 'high' water stress conditions due to mining projects being predominantly located in South Africa, whereas mining sites with more than 10 t of production output per year are mainly situated in Russian Siberia and are thus characterized by 'low–medium' water stress. Uranium mining, by contrast, is mainly performed in Kazakhstan under 'extremely high' water stress, resulting in comparatively high overall water scarcity effects as well as a high mine-site WSFP on average. In addition, on closer inspection, the production profiles and water scarcity profiles of platinum, palladium, cobalt, iron ore and uranium are significantly influenced by the water stress of only a few geographic regions hosting large production capacities which are of global relevance and characterized by major market shares. For instance, 68.6% of global platinum production in 2018 was supplied by South Africa, of which 99.4% was conducted in the Limpopo River Basin characterized by 'high' to 'extremely high' water stress.

The same effect can be depicted for palladium, 36.9% of the global production of which was also located in Limpopo River Basin. Additional 38.2% of palladium was produced at two mining sites adjacent to Norilsk in Yenisei Basin, Russia, although operating under 'low' water stress conditions. Thus, in terms of physical water scarcity, neglecting wastewater issues and other environmental or sustainability aspects, palladium mined in northern Siberia accounts for a significantly lower water scarcity impact than palladium mined in South Africa. Another mining commodity predominantly influenced by the water stress of one specific river basin is cobalt, 60.5% of the global production of which is situated in the southeast of DR Congo, in the Congo Basin. Further examples to be mentioned are iron ore and uranium. The production of iron ore fines in particular is significantly influenced by the water situation in the northern area of western Australia, hosting 60.2% of global production altogether, as well as Brazil, which supplied another 27.3% of global iron ore fine production in 2018. Furthermore, 41.3% of uranium mining was situated in the Issyk-Kul Basin and Aral Drainage, primarily located in the territory of Kazakhstan and Kyrgyzstan characterized by 'high' and 'extremely high' water stress. In contrast, gold and silver are mined all over the world in low volumes per mining site and are therefore globally allocated homogeneously over numerous areas characterized by a wide range of different water stress intensities.

Overall, Table 3 summarizes the comparative results of the commodity-wise water scarcity impacts as provided in Figure 3. It is indicated that the largest water scarcity impacts are related to copper, iron ore and gold. For instance, the global cumulative WSFP of these three commodities together accounted for 69.7% and 5536 Mm^3 in 2018. The WSFP of all 14 commodities considered accounted for almost 7950 Mm^3 in total.

Table 3. Global production volumes, number of mining properties, global water consumption and overall water scarcity footprint per commodity considered in this survey.

Mining Commodity	Global Production in 2018 (in t) (Except Bauxite in 2016) (Source: SNL [81])	Global Production Share (in Percent) (According to USGS [103])	Number of Mining Sites Considered (Source: SNL [81])	Averaged Water Consumption Factor (in m^3/t)	Global Water Consumption in 2018 (in Mm^3)	Global WSFP (in Mm^3) (=Water Consumption Multiplied with Regional WSI)	Mine-Site ØWSFP (in Mm^3) (=Global WSFP Divided by Number of Mining Properties)
Preprocessed Ores							
Bauxite	198,374,454	72.2	50	0.447	88.94	61.18	1.25
Iron ore	1,008,099,678	73.6	106	1.371	1382.30	1889.52	17.83
Concentrates							
Cobalt	117,332 *	79.3	45	208.4	24.45	16.25	0.36
Copper	19,876,739 *	97.0	439	43.235	859.38	2325.35	5.30
Lead	3,020,613 *	60.2	247	6.597	19.93	39.64	0.16
Manganese	44,637,552 *	73.0	28	1.404	62.67	149.88	5.92
Molybdenum	228,207 *	70.0	62	240.9	54.98	192.78	3.11
Silver	23,986.37 *	92.1	487	1713	41.08	97.64	0.20
Uranium	62,236.74 *	100.0	50	2746	170.91	617.02	12.34
Zinc	9,574,839 *	73.9	266	11.93	114.22	233.93	0.89
Refined Metals							
Gold	2,681.05	79.7	818	265,861	712.79	1320.98	1.61
Nickel	1,833,467	80.7	102	193.8	355.32	720.27	7.28
Palladium	217.51	98.9	40	210,713	45.83	77.89	1.95
Platinum	214.81	100.0	43	313,496	67.34	202.56	4.94
Summary							
Total	1,285,852,217.4		2783		4000.14	7944.89	

* The production numbers indicate the metal-eq. contained in the concentrates.

3.2. Mining's Influence on Regional Water Stress and Carrying Capacities

Due to the fact that production capacities in the case of certain mining commodities of global relevance are highly concentrated in few regions, e.g., cobalt, platinum, palladium, iron ore and uranium, the global water scarcity impact of these commodities is significantly influenced by the water stress conditions of the corresponding mining regions. Regarding these circumstances, and as the global demand for mining commodities is expected to grow in the future, this may also lead to increasing environmental impacts within the regions where mining is predominantly occurring. Hence, the question arises as to what impact industrial mining has on water resources on a regional scale taking into account mining's influence on the carrying capacity of regional hydrological systems to provide sufficient freshwater at regional and even local levels. To find an adequate answer to this question, a water impact assessment at the river basin level was conducted considering intra-annual changes in water demand and water availability affected by all mining activities operating in the particular basin.

Depending on economic structures and existing water-using sectors, the water consumption in every region varies in amount and over time, e.g., by seasonal irrigation activities in water-intensive agricultural landscapes or intense water use in heavily industrialized and thus energy-consumptive areas, affecting the availability and quality of region-specific water resources in different ways. Particularly in mining-intense regions, large-scale mining operations may also contribute different proportions of the overall water consumption in the given area, finally resulting in impacts on freshwater quantity and quality of aquifers, groundwater or river flow levels. As the water impact assessment of this survey focused on the influence of water use in the extractive industry on hydrological systems, this observation primarily addresses impacts at the watershed level, providing natural water resources for a wide range of different water-using sectors in the particular catchment.

Based on the relation between water availability and water use, each basin is characterized by individual water stress levels and volumes of metal mining directly affecting the basin's water sources and therefore competing with other water-consuming sectors, including environmental water requirements. This so-called watershed-specific impact assessment was conducted at the major river basin level as defined by the Global Runoff Data Centre (GRDC), providing GIS-based hydrological data for 405 major river basins and 687 associated river systems in total [83]. Hereafter the major river basins observed are termed 'GRDC-basins'. The GRDC-basins were implemented in the GIS model; correlated with mining coordinates, including mine-site data of commodity production, LCA-based water consumption data and water scarcity footprints for each mining property; and finally summed for each watershed.

Figure 4 shows the results of the basin-related observation. While Figure 4a illustrates the global coverage of all the GRDC-basins considered in this study, Figure 4b,c visualize the total water consumption of all mining activities considered in this study as well as the resulting water scarcity impacts according to each GRDC-basin.

According to Figure 4a, GRDC-basins cover approximately 65% of the global land area only; wide land surfaces are not included, such as Antarctica, Greenland and large areas of northern Canada. Mostly desert regions are also excluded, such as large areas of the Saharan desert, the Arabian Peninsula, the Afghan and Iranian deserts, the Gobi Desert in Asia, the Mojave Desert in North America and Australian deserts. Also excluded are areas of Oceania, the Indonesian archipelago and many minor fractions of land, often along the coasts, which are not part of major river basins. This also includes artificial land, which is mostly relevant to the Asian region. Due to these limitations, numerous mining areas are located outside officially defined GRDC-basins and were not considered in this regional assessment.

However, regarding all 14 commodities observed in this study, 1783 mining sites were identified in total, located in 147 out of 405 catchment areas and representing roughly two-thirds of all mining operations included in this survey, with a production volume of approximately 1.374 bn t of metal-eq. in 2018. All 147 GRDC-basins with mining activities were subjected to a brief analysis to determine the extent to which mining contributes to regional water stress. Overall, some basins are significantly characterized by a high concentration of mining activities simultaneously combined with high overall water consumption (Figure 4b) and low water availability, therefore leading to high annual mean water stress values and high physical water scarcity magnitudes (Figure 4c).

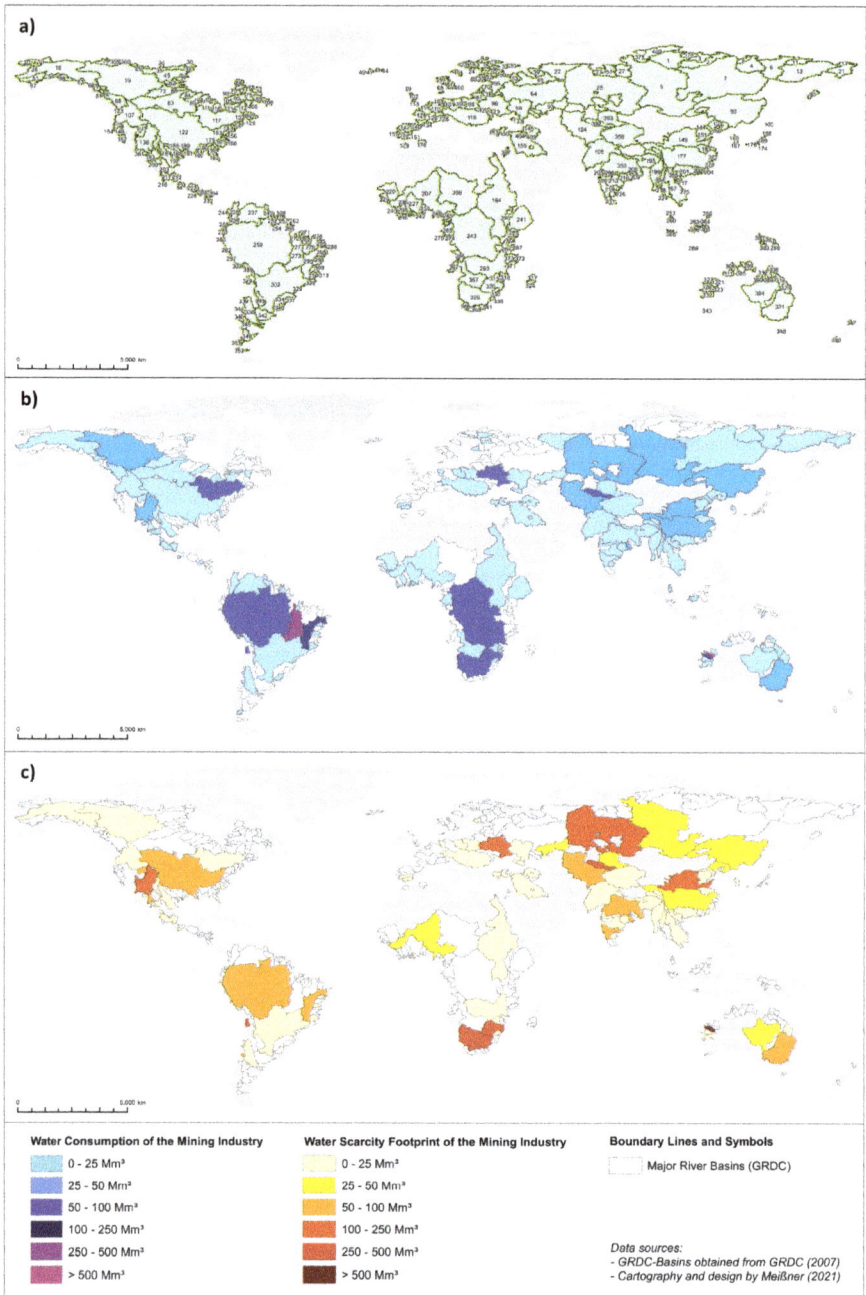

Figure 4. Total water consumption and water scarcity footprint as a result of mining according to major river basins as defined by the GRDC [83]: (**a**) global coverage of GRDC-basins; (**b**) total water consumption of mining; (**c**) total water scarcity footprint as a result of mining in each GRDC-river basin. (The description of the identification numbers of the GRDC catchment areas shown in Figure 4a as well as calculation data of Figure 4b,c is provided in the Supplementary Material spreadsheet, see Table S3).

In addition to Figure 4, Table 4 presents the basins with the highest annual water consumption levels observed in the mining sector as well as the resultant water scarcity effects. The basins with the highest global water consumption in the mining sector are mainly subject to large-scale iron ore mining, particularly Fortescue River (Australia), Tocantins (Brazil), Ashburton River (Australia) and Sao Francisco (Brazil), followed by Orange (South Africa), Amazonas (Brazil), Limpopo (South Africa), Loa (Chile) and others, where water consumption is predominantly driven by copper and gold mining. However, the total water scarcity impact resulted from mining-related water consumptions highly depends on the basin's overall water stress intensity—e.g., in the case of basins with lower water consumption for mining, such as Ob Basin, Colorado River or Huang He (Yellow River), the highest water scarcity effects can be observed due to the high water stress occurring in the respective basins. By contrast, Fortescue River Basin in Australia or Tocantins Basin in Brazil, both characterized by the highest water consumption rates in the mining industry, are only slightly affected by mining-related water scarcity impacts. This is due to the low average water stress conditions in the basins as well as partly being classified as 'arid with low water use'. Particularly in the case of higher water stress intensities, the 'water consumption/water scarcity impact ratio' is a meaningful indicator showing the magnitude of the water scarcity intensity resulting from a given volume of water used in a particular basin. According to this, the highest water scarcity effects were documented for Issyk-Kul Basin with a 'water consumption/water scarcity impact ratio' of 1:5, followed by Ashburton (~1:5), Huang He (1:4.7) and Colorado River Basin (1:4.4) which were characterized by high water stress intensities.

Table 4. GRDC-basins characterized by highest mining-related water consumption and water scarcity effects.

Basin	Basin-ID	Water Consumption (Mm^3)	Water Scarcity Impact (Mm^3)	Water Consumption/Water Scarcity Ratio	Most Relevant Mining Commodities (in Terms of Production Volumes)
Fortescue River (Australia)	No. 323	522.7	<1.0 [1]	>500:1	Iron ore
Tocantins (Brazil)	No. 273	259.9	<1.0	~260:1	Iron ore
Ashburton River (Australia)	No. 327	208.0	1040.0	~1:5	Iron ore
Sao Francisco (Brazil)	No. 290	166.0	63.2	2.6:1	Iron ore
Orange (South Africa)	No. 326	87.5	270.6	1:3.1	Gold, iron ore, manganese
Amazonas (Brazil)	No. 259	85.6	71.2	1:0.8	Copper, gold, bauxite, zinc, silver
Limpopo (South Africa)	No. 320	76.8	256.0	1:3.3	Platinum, palladium, nickel, gold, copper
Loa (Chile)	No. 319	73.6	270.6	1:3.7	Copper, molybdenum, silver
Congo (Central Africa)	No. 243	69.1	<1.0	~70:1	Copper, cobalt, gold
St. Lawrence (USA, Canada)	No. 117	62.5	<1.0	~100:1	Nickel, cobalt, copper, gold
Zambezi (Central Africa)	No. 293	57.5	1.4	~40:1	Copper, nickel
Dnieper (Ukraine, Belarus, Russia)	No. 96	55.3	114.8	1:2.1	Iron ore
Issyk-Kul (Kazakhstan, Kyrgyzstan)	No. 392	51.2	255.9	1:5	Uranium
Ob (Russia)	No. 25	46.2	132.0	1:2.9	Gold, copper, zinc, iron ore, bauxite, lead, silver, uranium
Colorado River (USA, Mexico)	No. 138	43.9	192.5	1:4.4	Copper, molybdenum, gold
Huang He (Yellow River) (China)	No. 149	43.1	200.8	1:4.7	Nickel, molybdenum, gold, copper, zinc

[1] Low water stress but classified as 'arid and low water use' according to Gassert et al. [69]. (Identification numbers of the GRDC catchment areas as well as basin-specific mining data, water consumption and water scarcity calculations are provided in the Supplementary Materials spreadsheet, see Table S3).

However, as water availability and water demand can significantly vary between basins and over the year, it may also be important to consider the intra-annual variability of water stress as well as water scarcity impacts caused by mining activities. By conducting a water stress assessment regarding the seasonal variability of both natural water supplies and water consumption, it could be shown that numerous basins are affected by mining-related water stress for limited time periods over the year. In order to quantify mining's contribution to the water stress of each individual major river basin, particularly taking into account intra-annual variability, the cumulative water consumption of all mining activities was compared to the monthly water stress of each basin. As industrial, large-scale mining commonly takes place under continuous operation, 24 h a day, 7 days per week [3], it was assumed that there is constant mining and commodity production throughout the entire year without being affected by significant interruptions of the operations. To conduct a monthly based water impact assessment at the basin level, runoff data were taken from Hoekstra and Mekonnen [84], who obtained basin-specific runoff data from the Composite Runoff Database (based on Fekete et al. [104]) referring to the average over the time period of 1996–2005. Furthermore, as proposed by Sood et al. [73], the water availability of each basin was calculated as the natural runoff minus the environmental flow requirements (EWRs). According to Vanham et al. [52], the consideration of region-specific EWRs is highly recommended in order to align water scarcity assessments with the framework of water-related sustainability goal SDG 6. In conclusion, this regional water impact assessment comprises several aspects, including seasonal variability of water availability and water consumption, particularly water consumption in the mining sector, taking into account the volumes of freshwater needed to sustain basin-related ecosystems and eco-services. By considering EWRs it is thus assumed that the amount of water available in a particular basin can be fully used without affecting the environmental integrity of ecosystems. This EWR-specific water availability is also determined as the 'economic carrying capacity' of water resources of each basin. Hence, in the case of the water consumption exceeding the water availability and the regional 'economic carrying capacity', it has to be assumed that there is significant pressure on regional ecosystems.

The results of this intra-annual water impact assessment are illustrated in Figures 5 and 6, showing the basin-wise cumulative water consumption of all mining operations on a monthly basis in relation to regional water availability, taking into account EWRs. Based on the relation between water availability and water use in the mining sector, the mining-related intra-annual water stress levels within all 147 GRDC-basins were calculated.

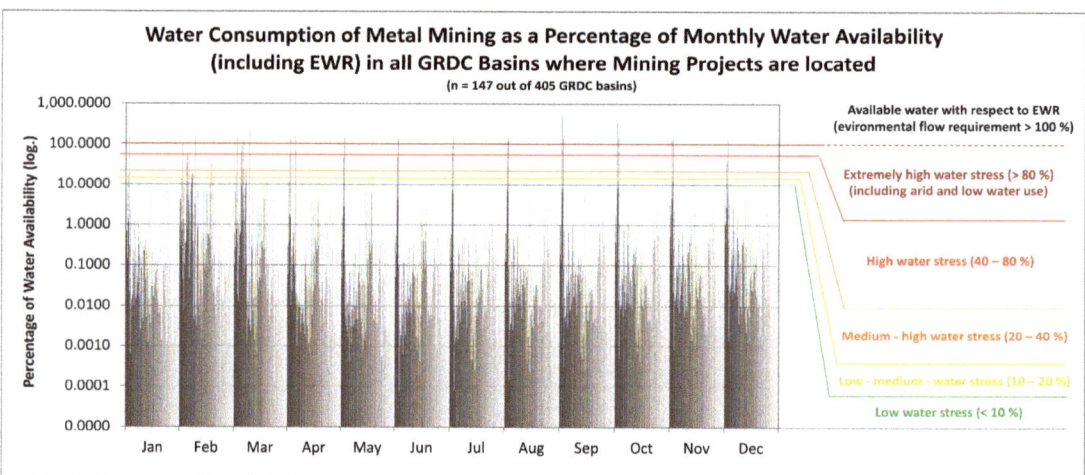

Figure 5. Monthly water consumption of mining as a percentage of water availability in 147 major GRDC-basins hosting at least one mining operation of all mining projects considered in this survey (available water is defined as basin-specific runoff minus environmental water requirements); depending on the monthly proportion of available water consumed, mining results in intra-annually varying water stress levels. Water stress was determined by the proportion of water consumed in relation to the water available in the corresponding time period. (Calculation data of Figure 5 is provided in the Supplementary Materials spreadsheet, see Table S3).

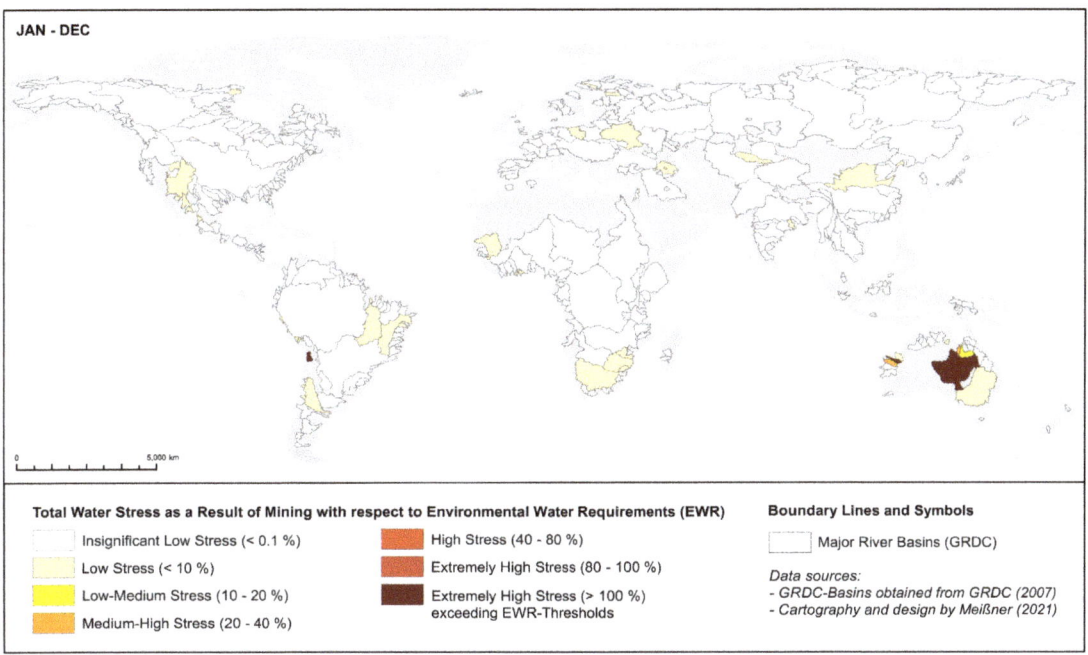

Figure 6. Mean annual water stress in 147 GRDC-basins as a result of water consumption in mining. Water stress was determined by the proportion of the overall water consumed in mining within a basin in relation to basin-specific water availability. Water availability is defined as natural runoff minus environmental water requirement (EWR). Depending on the monthly proportion of available water consumed, mining results in intra-annually varying water stress levels; particularly in the case of the water consumption exceeding 100% of the regional water availability (referred to as regional EWR threshold or 'economic carrying capacity'), mining significantly influences the carrying capacity of the corresponding basin within the given monthly time period. (GRDC-basins are obtained from GRDC [83]; a time series of detailed maps illustrating changes and intensity of monthly mining-induced water stress over the year is provided in the Supplementary Materials, see Figure S6a–m) calculation data of Figure 6 and Figure S6a–m is provided in the Supplementary Materials spreadsheet, see Table S3).

As a result, it could be validated that water stress caused by mining is predominantly limited to 'low' water stress over the year in many of the catchments, not surpassing 10% of the basins' available water used in total. In most cases, mining had only an insignificant influence on the basins' overall water stress, accounting for less than 0.1% of the total water consumption. However, there were also exceptional cases where the environmental flow requirements were completely surpassed by the consumption of freshwater by the mining industry during the entire year. This was the case when water consumption exceeded 100% of the available water. Thus, in some basins, mining was the overall dominant water user, mostly due to large-scale mining projects significantly influencing the regional water availability. However, this situation also offers notable potential to reduce water stress in the entire basin by improving water use efficiency, implementing further water management measures or increasing use of seawater and recycled mine waters at one particular mining location.

According to this intra-annual and watershed-specific water impact assessment, the following basins are amongst the river systems characterized by the highest mining-related water impacts in terms of water scarcity effects and water stress, surpassing the EWR thresholds which indicate the amount of water required to maintain the basin-specific ecosystems:

- Loa Basin (Figure 4a, basin no. 319), located in the northern part of the Antofagasta region in Chile, is one of the world's leading copper-mining areas, hosting 7.9% of global copper and 8.2% of global molybdenum production (according to production in 2018). Mining alone is the main driving factor causing the 'extremely high' water stress due to exceeding the available water limits during the entire year as a result of hyper-arid conditions simultaneously paired with the highest rates of water consumption in the mining sector. Owing to the fact that copper mining is ranked as one of the largest water-consuming sectors in the global mining industry, many large-scale production capacities are primarily located in Chile, therefore resulting in intense water scarcity impacts in the Chilean Loa Basin. As a consequence, the total annual water scarcity impact caused by mining operations in Loa Basin accounts for 368.17 Mm^3, of which 338.3 Mm^3 is associated with copper mining, which represents 14.5% of the global copper-related water scarcity effect. Another large water scarcity impact of 22.5 Mm^3 results from molybdenum mining, which is often jointly performed with copper production.
- Similar to Loa Basin, the Pilbara region in northwestern Australia is also characterized by high water scarcity impacts due to large-scale mining under arid conditions. Pilbara comprises three river basins, namely De Grey River (Figure 4a, basin no. 321), Fortescue (basin no. 323) and Ashburton (basin no. 327), altogether providing 53.3% of the global iron ore fine supply—i.e., the overall results in Pilbara are significantly influenced by iron ore mining. However, the percentage of water consumption for mining in Pilbara varies significantly between the basins. For instance, while mining's influence on the basin's water stress in De Grey River is basically low (usually below 10% of the basin's total water consumption per month), its contribution to the overall water stress in Ashburton is slightly above 10% in the period from January to March but exceeds the EWR limits significantly from September to October, thus causing a range of 'low–medium' to 'extremely high' water stress throughout the year. By contrast, in Fortescue River Basin, mining alone is responsible for the 'extremely high' water stress during most of the year, particularly surpassing EWR thresholds from April to December. Overall, as the production of iron ore fines is responsible for the largest global water consumption amongst all mining commodities observed in this study, Pilbara is, after Loa Basin, the most prominent area affected by high mining-related water stress as well as a water scarcity impact accounting for 1039.91 Mm^3 in total. This represents roughly 55% of the global water scarcity effect resulting from iron ore mining, particularly regarding iron ore fines.
- Orange Basin (Figure 4a, basin no. 326) and Limpopo Basin (basin no. 320) are further prominent examples of river catchments affected by high water stress and water scarcity effects caused by the mining industry. Both basins are located in South Africa, covering areas of the Republic of South Africa, Lesotho, Namibia, Botswana, Zimbabwe and Mozambique. While Limpopo Basin supplied 68.3% (146.7 t) of global platinum and 36.9% (80.3 t) of palladium production in 2018, mining operations situated in Orange Basin contributed to 4.3% (114.4 t) of global gold production, which is the largest gold production capacity in the basins observed. However, even relatively low quantities of metal production for gold or platinum group metals, for example, may cause high water stress and water scarcity impacts, particularly due to the relatively high specific water demand per t refined metal. Consequently, both basins are highly influenced by water consumption for the precious metals mining and refining industry. Overall, this mining sector is mostly responsible for 'low' water stress in both basins, primarily averaging between 5 and 9% from June and December and peaking at 13.5% in November, which is classified as 'low–medium' water stress. Besides gold and PGM production, mining of copper, iron ore and manganese also has a significant influence on the hydrological system of both basins. Manganese production in Orange Basin represents 35.6% of global manganese production, resulting in a global water scarcity footprint (WSFP) of 74.4% for manganese mining. While

the total WSFP of mining in Orange Basin is 270 Mm^3, the annual mining-related WSFP in Limpopo Basin is estimated to be 256 Mm^3, mainly caused by the mining and refining of platinum group metals. For instance, palladium production in Limpopo accounts for 72.6% of the global palladium WSFP and platinum production accounts for approximately 81% of the global platinum WSFP.

Besides Loa, Pilbara, Orange and Limpopo, other basins have to be mentioned due to the significant water scarcity impacts from mining. Firstly, the transboundary Issyk-Kul Basin (Figure 4, basin no. 392) shared by Kazakhstan and Kyrgyzstan holds a water scarcity impact of 255.89 Mm^3, which results predominantly from the production of uranium concentrate (U_3O_8), contributing to 40.3% of the global uranium WSFP. Secondly, in China's Huang He Basin (Yellow River, basin no. 149), mining-related water scarcity effects account for 200.82 Mm^3, mainly caused by lead, molybdenum and nickel production. Thirdly, Colorado River Basin (basin no. 138) and Great Salt Lake Basin (basin no. 401), both situated in the southwest of the United States, are also characterized by high mining-related WSFPs of 192.46 and 59.31 Mm^3, respectively. Both basins hold roughly 5.4% of global copper and 9.5% of global molybdenum mining production. Ob Basin is also considered a catchment with high water scarcity effects of mining, accounting for 131.99 Mm^3, mainly as it contributes to 3.4% of global bauxite mining as well as 4.7% of zinc and 5.4% of lead mining. However, mining's influence on water stress in Ob Basin accounts for less than 0.1%, which is almost negligible compared to other water-consuming sectors in the basin. Finally, Dnieper Basin is also ranked amongst the basins with the highest mining-related WSFPs at 114.75 Mm^3, with mining-induced 'low' water stress.

In addition to all the basins mentioned above, characterized by high water consumption in the mining industry, Eyre Lake Basin in Australia (basin no. 394) is an exceptional case with a moderate water scarcity impact accounting for 29.4 Mm^3 in total. However, due to the fact that the basin's climate is predominantly classified as arid to hyper-arid, mining is responsible for 'extremely high' water stress, exceeding EWR thresholds during the entire year.

Overall, the brief examinations of the selected basins clearly showed the wide range of region-specific influences under which water is used and consumed in the mining sector and its respective influence on water availability and water stress on both a monthly and an annual basis. It was also shown that in the case of high water availability, high water scarcity impacts of mining do not necessarily result in high water stress levels significantly affecting regional carrying capacities. However, in some basins mining is clearly the dominant factor in regional water consumption during the entire year, thus highly influencing local water stress conditions.

4. Discussion and Conclusions

The methodological approach provided in this article, combining a geographical information system (GIS) with LCA-based water use inventories and finally providing information about water stress and the water scarcity effects of the mining industry, not only enables conducting water impact assessments at the explicit regional level but also provides a comprehensive view on how mining activities are distributed all over the world in relation to regional water stress contexts. Consequently, this provides an advanced understanding of the complexity and wide range of water footprints occurring in the mining industry.

The results of this approach may also support decision making in both integrated water resource management processes at a regional level and raw materials criticality assessments at a global level, taking into account environmental risks caused by metal production. Raw materials criticality assessments primarily observe supply risks and environmental impacts as a result of a single raw material production; however, this GIS-based approach also provides data on cumulative water impacts caused by the production of multiple mining commodities at a particular place or region. Owing to the fact that numerous mining projects jointly produce various metals and their concentrates simultaneously, cumulative assessment considering all commodities produced at a particular mining location and within the entire basin should be the standard to cover the entire range of impacts instead of assessing the impacts of individual commodities separately—particularly with respect to SDG 6.4 [52], which recommends a comprehensive water impact assessment at the catchment level. As risk analyses have been carried out exclusively from the raw material point of view thus far, future impact assessments should also be carried out from the perspective of the region in which the entire production and the associated environmental impacts occur. This approach increasingly reflects the circumstances under which raw materials are produced, providing important information to address region-specific challenges to overcome in order to advance the mining sector towards sustainable supply of natural resources, particularly with respect to local environmental water requirements. Consequently, this transdisciplinary approach helps to significantly specify and improve the criticality assessment of raw materials with respect to water-related risks. In terms of future development, this approach could also be further expanded by increasing the scope by incorporating additional environmental impact categories, particularly those related to the SDG framework. Broadening this scope of sustainability assessment is necessary to gain a more holistic view of the complexity and wide range of impacts related to global mining and raw materials supply in general.

Despite the promising outcomes of this GIS-based approach, there are notable limits that were also identified in this study, particularly in relation to the quality and availability of water-specific LCA inventory data as well as the representativeness of the WSI characterization factor used in the GIS. These limitations have to be considered for future research on water footprints in the mining sector. The most relevant is the fact that the number of publications observing the water consumption of metal mining is still very limited, and the available LCA data show significant variability for individual mining operations and processing methods throughout the entire literature reviewed. This variation is due to the different mine types, ore types and given mineral grades and thus the different processing pathways applied to produce mining commodities. In addition, the LCA data provided by established LCA databases, e.g., Ecoinvent [86] and PROBAS [90], are not up to date consistently over all datasets. For example, some of the data records still refer to a time period before 2000 and are thus increasingly outdated or particularly addressing traditional mining and refining methods only, almost neglecting the latest developments in mining technology. Furthermore, depending on the geological setting of mining locations, different extraction methods are applied, e.g., leaching methods, or combined and modified as mining projects move from open pit to underground mining as a result of declining ore grades in the minerals mined. Correspondingly, each extraction technique is characterized by a wide range of water consumption volumes, which is not fully covered by LCA data. Furthermore, different system boundaries were determined by the consulted studies, conducted for different mine sites and time periods. As a consequence, the comparison of data within the literature was very limited, particularly between all commodities observed. As the water impact assessment of this survey is related to different processing stages of the particular commodities, such as preprocessed ore, concentrates or refined metals, even in this study comparison of the results of each commodity is only useful to a limited extent. Moreover, there is notable variability and substantial uncertainty inherent in the existing inventory data for mining commodities, particularly with regard to the origin of the water sources obtained for individual mining operations, and the shares of reused mine water are mostly unknown or neglected. Additionally, water treatment and desalination were not considered but play an important role in mining's current and future water supply. These measures in the field of water management might also help to reduce pressures on local water sources and thus minimize mining's influence on local water conflicts. However, these efforts are not measured as part of many LCA studies and inventory datasets, thus making interpretation of the water stress resulting from mining extremely difficult.

Besides LCA-related limits, there are also challenges and restrictions to be mentioned concerning the WSI characterization factor as well as the EWR thresholds applied in the

GIS. Even though EWR was considered in the water impact assessment, only average EWR values at the continental level were used in the GIS model. Thus, while working on a global scale with such a complex issue, numerous assumptions in defining the EWR values for both river water and groundwater had to be made. Based on these limitations, there is an urgent need for catchment-specific EWR quantifications because there are manifold ecological differences to be observed across a continent-wide area which are not representative for each particular mining location. Consequently, more recent, detailed and locally sourced data on water availability, water demand and environmental water requirements would be helpful to assess water-related risks more accurately. Finally, depending on the commodities observed, all the mining locations taken into account represent a range of approximately 60–100% of the global production volumes. Thus, uncertainty still remains with regard to the water impact intensities of mining locations which were neglected in this study.

However, as the findings of this study notably advance the understanding of the relationship between the increasing global demand of mining commodities and simultaneously rising water shortages on a regional scale, this GIS-based assessment approach does not aim to replace detailed water impact assessment at the local level. It is, rather, intended to enhance global impact assessments increasingly based on data derived from explicit locations where mining and metal production sites are situated. Hence, to understand the complex and individual interactions between mining and other water-consuming sectors, further case studies are needed to complement both the quantitative and qualitative aspects of water-related impacts as a result of a global mining industry that is still expected to grow in the future.

Supplementary Materials: The following are available online at https://www.mdpi.com/article/10.3390/resources10120120/s1; Figure S2a–n: Locations of mining properties of all 14 individual mining commodities in relation to water stress at sub-basin level; Figure S6a–m: World maps of monthly basin-related water stress as a result of mining. Tables S1 and S2 in the calculation spreadsheet: Commodity-wise data on annual production, water consumption and water scarcity impact related to each mining operation (data for Figure 2a–c, Figure S2a–n, and Figure 3a,b); Table S3 in the calculation spreadsheet: Commodity-wise data on annual production, water consumption and water scarcity impact related to GRDC-basins (data for Figure 4a–c, Figure 5, Figure 6, Figure S6a–m and Table 4). In addition, Figures 1–6 are included in the Supplementary Materials as high-resolution pdf files as well.

Funding: The APC was funded by the Environmental Science Center at the University of Augsburg and the license fee of SNL Metals & Mining Database was funded by the Chair of Resource Strategy at the University of Augsburg. Apart from this, the research received no external funding.

Institutional Review Board Statement: Not applicable.

Informed Consent Statement: Not applicable.

Data Availability Statement: The data presented in this study were predominantly obtained from publicly available literature sources and databases. The respective references are mentioned in the article. The only exceptions are selected datasets from the fee-based "SNL Metals & Mining database" (geographic coordinates and mine-site production volumes of all mining operations considered) and "Ecoinvent database" (commodity-specific water consumption values). The particular datasets are not publicly available due to the terms of use given by the publishers.

Conflicts of Interest: The author declares no conflict of interest.

Abbreviations

BWS	Baseline water stress
CTA	Consumption-to-availability ratio
EFR	Environmental flow requirement

EWR	Environmental water requirement
GIS	Geographic information system
GRDC	Global Runoff Data Centre
ISO	International Organization for Standardization
LCA	Life cycle assessment
MAR	Mean annual runoff
RCP	Representative concentration pathway
SDG	Sustainable development goal
SSP	Shared socioeconomic pathway
WA	Water availability
WS	Water stress
WSFP	Water scarcity footprint
WSI	Water stress index
WTA	Withdrawal-to-availability ratio
WU	Water use

References

1. Helbig, C.; Bruckler, M.; Thorenz, A.; Tuma, A. An Overview of Indicator Choice and Normalization in Raw Material Supply Risk Assessments. *Resources* **2021**, *10*, 79. [CrossRef]
2. Sonderegger, T.; Berger, M.; Alvarenga, R.; Bach, V.; Cimprich, A.; Dewulf, J.; Frischknecht, R.; Guinée, J.; Helbig, C.; Huppertz, T.; et al. Mineral resources in life cycle impact assessment—Part I: A critical review of existing methods. *Int. J. Life Cycle Assess.* **2020**, *25*, 784–797. [CrossRef]
3. Spitz, K.; Trudinger, J. *Mining and the Environment. From Ore to Metal*; CRC Press Taylor & Francis Group: London, UK, 2009.
4. Schrijvers, D.; Hool, A.; Blengini, G.A.; Chen, W.-Q.; Dewulf, J.; Eggert, R.; van Ellen, L.; Gauss, R.; Goddin, J.; Habib, K.; et al. A review of methods and data to determine raw material criticality. *Resour. Conserv. Recycl.* **2020**, *155*, 104617. [CrossRef]
5. Gunson, A.J.; Klein, B.; Veiga, M.; Dunbar, S. Reducing mine water requirements. *J. Clean. Prod.* **2012**, *21*, 71–82. [CrossRef]
6. Brown, E. Water for a sustainable minerals industry—A review. In Proceedings of the Water in Mining 2003, Brisbane, Australia, 13–15 October 2003; The Australasian Institute of Mining and Metallurgy (AusIMM): Brisbane, Australia, 2003.
7. Bangerter, P.; Dixon, R.; Villegas, M. Improving overall usage of water in mining-A sustainable development approach. In Proceedings of the 2nd International Congress on Water Management in the Mining Industry, Water in Mining, Antofagasta, Chile, 2–4 June 2010; Wiertz, J., Ed.; p. 403.
8. Nedved, M.; Jansz, J. Waste water pollution control in the Australian mining industry. *J. Clean. Prod.* **2006**, *14*, 1118–1120. [CrossRef]
9. Akcil, A.; Koldas, S. Acid mine drainage (AMD): Causes, treatment and case studies. *J. Clean. Prod.* **2006**, *14*, 1139–1145. [CrossRef]
10. Cohen, R. Use of microbes for cost reduction of metal removal from metals and mining industry waste streams. *J. Clean. Prod.* **2006**, *14*, 1146–1157. [CrossRef]
11. Santana, C.S.; Montalván, D.M.; Vinnícius, O.; Francisco, H.C.S.; Fermin, H.M.L.; Raildo, G.V.; de Jesus, M. Assessment of water resources pollution associated with mining activity in a semi-arid region. *J. Environ. Manag.* **2020**, *273*, 111148. [CrossRef]
12. Ugya, A.; Ajibade, F.; Ajibade, T. Water pollution resulting from mining activity: An overview. In Proceedings of the 2018 Annual Conference of the School of Engineering & Engineering Technology, The Federal University of Technology (FUTA), Akure, Nigeria, 17–19 July 2018. Available online: https://www.researchgate.net/publication/326925600_Water_Pollution_Resulting_From_Mining_Activity_An_Overview (accessed on 6 November 2021).
13. Mudd, G.M. Sustainability Reporting and Water Resources: A Preliminary Assessment of Embodied Water and Sustainable Mining. *Mine Water Environ.* **2008**, *27*, 136–144. [CrossRef]
14. Northey, S.A.; Mudd, G.M.; Saarivuori, E.; Wessman-Jääskeläinen, H.; Haque, N. Water footprinting and mining: Where are the limitations and opportunities? *J. Clean. Prod.* **2016**, *135*, 1098–1116. [CrossRef]
15. Meißner, S. Global Metal Mining and Physical Water Stress–Water Impact Assessment of the Mining Industry within the Raw Materials Criticality Methodology. Habilitation-Thesis, University of Augsburg, Augsburg, Germany, 20 June 2021.
16. Northey, S.A.; Mudd, G.M.; Werner, T.T.; Jowitt, S.M.; Haque, N.; Yellishetty, M.; Weng, Z. The exposure of global base metal resources to water criticality, scarcity and climate change. *Glob. Environ. Chang.* **2017**, *44*, 109–124. [CrossRef]
17. International Council on Mining & Metals (ICMM) (Ed.) *Where and How does Mining Take Place?* London, UK. 2019. Available online: https://www.icmm.com/en-gb/metals-and-minerals/producing-metals/where-and-how-does-mining-take-place (accessed on 20 December 2019).
18. Gilsbach, L.; Schütte, P.; Franken, G. Applying water risk assessment methods in mining: Current challenges and opportunities. *Water Resour. Ind.* **2019**, *22*, 100118. [CrossRef]
19. Kinnunen, P.; Obenaus-Emler, R.; Raatikainen, J.; Guignot, S.; Guimerà, G.; Ciroth, A.; Heiskanen, K. Review of closed water loops with ore sorting and tailings valorisation for a more sustainable mining industry. *J. Clean. Prod.* **2021**, *278*, 123237. [CrossRef]
20. Masood, N.; Hudson-Edwards, K.; Farooqi, A. True cost of coal: Coal mining industry and its associated environmental impacts on water resource development. *J. Sustain. Min.* **2020**, *19*, 135–149. [CrossRef]

21. Miller, K.D.; Bentley, M.J.; Joseph, N.; Linden, R.K.G.; Larison, C.; Kienzle, B.A.; Katz, L.E.; Wilson, A.M.; Cox, J.T.; Kurup, P.; et al. Mine Water Use, Treatment, and Reuse in the United States: A Look at Current Industry Practices and Select Case Studies. *ACS ES&T Eng.* **2021**. [CrossRef]
22. Le, T.M.K.; Miettinen, H.; Bomberg, M.; Schreithofer, N.; Dahl, O. Challenges in the Assessment of Mining Process Water Quality. *Minerals* **2020**, *10*, 940. [CrossRef]
23. International Aluminium Institute (IAI). *Life Cycle Inventory Data and Environmental Metrics for the Primary Aluminium Industry, 2015 Data*; International Aluminium Institute: London, UK, 2017.
24. Cote, C.; Moran, C.; Cummings, J.; Ringwood, K. Developing a water accounting framework for the Australian minerals industry. *Min. Technol.* **2009**, *118*, 162–176. [CrossRef]
25. Cote, C.; Moran, C.; Hedemann, C.; Koch, C. Systems modelling for effective mine water management. *Environ. Model. Softw.* **2010**, *25*, 1664–1671. [CrossRef]
26. Ferreira, H.; Leite, M.G.P. A Life Cycle Assessment study of iron ore mining. *J. Clean. Prod.* **2015**, *108*, 1081–1091. [CrossRef]
27. Gunson, A.J. Quantifying, Reducing and Improving Mine Water Use. Ph.D. Thesis, University of British Columbia, Vancouver, BC, Canada, 2013. [CrossRef]
28. Lutter, S.; Giljum, S. *Copper Production in Chile Requires 500 Million Cubic Metres of Water. An Assessment of the Water Use by Chile's Copper Mining Industry*; FINEPRINT Brief No. 9; Vienna University of Economics and Business: Vienna, Austria, 2019. Available online: https://www.fineprint.global/publications/briefs/chile-copper-water/ (accessed on 26 October 2020).
29. Mudd, G.M. An analysis of historic production trends in Australian base metal mining. *Ore Geol. Rev.* **2007**, *32*, 227–261. [CrossRef]
30. Mudd, G.M. Gold mining in Australia: Linking historical trends and environmental and resource sustainability. *Environ. Sci. Policy* **2007**, *10*, 629–644. [CrossRef]
31. Mudd, G.M. *The Sustainability of Mining in Australia: Key Production Trends and Their Environmental Implications for the Future*; Research Report No. RR5; Department of Civil Engineering, Monash University and Mineral Policy Institute: Clayton, Australia, 2009.
32. Norgate, T.; Haque, N. Energy and greenhouse gas impacts of mining and mineral processing operations. *J. Clean. Prod.* **2010**, *18*, 266–274. [CrossRef]
33. Northey, S.A.; Haque, N.; Mudd, G. Using sustainability reporting to assess the environmental footprint of copper mining. *J. Clean. Prod.* **2013**, *40*, 118–128. [CrossRef]
34. Northey, S.A. Assessing Water Risks in the Mining Industry Using Life Cycle Assessment Based Approaches. Ph.D. Thesis, Monash University, Clayton, Australia, 2018.
35. Northey, S.A.; Mudd, G.M.; Werner, T.T.; Haque, N.; Yellishetty, M. Sustainable water management and improved corporate reporting in mining. *Water Resour. Ind.* **2019**, *21*, 100104. [CrossRef]
36. Suppen, N.; Carranza, M.; Huerta, M.; Hernández, M.A. Environmental management and life cycle approaches in the Mexican mining industry. *J. Clean. Prod.* **2006**, *14*, 1101–1115. [CrossRef]
37. Worrall, R.; Neil, D.; Brereton, D.; Mulligan, D. Towards a sustainability criteria and indicators framework for legacy mine land. *J. Clean. Prod.* **2009**, *17*, 1426–1434. [CrossRef]
38. Damkjaer, S.; Taylor, R. The measurement of water scarcity: Defining a meaningful indicator. *Ambio* **2017**, *46*, 513–531. [CrossRef]
39. International Organization for Standardization (ISO). *ISO 14040:2006. Environmental Management-Life Cycle Assessment—Principles and Framework*; International Organization for Standardization (ISO): Geneva, Switzerland, 2006.
40. International Organization for Standardization (ISO). *ISO 14046:2014. Environmental Management-Water Footprint-Principles, Requirements and Guidelines*; International Organization for Standardization (ISO): Geneva, Switzerland, 2014.
41. Allan, J.A. Fortunately there are Substitutes for Water Otherwise our Hydro-political Futures would be Impossible, Priorities for Water Resources Allocation and Management. In *Priorities for water resources allocation and management*; ODA: London, UK, 1993; pp. 13–26.
42. Allan, J.A. Overall Perspectives on Countries and Regions. In *Water in the Arab World: Perspectives and Prognoses*; Rogers, P., Lydon, P., Eds.; Harvard University Press: Cambridge, UK, 1994; pp. 65–100.
43. Hoekstra, A.Y. (Ed.) Virtual Water Trade. In *Proceedings of the International Expert Meeting on Virtual Water Trade*; Value of Water Research Report Series No. 12; IHE Delft: Delft, The Netherlands, 2003.
44. Hoekstra, A.Y.; Chapagain, A.K.; Aldaya, M.M.; Mekonnen, M.M. *Water Footprint Manual, State of the Art 2009*; Water Footprint Network, Earthscan: London, UK, 2009.
45. Hoekstra, A.Y.; Chapagain, A.K.; Aldaya, M.M.; Mekonnen, M.M. *The Water Footprint Assessment Manual: Setting the Global Standard*; Water Footprint Network, Earthscan: London, UK, 2011.
46. Pfister, S. Environmental Evaluation of Freshwater Consumption within the Framework of Life Cycle Assessment. Ph.D. Thesis, Diss. ETH no. 19490. ETH Zurich, Zurich, Switzerland, 2011. [CrossRef]
47. Kasbohm, J.; Grothe, S.; Steingrube, W.; Lài, L.; Ngân, L.; Hồng, N.T.; Oanh, L.T.K.; Hương, N.Q. Integrated water resources management (iwrm)–An introduction. *J. Geol.* **2009**, *33*, 3–14. Available online: https://www.researchgate.net/publication/237062491 (accessed on 26 October 2020).
48. Falkenmark, M.; Lundqvist, J.; Widstrand, C. Macro-scale water scarcity requires micro-scale approaches. *Natl. Resour. Forum* **1989**, *13*, 258–267. [CrossRef] [PubMed]

49. Brown, A.; Matlock, M.D. *A Review of Water Scarcity Indices and Methodologies*; White Paper; Sustainability Consortium: Scottsdale, AZ, USA; Fayetteville, NC, USA; Den Haag, The Netherlands, 2011; Volume 106. Available online: https://www.sustainabilityconsortium.org/tsc-downloads/a-review-of-water-scarcity-indices-and-methodologies/ (accessed on 24 October 2020).
50. Liu, J.; Yang, H.; Gosling, S.N.; Kummu, M.; Flörke, M.; Pfister, S.; Hanasaki, N.; Wada, Y.; Zhang, X.; Zheng, C. Water scarcity assessments in the past, present, and future. *Earth's Future* **2017**, *5*, 545–559. [CrossRef]
51. Rijsberman, F.R. Water scarcity: Fact or fiction? *Agric. Water Manag.* **2006**, *80*, 5–22. [CrossRef]
52. Vanham, D.; Hoekstra, A.Y.; Wada, Y.; Bouraoui, F.; de Roo, A.; Mekonnen, M.; van de Bund, W.; Batelaan, O.; Pavelic, P.; Bastiaanssen, W.; et al. Physical water scarcity metrics for monitoring progress towards SDG target 6.4: An evaluation of indicator 6.4.2 "Level of water stress". *Sci. Total Environ.* **2018**, *613–614*, 218–232. [CrossRef]
53. Falkenmark, M.; Gunnar, L. How can we cope with the water resources situation by the year 2015? *Ambio* **1974**, *3*, 114–122. Available online: https://www.jstor.org/stable/4312063 (accessed on 26 October 2020).
54. Raskin, P.; Gleick, P.; Kirshen, P.; Pontius, G.; Strzepek, K. *Water Futures: Assessment of long-range patterns and problems. Comprehensive Assessment of the Freshwater Resources of the World*; SEI report; Stockholm Environment Institute: Stockholm, Sweden, 1997.
55. Rockström, J.; Falkenmark, M.; Karlberg, L.; Hoff, H.; Rost, S.; Gerten, D. Future water availability for global food production: The potential of green water for increasing resilience to global change. *Water Resour. Res.* **2009**, *45*, W00A12. [CrossRef]
56. Swain, A.; Wallensteen, P. *Comprehensive Assessment of the Freshwater Resources of the World, International Fresh Water Resources: Conflict or Cooperation?* Stockholm Environment Institute: Stockholm, Sweden, 1997.
57. Alcamo, J.; Henrich, T.; Rosch, T. *World Water in 2025-Global Modelling and Scenario Analysis for the World Commission on Water for the 21st Century*; Report A0002; Centre for Environmental System Research, University of Kassel: Kassel, Germany, 2000.
58. Arnell, N.W. Climate change and global water resources. *Glob. Environ. Chang.* **1999**, *9* (Suppl. 1), S31–S49. [CrossRef]
59. Arnell, N.W. Climate change and global water resources: SRES emissions and socioeconomic scenarios. *Glob. Environ. Chang.* **2004**, *14*, 31–52. [CrossRef]
60. Oki, T.; Agata, Y.; Kanae, S.; Saruhashi, T.; Yang, D.; Musiake, K. Global assessment of current water resources using total runoff integrating pathways. *Hydrol. Sci. J.* **2001**, *46*, 983–995. [CrossRef]
61. Seckler, D.; Barker, R.; Amarasinghe, U. Water scarcity in the twenty-first century. *Int. J. Water Resour. Dev.* **1999**, *15*, 29–42. [CrossRef]
62. Vörösmarty, C.J.; Green, P.; Salisbury, J.; Lammers, R.B. Global water resources: Vulnerability from climate change and population growth. *Science* **2000**, *289*, 284–288. [CrossRef] [PubMed]
63. European Commission (EC). *Ecological Flows in the Implementation of the Water Framework Directive*; CIS Guidance Document n°31; European Commission, DG Environment: Bruxelles, Belgium, 2015. [CrossRef]
64. European Environmental Agency (EEA). Water Exploitation Index. Available online: https://www.eea.europa.eu/data-and-maps/indicators/use-of-freshwater-resources-3/assessment-4 (accessed on 25 October 2020).
65. Pfister, S.; Koehler, A.; Hellweg, S. Assessing the environmental impacts of freshwater consumption in LCA. *Environ. Sci. Technol.* **2009**, *43*, 4098–4104. [CrossRef] [PubMed]
66. Ridoutt, B.G.; Pfister, S. A revised approach to water footprinting to make transparent the impacts of consumption and production on global freshwater scarcity. *Glob. Environ. Chang.* **2010**, *20*, 113–120. [CrossRef]
67. Hoekstra, A.Y. A critique on the water-scarcity weighted water footprint in LCA. *Ecol. Indic.* **2016**, *66*, 564–573. [CrossRef]
68. Pfister, S.; Boulay, A.-M.; Berger, M.; Hadjikakou, M.; Motoshita, M.; Hess, T.; Ridoutt, B.; Weinzettel, J.; Scherer, L.; Döll, P.; et al. Understanding the LCA and ISO water footprint: A response to Hoekstra (2016) "A critique on the water-scarcity weighted water footprint in LCA". *Ecol. Indic.* **2017**, *72*, 352–359. [CrossRef]
69. Gassert, F.; Luck, M.; Landis, M.; Reig, P.; Shiao, T. *Aqueduct Global Maps 2.1: Constructing Decision-Relevant Global Water Risk Indicators*; Working Paper; World Resources Institute: Washington, DC, USA, 2014. Available online: https://www.wri.org/research/aqueduct-global-maps-21-indicators (accessed on 23 September 2021).
70. Gassert, F.; Reig, P.; Luo, T.; Maddocks, A. *Aqueduct Country and River Basin Rankings: A Weighted Aggregation of Spatially Distinct Hydrological Indicators*; Working paper; World Resources Institute: Washington, DC, USA, 2013. Available online: https://www.wri.org/research/aqueduct-country-and-river-basin-rankings (accessed on 23 September 2021).
71. Commonwealth of Australia (CoA). *Leading Practice Sustainable Development Program for the Mining Industry, Water Stewardship*; Australian Government, Department of Industry, Science, Energy and Resources: Canberra, Australia, 2016. Available online: https://www.industry.gov.au/data-and-publications/leading-practice-handbook-water-stewardship (accessed on 23 September 2021).
72. Sullivan, C. Calculating a Water Poverty Index. *World Dev.* **2002**, *30*, 1195–1210. [CrossRef]
73. Sood, A.; Smakhtin, V.; Eriyagama, N.; Villholth, K.G.; Liyanage, N.; Wada, Y.; Ebrahim, G.; Dickens, C. *Global Environmental Flow Information for the Sustainable Development Goals*; IWMI Research Report 168; International Water Management Institute (IWMI): Colombo, Sri Lanka, 2017. [CrossRef]
74. Zeiringer, B.; Seliger, C.; Greimel, F.; Schmutz, S. River Hydrology, Flow Alteration, and Environmental Flow. In *Riverine Ecosystem Management. Science for Governing towards a Sustainable Future*; Aquatic Ecology Series; Schmutz, S., Sendizimir, J., Eds.; Springer: Cham, Germany, 2018; Volume 8, pp. 67–89. [CrossRef]
75. Hirji, R.; Davis, R. *Environmental Flows in Water Resources Policies, Plans, and Projects. Findings and Recommendations*; World Bank: Washington, DC, USA, 2009. [CrossRef]

76. Smakhtin, V.; Revenga, C.; Döll, P. *Taking into Account Environmental Water Requirements in Global-Scale Water Resources Assessments*; Comprehensive Assessment Research Report 2; Comprehensive Assessment Secretariat: Colombo, Sri Lanka, 2004.
77. Falkenmark, M.; Berntell, A.; Jägerskog, A.; Lundqvist, J.; Matz, M.; Tropp, H. *On the Verge of a New Water Scarcity: A Call for Good Governance and Human Ingenuity*; SIWI Policy Brief; Stockholm International Water Institute (SIWI): Stockholm, Sweden, 2007.
78. Pastor, A.V.; Ludwig, F.; Biemans, H.; Hoff, H.; Kabat, P. Accounting for environmental flow requirements in global water assessments. *Hydrol. Earth Syst. Sci.* **2014**, *18*, 5041–5059. [CrossRef]
79. Richter, B.D.; Davis, M.M.; Apse, C.; Konrad, C. A presumptive standard for environmental flow protection. *River Res. Appl.* **2012**, *28*, 1312–1321. [CrossRef]
80. Tharme, R.E. A global perspective on environmental flow assessment: Emerging trends in the development and application of environmental flow methodologies for rivers. *River Res. Appl.* **2003**, *19*, 397–441. [CrossRef]
81. SNL Metals & Mining Database. *Essential Mining Industry Data with Actionable Insights*; S&P Global Market Intelligence (S&P): New York, NY, USA, 2019. Available online: https://www.spglobal.com/marketintelligence/en/solutions/metals-and-mining (accessed on 1 June 2019).
82. Luck, M.; Landis, M.; Gassert, F. *Aqueduct Water Stress Projections: Decadal Projections of Water Supply and Demand Using CMIP5 GCMs*; Technical Note; World Resources Institute: Washington, DC, USA, 2015. Available online: https://www.wri.org/research/aqueduct-water-stress-projections-decadal-projections-water-supply-and-demand-using-cmip5 (accessed on 23 September 2021).
83. Global Runoff Data Centre (GRDC). *Major River Basins of the World/Global Runoff Data Centre*; Federal Institute of Hydrology (BfG): Koblenz, Germany, 2007. Available online: https://www.bafg.de/GRDC/EN/02_srvcs/22_gslrs/221_MRB/riverbasins_node.html (accessed on 23 September 2021).
84. Hoekstra, A.Y.; Mekonnen, M.M. *Global Water Scarcity: Monthly Blue Water Footprint Compared to Blue Water Availability for the World's Major River Basins*; Value of Water Research Report Series No. 53; UNESCO-IHE: Delft, The Netherlands, 2011.
85. Giurco, D.; Prior, T.; Mudd, G.; Mason, L.; Behrisch, J. *Peak Minerals in Australia: A Review of Changing IMPACTS and Benefits*; Prepared for CSIRO Minerals Down Under Flagship; Institute for Sustainable Futures (University of Technology, Sydney) and Department of Civil Engineering (Monash University): Sydney/Clayton, Australia, 2010.
86. Frischknecht, R.; Jungbluth, N.; Althaus, H.-J.; Doka, G.; Dones, R.; Heck, T.; Hellweg, S.; Hischier, R.; Nemecek, T.; Rebitzer, G.; et al. The ecoinvent database: Overview and methodological framework. *Int. J. Life Cycle Assess.* **2005**, *10*, 3–9. [CrossRef]
87. Buxmann, K.; Koehler, A.; Thylmann, D. Water scarcity footprint of primary aluminium. *Int. J. Life Cycle Assess.* **2016**, *21*, 1605–1615. [CrossRef]
88. Dai, Q.; Kelly, J.C.; Elgowainy, A. *Cobalt Life Cycle Analysis Update for the GREET Model*; Argonne National Laboratory: Argonne, IL, USA, 2018. Available online: https://greet.es.anl.gov/publication-Li_battery_update_2017 (accessed on 25 October 2020).
89. Shahjadi, H.F.; Nazmul Huda, M.A.; Parvez, M. Life cycle assessment of cobalt extraction process. *J. Sustain. Min.* **2019**, *18*, 150–161. [CrossRef]
90. Fritsche, U. Process-oriented Basic Data for Environmental Management Instruments-www.probas.umweltbundesamt.de. *Int. J. Life Cycle Assess.* **2005**, *10*, 225. [CrossRef]
91. Pena, C.; Huijbregts, M. The Blue Water Footprint of Primary Copper Production in Northern Chile. *J. Ind. Ecol.* **2013**, *18*, 49–58. [CrossRef]
92. Haque, N.; Norgate, T. Life cycle assessment of iron ore mining and processing. In *Iron Ore. Mineralogy, Processing and Environmental Sustainability*; Lu, L., Ed.; Woodhead Publishing: Cambridge, UK, 2015; pp. 615–630. [CrossRef]
93. Tost, M.; Bayer, B.; Hitch, M.; Lutter, S.; Moser, P.; Feiel, S. Metal Mining's Environmental Pressures: A Review and Updated Estimates on CO_2 Emissions, Water Use, and Land Requirements. *Sustainability* **2018**, *10*, 2881. [CrossRef]
94. Mudd, G.M.; Diesendorf, M. Sustainability of Uranium Mining and Milling: Toward Quantifying Resources and Eco-Efficiency. *Environ. Sci. Technol.* **2008**, *42*, 2624–2630. [CrossRef]
95. Norgate, T.; Haque, N. Using life cycle assessment to evaluate some environmental impacts of gold production. *J. Clean. Prod.* **2012**, *29–30*, 53–63. [CrossRef]
96. Mudd, G.M. Resource Consumption Intensity and the Sustainability of Gold Mining. In Proceedings of the 2nd International Conference on Sustainability Engineering & Science, Auckland, New Zealand, 20–23 February 2007.
97. Talbot, D.; Barbat, G. Water disclosure in the mining sector: An assessment of the credibility of sustainability reports. *Corp. Soc. Responsib. Environ. Manag.* **2020**, *27*, 1241–1251. [CrossRef]
98. World Water Assessment Programme (WWAP). *The United Nations World Water Development Report 2019: Leaving No One Behind*; UNESCO: Paris, France, 2019.
99. Mindat.org dba. Hudson Institute of Mineralogy, Ed.: Keswick, USA. 2019. Available online: https://www.mindat.org/ (accessed on 21 September 2021).
100. United States Geological Survey (USGS). USGD, Ed.: Reston, USA. 2021. Available online: https://www.usgs.gov/ (accessed on 21 September 2021).
101. British Geological Survey (BGS). BGS, Ed.: Nottingham, UK. 2021. Available online: https://www.bgs.ac.uk/ (accessed on 21 September 2021).
102. Australian Geoscience Information Network (AUSGIN)-Geoscience Portal. Australian Government's Department of Industry, Innovation and Science, Ed.: Canberra, Australia. 2021. Available online: https://portal.geoscience.gov.au/ (accessed on 21 September 2021).

103. United States Geological Survey (USGS). *Mineral Commodity Summaries 2020*; U.S. Geological Survey: Reston, VA, USA, 2020. [CrossRef]
104. Fekete, B.M.; Vörösmarty, C.J.; Grabs, W. High-resolution fields of global runoff combining observed river discharge and simulated water balances. *Glob. Biogeochem. Cycles* **2002**, *16*, 15/1–15/10. [CrossRef]

Article

Assessing the Application-Specific Substitutability of Lithium-Ion Battery Cathode Chemistries Based on Material Criticality, Performance, and Price

Steffen Kiemel [1,*], Simon Glöser-Chahoud [2], Lara Waltersmann [1], Maximilian Schutzbach [1,3], Alexander Sauer [1,3] and Robert Miehe [1]

[1] Fraunhofer Institute for Manufacturing Engineering and Automation IPA, Nobelstraße 12, 70569 Stuttgart, Germany; lara.waltersmann@ipa.fraunhofer.de (L.W.); maximilian.schutzbach@ipa-extern.fraunhofer.de (M.S.); alexander.sauer@ipa.fraunhofer.de (A.S.); robert.miehe@ipa.fraunhofer.de (R.M.)
[2] Institute for Industrial Production IIP, Karlsruhe Institute of Technology, Hertzstraße 16, 76187 Karlsruhe, Germany; simon.gloeser-chahoud@kit.edu
[3] Institute for Energy Efficiency in Production EEP, University of Stuttgart, Nobelstraße 12, 70569 Stuttgart, Germany
* Correspondence: steffen.kiemel@ipa.fraunhofer.de

Citation: Kiemel, S.; Glöser-Chahoud, S.; Waltersmann, L.; Schutzbach, M.; Sauer, A.; Miehe, R. Assessing the Application-Specific Substitutability of Lithium-Ion Battery Cathode Chemistries Based on Material Criticality, Performance, and Price. *Resources* **2021**, *10*, 87. https://doi.org/10.3390/resources10090087

Academic Editors: Andrea Thorenz and Armin Reller

Received: 30 July 2021
Accepted: 19 August 2021
Published: 25 August 2021

Publisher's Note: MDPI stays neutral with regard to jurisdictional claims in published maps and institutional affiliations.

Copyright: © 2021 by the authors. Licensee MDPI, Basel, Switzerland. This article is an open access article distributed under the terms and conditions of the Creative Commons Attribution (CC BY) license (https://creativecommons.org/licenses/by/4.0/).

Abstract: The material use of lithium-ion batteries (LIBs) is widely discussed in public and scientific discourse. Cathodes of state-of-the-art LIBs are partially comprised of high-priced raw materials mined under alarming ecological and social circumstances. Moreover, battery manufacturers are searching for cathode chemistries that represent a trade-off between low costs and an acceptable material criticality of the comprised elements while fulfilling the performance requirements for the respective application of the LIB. This article provides an assessment of the substitutability of common LIB cathode chemistries (NMC 111, −532, −622, −811, NCA 3%, −9%, LMO, LFP, and LCO) for five major fields of application (traction batteries, stationary energy storage systems, consumer electronics, power-/garden tools, and domestic appliances). Therefore, we provide a tailored methodology for evaluating the substitutability of products or components and critically reflect on the results. Outcomes show that LFP is the preferable cathode chemistry while LCO obtains the worst rating for all fields of application under the assumptions made (as well as the weighting of the considered categories derived from an expert survey). The ranking based on the substitutability score of the other cathode chemistries varies per field of application. NMC 532, −811, −111, and LMO are named recommendable types of cathodes.

Keywords: lithium-ion battery; LIB; raw material criticality; substitutability; cathode chemistries; traction batteries; stationary energy storage systems; consumer electronics; power-/garden tools; domestic appliances

1. Introduction

Tackling climate change and decarbonizing the economy and society may be considered as some of the greatest challenges of this century. The Paris Agreement limits global warming to 1.5 degrees [1]. The European Union aims to achieve climate neutrality by 2050 [2]. This results in massive pressure for technology development and applications in industries. One of the most promising solutions for achieving climate neutrality is electrification in combination with the expansion of renewable energies. With the recent proposal "Fit-for-55" within its Green Deal Framework, the European Commission has significantly increased the need for electrification, especially in the transport sector [3]. Emissions trading is to be introduced for road traffic and only new registrations of zero-emission cars are to be allowed from 2035 onwards. Therefore, electrification will be the key measure for reaching the climate targets. This results in a significantly increasing demand for batteries. In this context, lithium-ion batteries (LIBs) are of utmost importance due to their rechargeability and favorable performance properties. LIBs cover a wide range of applications, such as traction batteries, stationary energy storage systems, consumer electronics, power/garden tools, and domestic appliances. Traction batteries for electric vehicles represent a huge amount of potential for achieving climate goals, as road transport is responsible for approximately 15% of global CO_2 emissions [4]. Electrified transport in combination with a low-carbon energy supply can reduce these CO_2 emissions significantly. According to Zhang and Fujimori, the CO_2 emissions of this sector can be reduced by 50% until 2050 only via electrification [5]. The International Energy Agency (IEA) even states that the CO_2 emissions from road transport may be reduced to zero for light commercial vehicles, passenger cars, and buses by 2070 [6]. In addition, batteries can be used as stationary energy storage by private, commercial, or utility users, and thus balance wind or solar energy shortages and contribute to energy flexibility [7].

To cover the huge range of applications of LIBs, different cathode chemistries are used. These include "Lithium Nickel Manganese Cobalt Oxide" (NMC 111, NMC 532, NMC 622, and NMC 811), "Lithium Nickel Cobalt Aluminum Oxide" (NCA 3% Co and NCA 9% Co), "Lithium Iron Phosphate" (LFP), "Lithium Cobalt Oxide" (LCO), and "Lithium Manganese Oxide" (LMO) [8]. The well-established alphabetic abbreviations denote the elements the cathodes are comprised of. The stated numbers define the mass percentages of the respective materials. For example, the label NMC 532 indicates that the cathode comprises five parts nickel, three parts manganese, and two parts cobalt. The percentage after the code NCA (nickel cobalt aluminum oxide) describes the material share of cobalt within the cathode [9].

As shown by Merriman, the market share of the different cathode chemistries will change in favor of higher energy density material compositions with the additional advantages of lower costs and higher social acceptance [8]. This means that cathode chemistries with higher nickel content are likely to increase their market share significantly, while cathode chemistries with higher cobalt content are predicted to decline to a market share of less than 10% in 2025 [8].

This paper focuses on cathode chemistries that either already have a high market share or will be important in the near future. Thus, solid-state batteries and other innovative cell generations, such as lithium air, are neglected, although they are expected to gain significant market shares in the future [9]. This is because no market-ready battery has been developed yet and, according to forecasts, they will initially become relevant in 2030 [9]. For the battery industry, such forecasts are very important for strategically planning the production infrastructure and portfolio. This article will provide an additional approach to such considerations. By assessing the substitutability of cathode chemistries, decision-makers in industries receive a detailed overview of the potential benefits and disadvantages of staying with the existing product portfolio or changing to another.

One of the critical factors for the market penetration of LIBs is the low security of cathode materials [10]. In order to assess impact factors for a sustainable supply of resources, the holistic methodology of resource criticality assessments was developed a few years ago. The first criticality assessments were conducted in the context of both world wars to evaluate strategically valuable resources [11]. While these studies usually concentrated on minerals, later analyses have significantly broadened the scope.

According to the standard denotation by Erdmann and Graedel, a critical raw material is defined as one for which the medium to long-term supply situation might turn out to be critical from the point of view of various systems (e.g., companies, industries, national economies) [12]. This article focuses on the company perspective, for which various authors have developed respective models and methodologies [13–15]. Schrijvers et al. provide a comprehensive overview [16]. In the case of an indicator-based criticality assessment, this is carried out based on different dimensions such as vulnerability, supply risk, environmental impact, and social implications [13,14]. Each dimension is assessed quantitatively using various indicators. Qualitative information (e.g., concerning recyclability) is transformed to a quantitative scale where possible. There are various approaches to interpreting the results. While it is possible to aggregate the individual results of the considered indicators to calculate an overall criticality value, some studies remain on the indicator level, since this allows for detailed analyses of potential problems in the supply chain of raw materials.

Due to the limited nature of raw materials, criticality assessments have been extensively investigated in the recent literature. Among others, numerous commodity-specific analyses have been performed for metals and metalloids [17–20], such as copper [21], iron [22], and rare earth elements [23]. However, the analysis of raw material criticality is an ongoing process since the respective methodologies are continually being improved and further possible indicators integrated. Furthermore, results may vary significantly depending on the underlying base year of the utilized data. A prominent example of such variations in results for individual indicators is the significant increase in the static reach of lithium due to the exploitation of wider reserves within the past few years [24,25]. Furthermore, demand impulses through new emerging technologies should be mentioned here. In addition to the consideration of individual elements, an aggregated consideration of criticality at the product or technology level is also possible. Exemplarily for this context, specific use cases for clean energy technologies [26], such as photovoltaics [27] and water electrolyzers [28], and other emerging technologies [29] have been described. A few criticality analyses have also been carried out for different battery technologies [9,30,31] and energy storage systems [32].

Nevertheless, the results from previous studies cannot be adopted, since in this paper, substitutability is not implicitly analyzed in the vulnerability dimension, but explicitly as the main finding quantified with a separate set of indicators and criteria. Furthermore, in addition to supply risk, the other dimensions of environmental impact and social implications, according to Kolotzek et al. [14], are included to generate an overview of resource criticality that is as comprehensive as possible [33]. The combination of classic approaches to criticality assessments and methods from the fields of life cycle analysis as well as social life cycle analysis [14] results in an evaluation of the effects of resource use that involves all three pillars of sustainability (economic, environmental, social). To guarantee a consistent approach, all battery raw materials must be evaluated again according to the corresponding methodology.

In general, the substitution of single raw materials can lead to a significant decrease in the overall criticality of a material compound or a product/technology, respectively. However, this is not possible for complex technology components such as those of LIBs. Pivotal examples are the cathodes of LIBs, as they are comprised of various materials that cannot be individually substituted. Hence, the substitutability has to be analyzed on a composite, rather than on an elemental, level. As individual fields of applications are defined by specific requirements, this evaluation needs to be conducted in an application-specific manner. The present article provides a respective methodology. Consequently, the central research question is defined as: "Which LIB cathode chemistry is to be suggested for the considered fields of application, and what are the respective benefits when switching from the status quo to another cathode chemistry?".

This paper's contribution to the existing literature on raw material criticality assessment, substitutability, and, consequently, sustainability is twofold. First, we provide a comprehensive approach to how the substitutability of materials at a technology level could be quantified as an enhancement of basic criticality assessments. Second, we apply the methodology to current LIB chemistries as one of the most relevant emerging technologies in the debate about future material availability from a company perspective. Substitutability in many cases is included in the vulnerability score as a single sub-indicator for each raw material [20,34,35]. However, this is not sufficient to provide decision support at the firm level, where individual alternative technologies and material systems need to be compared.

2. Materials and Methods

The methodology is separated into two main modules: (1) the assessment of material criticality; and (2) the determination of substitutability. For the latter, the two considered categories of price and performance are described, and the methodological approach in merging the individual results is outlined (Figure 1).

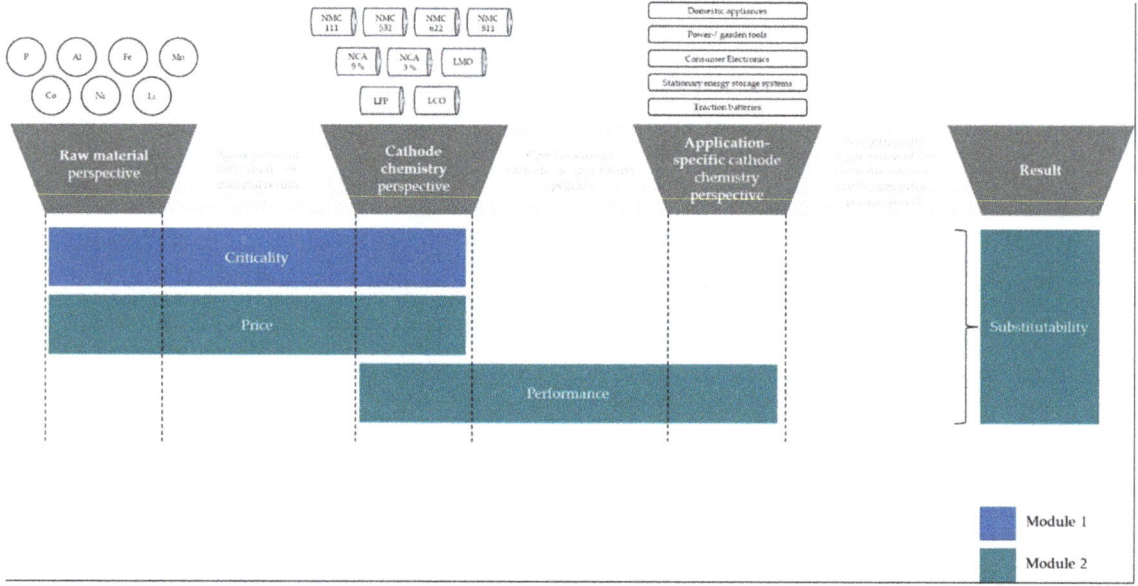

Figure 1. Visualization of the methodological approach.

The scope of the presented research includes nine different LIB cathode compositions. The respective selection illustrated in Figure 1 is based on state-of-the-art LIBs that either had, currently have, or are expected to gain relevant shares of the global LIB market [8,9]. As this article focuses on LIBs, each considered cathode chemistry comprises derivates of lithium, regardless of whether it is indicated in the abbreviation (L) or not. Thereby, the share of lithium in the cathode composition remains constant for all versions of NMC and NCA LIBs [8]. For assessing the raw material criticality of the considered cathode chemistries, the contained elements are individually assessed and later on aggregated by their material shares. The article focuses on cathode or so-called active materials and neglects other materials a LIB cell is comprised of.

The substitutability is assessed for five product families or fields of application (electric vehicles, stationary energy storage, consumer electronics, power-/garden tools, and domestic appliances). The respective choice is adopted from Full et al., who define the mentioned product families as the most popular use cases for LIBs regarding market shares based on available market analytics [36].

2.1. Methodology of Criticality Assessment

One of the most prominent approaches to criticality assessment is based on the work of Graedel et al. Thereby, material criticality is expressed by the three dimensions of supply risk, environmental implications, and vulnerability [13]. This article, however, neglects the dimension of vulnerability to supply restrictions. This is done in order to prevent double accounting of certain aspects. The substitutability as one of three criteria for assessing the vulnerability to supply restrictions is analyzed in detail in the subsequent steps. The inclusion of the dimension of vulnerability to supply restrictions would result in a circular reference between the evaluation of substitutability, which is based on the preceding assessment of material criticality (compare Section 2.2). The presented approach adopts the so-called "company-oriented model" developed by Kolotzek et al., which can be adapted to a product or technology level. The underlying three dimensions are supply risk, environmental impact, and social implications of raw material supply [14]. Incorporating the social aspect represents a significant added value to classic approaches to raw material criticality assessment. This is due to the fact that corporate social responsibility is expected both from customers as well as public authorities. The respective applied set of indicators provides a comprehensive overview of potential social problems in the upstream supply chain. The considered indicators for each dimension are documented in Table 1. For a short description of each utilized indicator, the reader is referred to the Supplementary Materials of [14]. The underlying base year of the obtained data for each indicator is 2019.

The selection of indicators for the supply risk dimension is equivalent to the eight most frequently applied indicators in related studies [37]. Thereby, the original indicator "country risk" is translated into the category "political risk", which is comprised of three individual indicators ("policy perception", "political stability", and "regulation"). In addition, the original indicator "depletion time" [37] is split into two indicators: "static reach of reserves" and "static reach of resources".

In turn, the dimension "environmental impact" is based on the methodology by Graedel et al., who apply the two LCIA endpoint categories "human health" and "ecosystem quality" by allocating various midpoint indicators (see Table 1) [13]. The "hierarchist" perspective of ReCiPe 2008 is chosen as the underlying LCIA method [38]. As ReCiPe already includes a weighting to transform mid- to endpoint indicators, the definition of an individual weighting for the present work is neglected.

The dimension "social implications" is defined by a selection of indicators based on the "research field of social life cycle assessments" [14]. Standardization and weighting of individual indicators for each dimension are adopted from Kolotzek et al., who applied an analytical hierarchy process (AHP) for the respective determination. Each indicator is normalized on a scale from 0 to 100, where zero equals the best and 100 the worst performance. For detailed information concerning the selection process of dimensions, indicators, and further methodological aspects, the reader is referred to the work of Kolotzek et al. [14]. In accordance with other recent studies [22,30], the presented work does not define criticality thresholds, since criticality itself generally represents a subjective concept that is largely based on the respective perception of stakeholders. However, the comparison between the single raw materials and cathode chemistries allows for classification in terms of "less and more critical". The circular form of visual depiction of the results of the raw material assessment (adapted from Kolotzek et al. [14]) allows for an indicator-specific interpretation and analysis (compare Section 3.1). Hence, potential problems in the supply chain of the evaluated raw materials can be identified, and approaches to improvement can be derived for further analyses.

Table 1. Applied indicators for criticality assessment. The selection and weighting are based on [5], and the origin of data is stated in the column "reference".

Dimension	Category		Indicator	Abbreviation	Reference
Supply Risk	Concentration Risk		Company Concentration	CompC	[24]
			Country Concentration	CountC	[39]
	Political Risk		Policy Perception Index	PPI	[40]
			WGI: Political Stability and Absence of Violence/Terrorism	WGI-PV	[41]
			Human Development Index: Regulation	HDI	[42]
	Risk of Demand Increase		Companion Metal Fraction	CMF	[43]
			Future Technology Demand	FTD	[44]
			Substitutability (raw material)	Subs	[45]
	Risk of Supply Reduction		Recycling Rate	RR	[46]
			Static Reach Reserves	SRRV	[24]
			Static Reach Resources	SRRC	[24]
Environmental Impact	Ecosystem Quality		Agricultural Land Occupation	ALO	[47]
			Climate Change, Ecosystem	CCE	[47]
			Freshwater Ecotoxicity	FEuc	[47]
			Freshwater Eutrophication	FEut	[47]
			Marine Ecotoxicity	MEct	[47]
			Natural Land Transformation	NLT	[47]
			Terrestrial Acidification	Tacd	[47]
			Terrestrial Ecotoxicity	Tect	[47]
			Urban Land Occupation	ULO	[47]
	Human Health		Climate Change, Human Health	CCHH	[47]
			Human Toxicity	HAT	[47]
			Ionising Radiation	IR	[47]
			Ozone Depletion	OD	[47]
			Particulate Matter Formation	PMF	[47]
			Photochemical Oxidant Formation	POF	[47]
	Local Community	Access to Immaterial Resources	WGI: Voice and Accountability	AIR	[48]
			Global Competitiveness Report: FDI and technology transfer		[49]
		Access to Material Resources	Environmental Performance Index: Water and Sanitation	EPI-WS	[50]
		Community Engagement	GCR: Public Trust of Politicians	CE	[49]
			GCR: Transparency of Government Policymaking		[49]
		Cultural Heritage	Fragile State Index: Group Grievance	FSI-GG	[51]
		Delocalization and Migration	Fragile State Index: Refugees and IDPs	FSI-R	[51]
		Local Employment	Risk of Unemployment	LE	[52]

Table 1. Cont.

Dimension	Category		Indicator	Abbreviation	Reference
Social Implications	Society	Respect of Indigenous Rights	Risk That a Country Does not Provide Laws to Protect Indigenous People	RIR	[52]
			Risk that Indigenous People are Negatively Impacted		
		Safe and Healthy Living Conditions	WHO: Age-standardized DALY rates	DALY	[53]
		Secure Living Conditions	GCR: Security of Public Institutions	SLC	[49]
			GCR: Reliability of Police Services		[49]
		Corruption	WGI: Control of Corruption	WGI-CC	[48]
		Prevention and Mitigation of Armed Conflicts	HIIK Conflict Barometer	HIIK	[54]
	Worker	Child Labor	Risk of Child Labor	CL	[52]
		Equal Opportunities/Discrimination	Gender Inequality Index	GII	[55]
		Fair Salary	Risk of Average Wage Being Lower Than Non-Poverty Guideline	FS	[52]
		Forced Labor	Risk of Forced Labor	FL	[52]
		Freedom of Association and Bargaining	Risk of Not to Enforce the Right to Strike	FA&B	[52]
			Risk of Not to Enforce Freedom of Association Rights		
			Risk of Not to Enforce Collective Bargaining Rights		
		Health and Safety	Risk of Non-Fatal Injuries	H&S	[52]
			Risk of Fatal Injuries		
		Working Hours	Risk of Excessive Working Time	WH	[52]

The results on the elemental level (supply risk, environmental impact, social implications) are aggregated to the cathode-specific technology level. This is done by weighing the obtained values by mass shares of the respective raw material (or precursor material) within the considered cathode chemistries. This approach to calculating the criticality of material compounds or products is accepted in the scientific community [30–32]. The material compositions of the individual cathode chemistries can be obtained from Table 2. Thereby, lithium and manganese are applied as different derivates for some cathode chemistries. While lithium carbonate is used for NMC 111, NMC 532, NMC 622, LMO, and LCO [56,57], lithium hydroxide is required for the nickel-rich cathode NMC 811, LFP, and the NCA technologies [57–59]. The NMC-active material is comprised of manganese sulfate [57], while for LMO manganese dioxide is applied [60]. The respective information is involved for the environmental dimension. The indicators of the considered LCIA endpoint categories are assessed in both a raw-material-specific and derivate-specific manner. As no dataset exists for cobalt sulfate hexahydrate in the ecoinvent database [47], a mixed calculation from sulfuric acid and cobalt based on stoichiometric calculations is conducted.

2.2. Evaluation of the Substitutability of LIB Cathode Chemistries

The applied methodology for evaluating the substitutability of cathode chemistries is partially based on the approach to vulnerability assessment by Graedel et al. [13]. The respective components and context are illustrated in Figure 2. The substitutability is calculated by considering three categories (highlighted in cyan: the performance, the criticality, and the price of the substitute), which in turn are comprised of various indicators (highlighted in dark grey for the performance of the substitute, in green for the criticality of the substitute, and in blue for the price of the substitute). Potential substitutes in this regard are the mentioned cathode chemistries (compare Table 2). The weightings of the individual categories and indicators are based on an expert survey (cyan, green, and blue) or available assessments from the literature (dark grey). The respective approach and results are described at the end of this paragraph. The category "criticality of the substitute" is based on the results described in Section 2.1. The other two categories are described in detail in the following. The levels and categories that are highlighted in light grey are part of the approach to assessing the vulnerability but are neglected in the presented approach to assessing the substitutability of LIB cathode chemistries.

Table 2. Material shares of evaluated cathode chemistries.

	Li_2CO_3	LiOH	$NiSO_4$	$MnSO_4$	MnO_2	$CoSO_4$	$Al_2(SO_4)_3$	$FeSO_4$	H_3PO_4	Reference
NMC 111	0.106		0.298	0.298		0.298				[8]
NMC 532	0.106		0.447	0.268		0.179				[8]
NMC 622	0.106		0.536	0.179		0.179				[8]
NMC 811		0.106	0.716	0.089		0.089				[8]
NCA 3%		0.106	0.824			0.045	0.025			[8]
NCA 9%		0.106	0.734			0.14	0.02			[8]
LMO	0.06				0.94					[8]
LFP		0.04						0.36	0.6	[8]
LCO	0.11					0.89				[61]

The category "performance of the substitute" is assessed by taking five indicators into account. The selection of these indicators is based on a comparison and linkage of the results of two scientific articles [36,62]. Zubi et al. evaluated the characteristics of various LIB cathode chemistries (LCO, LMO, LFP, NCA, and NMC) based on eight indicators [62]. In order to connect their chemistry-specific results to application-specific requirements for LIBs, the eight indicators from Zubi et al. are matched with the nine indicators applied by Full et al. [36]. In the course of their evaluation of different LIB cell formats, they assessed the importance of various technical criteria for the main applications of LIBs (traction batteries, stationary energy storage systems, consumer electronics, power/garden tools, and domestic appliances) [36]. The comparison of these technical criteria can be obtained from Table 3. The five adopted indicators enable the evaluation of the performance of cathode chemistries for the considered applications. The remaining indicators are either neglected due to consideration in other dimensions or the non-availability of an appropriate match (compare Table 3).

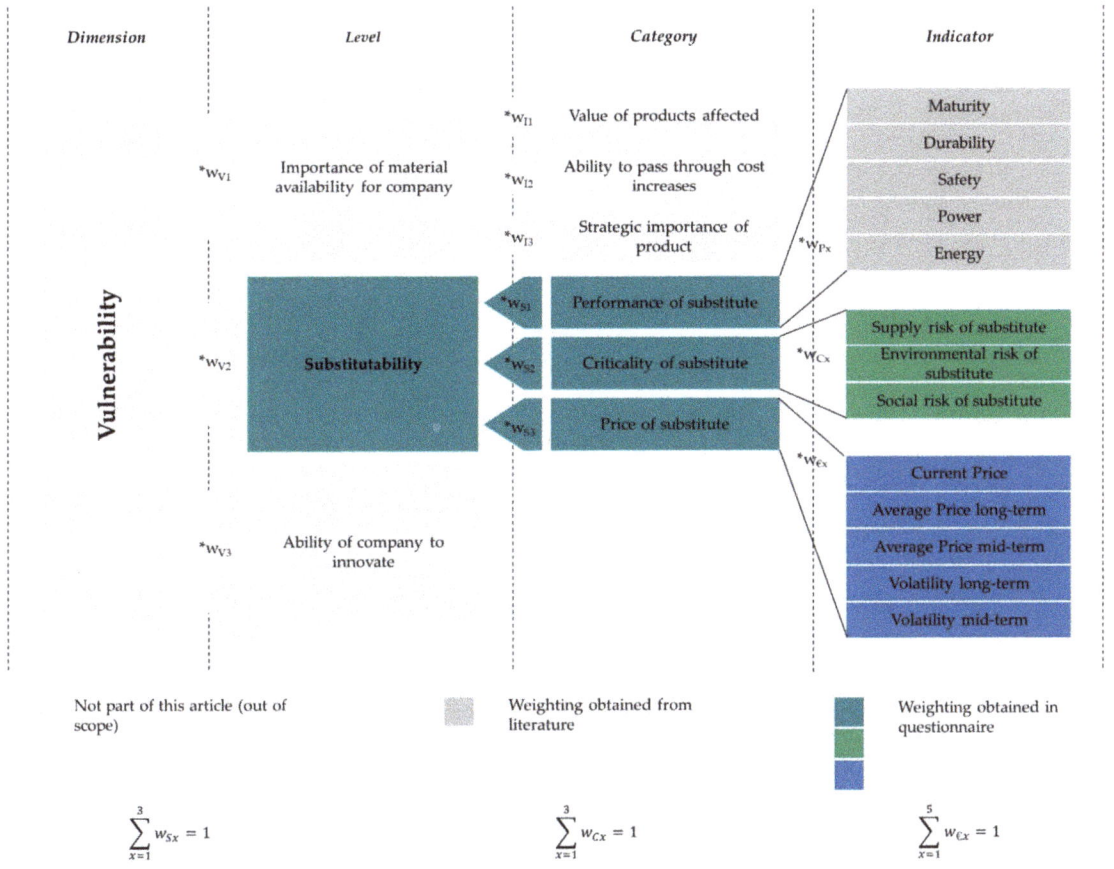

Figure 2. Structure of the substitutability assessment; partially based on Graedel et al. [13].

Table 3. Considered indicators for the category "performance of substitute".

Full et al. 2020 [36]	Zubi et al. 2018 [62]	Considered? Yes	Considered? No	Reason for Exclusion
Energy density	Energy	X		
Power density	Power	X		
Stability	Safety	X		
Lifespan	Durability	X		
Thermal properties			X	No matchable indicator used by Zubi et al.
Sustainability			X	The sustainability assessment of cathode chemistries is covered by the categories "price of substitute" and "criticality of substitute"
Degree of standardization	Maturity	X		
Shape flexibility			X	No matchable indicator used by Zubi et al.
Cost	Affordability		X	Covered by the category "price of substitute"
	Materials		X	The assessment of utilized materials is the focus of the present article ("criticality of substitute")
	Performance		X	Covered by the category "performance of substitute"

As Zubi et al. assess NMC and NCA batteries and do not further define the particular material composition (e.g., share of nickel, manganese, and cobalt), the supplied assessments are interpreted as averages for NMC 111–811 and NCA 3–9%. Consequently, the scoring published by Zubi et al. is extended to represent existing variances in performance between the different material shares in similar families of cathode chemistries. Hence, nickel-rich NMC cathode chemistries (NMC 622 and 811) are assessed with lower maturity [63], lower durability [64], lower safety [9,65], and higher energy density [9,66] than the previous generations (NMC 111 and 532). Similar to NMC cathodes, the maturity of low-cobalt NCA cathodes (NCA 3%) is defined to be lower than that of high-cobalt NCA cathodes (NCA 9%). Furthermore, the durability and safety are assessed to be lower, while the energy density is assessed to become higher with decreasing cobalt share [67]. The underlying scores are listed in the Supplementary Materials. The obtained values for the individual indicators concerning the performance of the cathode chemistries are multiplied by the respective importance for the considered applications. The assessment of importance obtained from Full et al. is transformed to a scale that ranges from 0 (not important) to 1 (very important) and the sum is normalized to 1. Consequently, the importance is used for indicator weighting. The sum of the weighted indicator values per application results in the desired application-specific performance rating of LIB cathode chemistries. The respective values can be found in the Supplementary Materials.

Identifying costs of raw materials represents a difficult task, as the raw material market is highly volatile and depends on numerous impact factors. In order to integrate the dimension "cost of the substitute", five indicators are introduced. These are expected to cover the current price of the substitute as well as the potential future development based on historical data. Historical data are provided on a medium (2000–2015) and long-term timescale (1990–2015) for the average price and the price volatility of the raw materials. Data for the introduced indicators are obtained on a raw material level, not for specific cathode materials, as the respective information is not publicly available. This applies especially to the indicators that are based on historic data. For the same reason, a differentiation between derivates of the utilized materials (e.g., lithium hydroxide vs. lithium carbonate) is not taken into consideration. Once more, the aggregation to the product level (cathode chemistry) is based on the material shares (compare Table 2). The current price of the commodities refers to the average price in 2019 [68]. The only exemption is iron, which is used in LFP cathodes. The price of iron is neither mentioned in the "DERA Preismonitor" nor in the "mineral commodity summaries" of the USGS [24,68]. The most current information concerning the price of iron in the literature was found for 2018 [69]. The historical price averages refer to prices in the United States [70]. The historical volatility is equated with the standard deviation of the logarithmic price changes from one year to another in the considered timespans [71]. The historical raw material prices in the United States were chosen as the underlying data [70]. The Supplementary Materials of the present article provides more detailed information concerning the utilized datasets.

For the final assessment (see Section 3.3), each indicator from each dimension is normalized on a scale that ranges from 0 to 1, where 0 is defined as the potential optimal value per indicator (e.g., the optimal value of the indicator "current price" in the dimension "price of the substitute" equals 0 US \$). The value 1, as the upper end of the scale, is equated with the worst value that is achieved by one of the considered cathode chemistries (e.g., LCO attains the highest current material costs; hence, the 30,295 US \$/t are defined as 1 on the introduced scaling). Consequently, cathode chemistries that attain the lowest score are qualified the most as potential substitutes in the respective field of application.

The weightings were obtained from an analytical hierarchy process (AHP). The respective methodology is based on Saaty et al. [72]. The weightings of the indicators of the categories "criticality of the substitute" (w_{cx}) and "price of the substitute" ($w_{\xi x}$) as well as of the three categories within the level "substitutability" (w_{sx}) were obtained by pairwise comparisons using a structured questionnaire. A sample of the utilized questionnaire can be found in the Supplementary Materials. As the importance of the individual indicators for the considered fields of application was obtained from Full et al., the respective weighting of the category "performance of substitute" was not requested in the questionnaire [36]. The sum of the individual weightings for the categories as well as for the level "substitutability" equals 1 (or 100%), respectively. Figure 2 visualizes the content of the questionnaire. Experts from battery-related research institutions and companies were asked to give their opinion about the importance of the mentioned indicators and categories from the viewpoint of a producer/distributor of LIBs. Information concerning the surveyed experts can be found in the Supplementary Materials. The final weightings were obtained by calculating the average of the valid expert opinions. The threshold value for the consistency check of the expert's answers, according to Meixner and Haas, was defined as 0.2 [73]. Only one returned questionnaire had to be neglected in the calculation of the average weightings due to the exceedance of the consistency value. The obtained weightings can be found in Table 4. The weightings for the indicators of the performance category (w_{px}), obtained from Full et al. [36], can be found in the Supplementary Materials.

Table 4. Weightings of categories/indicators obtained from the AHP process.

Level/Category	Category/Indicator	Abbreviation	Weighting
Substitutability	Performance of the substitute	w_{s1}	0.410
	Criticality of the substitute	w_{s2}	0.269
	Price of the substitute	w_{s3}	0.321
	Σ		1
Criticality of substitute	Supply risk of the substitute	w_{c1}	0.520
	Environmental risk of the substitute	w_{c2}	0.244
	Social risk of the substitute	w_{c3}	0.236
	Σ		1
Price of substitute	Current price of the substitute	$w_{\epsilon 1}$	0.161
	Average price long-term of the substitute	$w_{\epsilon 2}$	0.273
	Average price mid-term of the substitute	$w_{\epsilon 3}$	0.276
	Volatility long-term of the substitute	$w_{\epsilon 4}$	0.134
	Volatility mid-term of the substitute	$w_{\epsilon 5}$	0.157
	Σ		1

3. Results

The results are organized into three subsections. The first summarizes the findings from the criticality analysis on the raw material level as well as the aggregated perspective on the cathode-chemistry level. The second subsection describes the results concerning the price and performance category of the considered cathode chemistries. Finally, all individual results are merged in the last subsection, where the substitutability of the mentioned cathode chemistries for the considered fields of application of LIBs is described.

3.1. Criticality of Cathode Materials for Lithium-Ion Batteries

The criticality of raw materials is not aggregated to a single criticality value. The results are discussed on the indicator level to identify specific problems in the upstream supply chain. This shall create sensitization to potential difficulties in supply from a market and political perspective, but also under the circumstance of the expected transformation to stricter requirements concerning environmental and social standards from consumers as well as legislative entities. The depiction of raw material criticality is based on Kolotzek et al., who developed an expressive layout by using three superimposed circle diagrams [14]. The inner circle visualizes the accumulated results of the three dimensions. The middle circle is a little more specific in depicting the defined categories and the outer circle provides the most detailed information on the indicator level. The color-coded scale, as depicted in Figure 3, reaches from green (least critical = transformed value of 0 per indicator) to red (most critical = transformed value of 100 per indicator).

The results for the individual material criticality assessments are illustrated in Figure 4. Compared with the other considered elements, cobalt is assessed with the highest supply risk, directly followed by lithium. This is in line with the results of other studies [18,30,31].

In turn, iron is scored with the least critical supply risk of the considered elements, followed by aluminum. This is the other way round in the assessment of Wentker et al. [31]. As the same indicators are selected, the variation should be considered in the light of the more up-to-date data applied in this article (base year 2019). This results in less critical scores of iron for the indicators "company concentration", "country concentration", and "regulation".

Figure 3. Applied color-coded scale for raw material criticality.

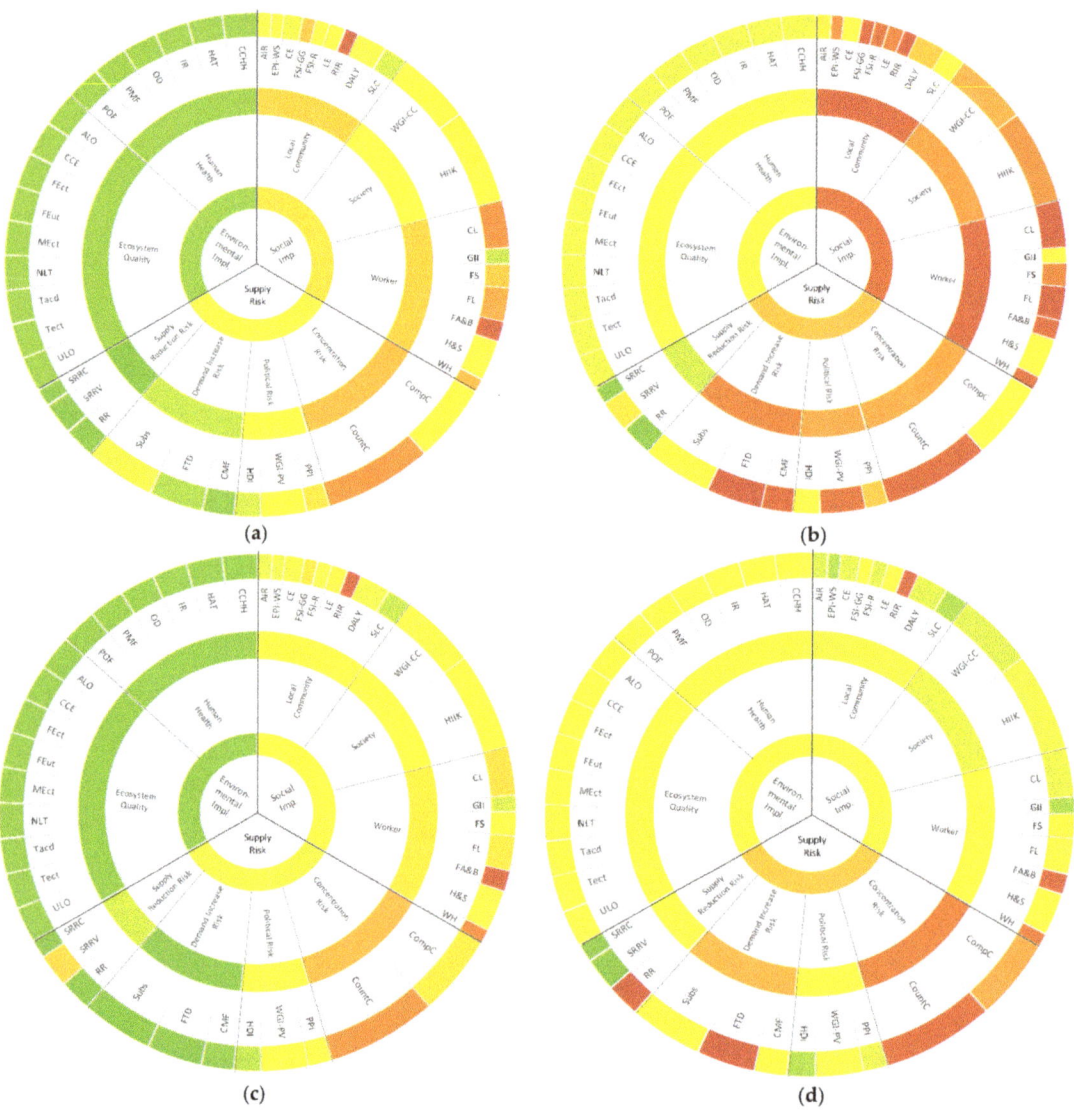

Figure 4. *Cont.*

Figure 4. Results of the raw material criticality assessments for: (**a**) aluminum; (**b**) cobalt; (**c**) iron; (**d**) lithium; (**e**) manganese; (**f**) nickel; and (**g**) phosphorus.

The lowest ecological implications occur from the mining of iron, followed by phosphorus and aluminum. Manganese, nickel, and cobalt follow after a considerable gap. The supply of lithium is accompanied by the highest ecological implications by far. Wentker et al. come to the same conclusions concerning cobalt, nickel, and lithium. However, according to their assessment, manganese and nickel are scored with an even lower environmental impact than phosphorus and aluminum [31]. The environmental impacts published by Graedel et al. are nearly the same for Co, Mn, Al, Li, and Fe. Only nickel is assessed as a little more critical [18]. These differences suggest that the dimension of ecological implications is highly dependent on the reference year of the applied data.

Compared with the other assessed elements, aluminum trends towards a rather uncritical assessment. The noticeable concentration risk (particularly the country concentration) is based on the fact that more than half of the world's aluminum production takes place in Australia and China [24]. Political circumstances in China, Brazil, Guinea, and India are the central reasons for the rather critical overall rating of the "policy perception index" (PPI) and the political stability and absence of violence indicator (WGI-PV) [40,41]. The implications for the environment are assessed as low (the third-lowest score of the considered elements) [47]. In the social dimension, the Fragile State Index: Group Grievance (FSI-GG) is scored as rather critical—as are Child Labor (CL) and Forced Labor (FL). Generally, it has to be stated that, for all elements, the indicators Respect of Indigenous Rights (RIR) and Freedom of Association and Bargaining (FA&B) are assessed as highly critical. Further research needs to analyze if this is a coincidence, or if the transformational rule obtained from Kolotzek et al. needs to be revised.

Cobalt mining is highly centralized. In 2019, around 65% of the world's production was mined in the Democratic Republic of Congo [24]. The Central African state is also the main reason for alarming political risks due to the critical ratings of the PPI, the WGI-PV, and the human development index (HDI) [40–42]. The supply risk is enhanced as cobalt is mainly extracted as a by-product metal of nickel and copper mining [43]. Furthermore, a significant increase in demand is expected for future technologies (mainly for LIBs) [44]. From the considered elements, the mining of cobalt is responsible for the second-highest environmental impact [47]. Concerning the social implications of cobalt mining, similar tendencies to the supply risk dimension can be observed. The strong concentration of mining operations in the Democratic Republic of Congo affects the social indicators negatively. Particularly highlighted in this regard are the scores for the FSI-GG and the Fragile State Index: Refugees and IDPs (FSI-R) [51]. Worth mentioning are also the rather critical assessments for the control of corruption indicator (WGI-CC), the HIIK conflict barometer (HIIK), and the water and sanitation indicator of the environmental performance index (EPI-WS) [48,50,54].

Except for the dimension of social implications (the second-lowest score after lithium), iron is assessed as the least critical of the considered elements. One of the aspects that increases the overall supply risk of the raw material is the high concentration of production. In 2019, more than half of the world's mining activity was located in Australia and Brazil. Taking China and India additionally into account, the production share of these four countries accounts for more than 80%. While the static reach of resources can be regarded as uncritical, the static reach of reserves is advised to be monitored as it currently only accounts for approximately 60 years [24].

The high supply risk of lithium is only surpassed by that of cobalt. The concentration risk of lithium is extremely high, as, in 2019, 60% of the world's lithium supply was produced in Australia (2019) [24]. However, it has to be stated that the political risk in the lithium-producing countries is rather low [40–42]. The future technology demand is scored with the highest possible value. This is mainly due to the expected increasing demand for LIBs [44]. Additionally, the recyclability of lithium is assessed as highly critical [46]. Nevertheless, it has to be stated that some processes exist that are capable of extracting derivates of lithium from end-of-life LIBs [74]. From a social point of view, lithium mining is rather uncritical compared with the other assessed elements.

Although the concentration risk of manganese is worth noting, it is low compared with the other considered elements (except for nickel) [24]. Manganese-producing countries are assessed with a medium political risk [40–42]. However, manganese is difficult to substitute on a material level [45]. Furthermore, the static reach of the identified reserves is comparatively low (44 years) [24]. In summary, this ranks manganese in fourth place (out of seven) concerning the supply risk dimension. The same ranking applies to the dimension of "environmental impact" [47]. Nearly all social indicators are assessed with a medium criticality. Thereby, the EPI-WS and the FSI-GG in particular increase the average value for the social implications dimension [50,51].

From the considered elements, the supply of nickel is the most diversified. However, significant mining activities in Colombia, Guatemala, and, most relevantly, China, Russia, Indonesia, and the Philippines (in ascending order by mined metric tons) impact the political risk negatively [40–42]. Worth mentioning is the short static reach of reserves (41 years), but even more the static reach of resources. A static reach of resources of approximately 60 years is by far the shortest of the considered elements (the second-shortest static reach of resources is that of cobalt with around 220 years) [24]. Although the color code suggests a rather uncritical environmental impact, that of nickel represents the third-highest score. Consequently, it is interpreted as worth mentioning and monitoring [47]. Similarly, for manganese, most of the social indicators are assessed with medium criticality. The deviations are the high scores for the indicators "cultural heritage" and "prevention and mitigation of armed conflicts" (both mainly caused by the respective situations in Brazil, Colombia, Guatemala, Indonesia, Russia, and the Philippines) [51,54].

Phosphorus is scored with the third-highest supply risk of the assessed elements. This is mainly due to the significant company and country concentration of phosphate mining activities [24], high risks concerning the policy potential and political stability in the respective countries [40,41], the non-substitutability in the main applications [45], and the low recycling rates [46]. The environmental impact of phosphorus supply is scored as very low (only undercut by iron) [47]. The social implications of phosphorus mining, in turn, are comparably high, which is mainly due to high scores for the FSI-GG, the HIIK, the WGI-CC, and the FSI-R [48,51,54].

The results concerning the criticality of the individual elements were merged according to the approach described in Sections 2.1 and 2.2. By multiplying the scores in the three dimensions per element with the respective material shares in the cathode chemistries (compare Table 2), absolute values per dimension and cathode chemistry were obtained. Figure 5a illustrates the corresponding results. By combining the weighting obtained from the AHP process, the dimensions were summed to one weighted criticality value per cathode chemistry (compare Figure 5b).

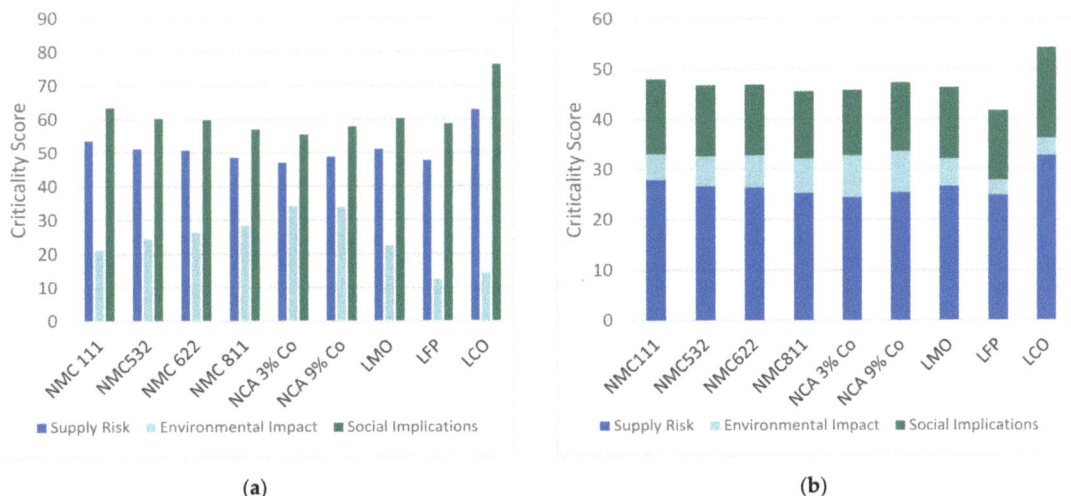

Figure 5. Criticality of LIB cathode chemistries: (**a**) absolute values of the three dimensions of criticality (supply risk, environmental impact, social implications); (**b**) weighted criticality value.

It is noticeable that the supply risk and the social implications seem to be correlated, with the social implications being more critical for all cathode chemistries. This is due to the fact that some of the supply risk indicators as well as all of the social implications indicators are dependent on the mining countries. Hence, there seems to be a causal relation between political and social circumstances. The high cobalt contents of NMC 111 and LCO result in significant supply risks as well as social implications. It is interesting to see that LMO has a high supply risk even though it contains no cobalt. This results from the supply risk of manganese, which is rated between cobalt and nickel. The NCA cathodes with a low cobalt share (NCA 3%) are scored with the lowest supply risk and social implications. For all of the considered elements, the environmental impact is assessed as rather low compared with the other two dimensions. Thereby, the NCA cathodes lead the order by far. Increasing the shares of nickel while decreasing the share of manganese from NMC 111 to NMC 811 results in higher environmental impacts. LFP obtains the lowest scores concerning environmental impact, which are due to the very low scores of phosphorus and iron in this dimension.

For the further analysis of substitutability, the normalized criticality per dimension and cathode chemistry is used. The normalization, which is described in Section 2.2, transforms the results on a scale from 0 (the best case) to 1 (defined as the maximum value one of the cathode chemistries achieves in the individual dimension). The respective values can be obtained from the Supplementary Materials.

3.2. Performance and Price of the Considered Cathode Chemistries

The performance of cathode chemistries is assessed on a scale from 0 to 2 (compare Section 2.2) [62]. The importance of the five analyzed indicators for the fields of application is based on the same scale [36]. However, as described in Section 2.2, for the purpose of normalization, the importance is transformed into a percentage scale (e.g., if all indicators are assessed with the highest importance of 2, they are transformed to be equally important: 0.2/0.2/0.2/0.2/0.2). Consequently, the maximum performance score per field of application and cathode chemistry equals 2. Figure 6 visualizes the results. It can be observed that, on average, the requirements for power- and garden tools are fulfilled the most with state-of-the-art cathode chemistries.

The LCO cathode chemistry is scored with the worst performance in every field of application. Even for consumer electronics, which is the main application for LCO LIBs, the preferred characteristics are not covered satisfactorily. The low-cobalt-containing NCA cathode chemistry follows with the second-worst rating. Between the other cathode chemistries, the one that is to be preferred varies depending on the field of application. The NMC 532 is top-rated in every field of application except for "stationary energy storage systems", which is dominated by NMC 111.

NMC cathodes are the technologies most favored for application in traction batteries. Following NMC 532, NMC 111, and NMC 811, LFP also achieves a high performance rating. LCO, NCA 3%, and LMO are not advised to be utilized for traction batteries based on the present assessment.

For stationary energy storage systems, NMC 111 and LFP are best suited. With a rating of 1.85 and 1.84, respectively, their performance score is nearly the same. Interesting to observe is the low performance rating of NMC 811, although it achieves mostly high scores for the other fields of application. LMO performs satisfactorily and nearly achieving the same performance rating as NMC 622.

The order of preference based on the performance rating is very similar for the remaining fields of application (consumer electronics, power-/garden tools, and domestic appliances). The second place varies between NMC 111 (power-/garden tools, domestic appliances) and NMC 811 (consumer electronics). LFP is also assessed with a high performance rating for all three fields of application. The sixth and seventh places alternate between cobalt-rich NCA (consumer electronics, domestic appliances) and LMO (power-

and garden tools). Again, the transformed values (scale: 0—maximum value) were used for further analysis and can be found in the Supplementary Materials.

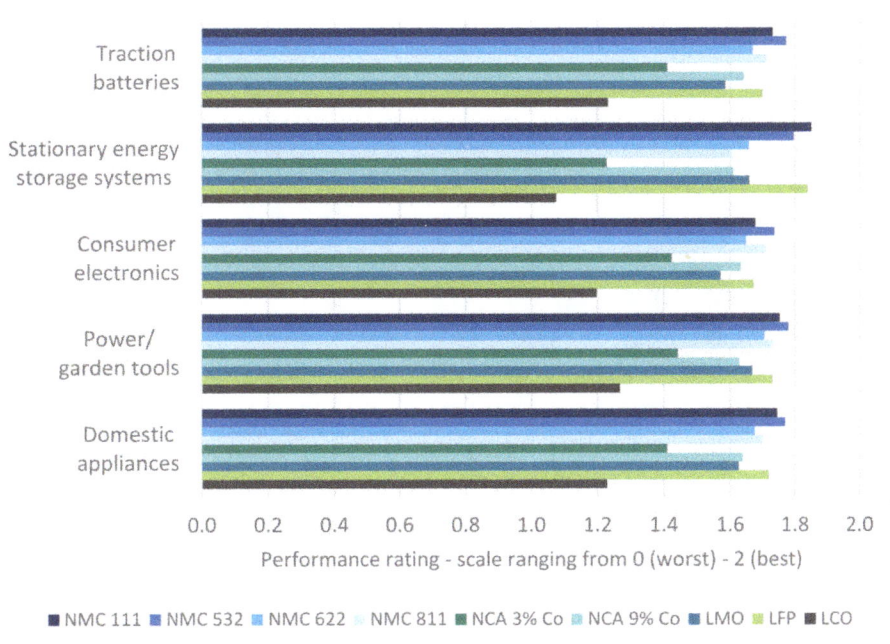

Figure 6. Performance of the considered cathode chemistries in different fields of application.

The price of the substitute is assessed by the use of five indicators (compare Section 2.2). Figure 7 illustrates the respective results for the nine considered cathode chemistries. The decrease in cobalt shares in cathodes results in significant price drops (on a raw material level). The price increase resulting from the evolution from NMC 532 to NMC 811 represents an intriguing finding. This is due to the fact that nickel is more expensive than manganese. Although the share of cobalt is decreased from 20% to 10%, the accompanying monetary benefit cannot compensate for the additional costs originating from the increase in the nickel share from 50% to 80% at the cost of a decreasing manganese share. LCO is by far the most expensive cathode chemistry due to the significant amounts of cobalt the cathode is comprised of. The cobalt-rich NCA (NCA 9%) and NMC (NMC 111) are the second and third most expensive cathode chemistries based on raw material prices. The derivates of NMC and NCA vary in price but mostly are on a similar level between 13,600 US$ and 16,000 US$ per ton of raw materials. As the identified costs for raw materials for the LFP cathode are extremely low, an additional approach was taken that is expected to be more realistic. This was done to allow for a discussion about the impact of the price dimension for the substitutability of cathode chemistries. According to the literature, LFP cathode material currently costs around 43% of the price of NMC 811 cathode material [75]. Consequently, the price indicators visualized on the right-hand side of Figure 7 equal 43% of the identified NMC 811 prices on a raw material level, which is also depicted in Figure 7 (fourth from the left). It appears that the price indicators are consequently assessed as significantly higher than the original score of LFP (Figure 7—third from the right). Although it surpasses LMO, the adjusted price of LFP is still lower than that of the other cathode chemistries by far. Once again, all obtained values were transformed to the introduced scale from 0 to the maximum value (see Supplementary Materials). The resulting scores were weighted by the weightings obtained from the AHP and merged to an overall price score.

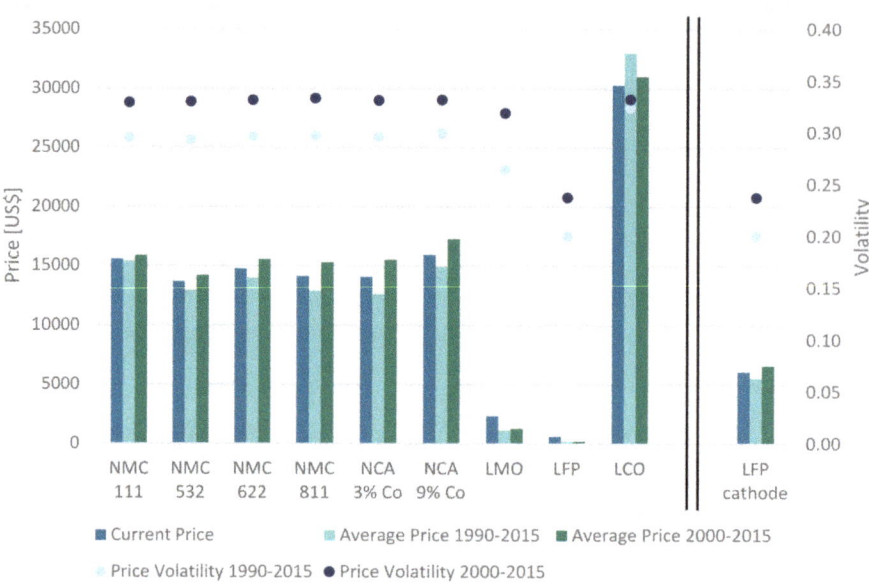

Figure 7. Price indicators for cathode chemistries.

3.3. Substitutability of Cathode Chemistries

The obtained scores for the criticality, performance, and price categories (compare Sections 3.1 and 3.2) are summed to a substitutability value per cathode chemistry and field of application by deploying the respective weightings from the AHP. Consequently, the substitutability score is in the range from 0 to 1, where 0 equals the highest potential for substitution and 1 equals the lowest. The results can be obtained from Figure 8. Generally, the impact of the criticality category is rather small, as the respective variations in criticality scores are minimal and the consulted experts assessed the importance of the category as low. Using the described transformation rules as well as the introduced weightings, noticeable differences can only be identified between high cobalt-containing cathode chemistries (LCO, NCA 9%, NMC 111) and LFP.

In contrast, the price dimension has a significant impact on the overall substitutability score due to the large spreads of the individual raw materials (especially phosphorus/iron vs. cobalt/nickel) as well as a higher weighting than the criticality dimension.

The dimension of performance, which is attributed as the most important from the viewpoint of a producer/distributor of LIBs, consequently has substantial effects on the substitutability of cathode chemistries. Various chemistries are more or less on a similar level of substitutability when only taking the criticality and price dimensions into account. The decisive factor in these cases is the dimension of performance.

LFP achieves the lowest and thus best scores for every field of application. This is not even changed by taking the higher price of the cathode material into account as described in Section 3.2. The respective score of LFP with the adjusted price of the cathode material is depicted on the right-hand side of Figure 8a–e (labeled 'LFP cathode').

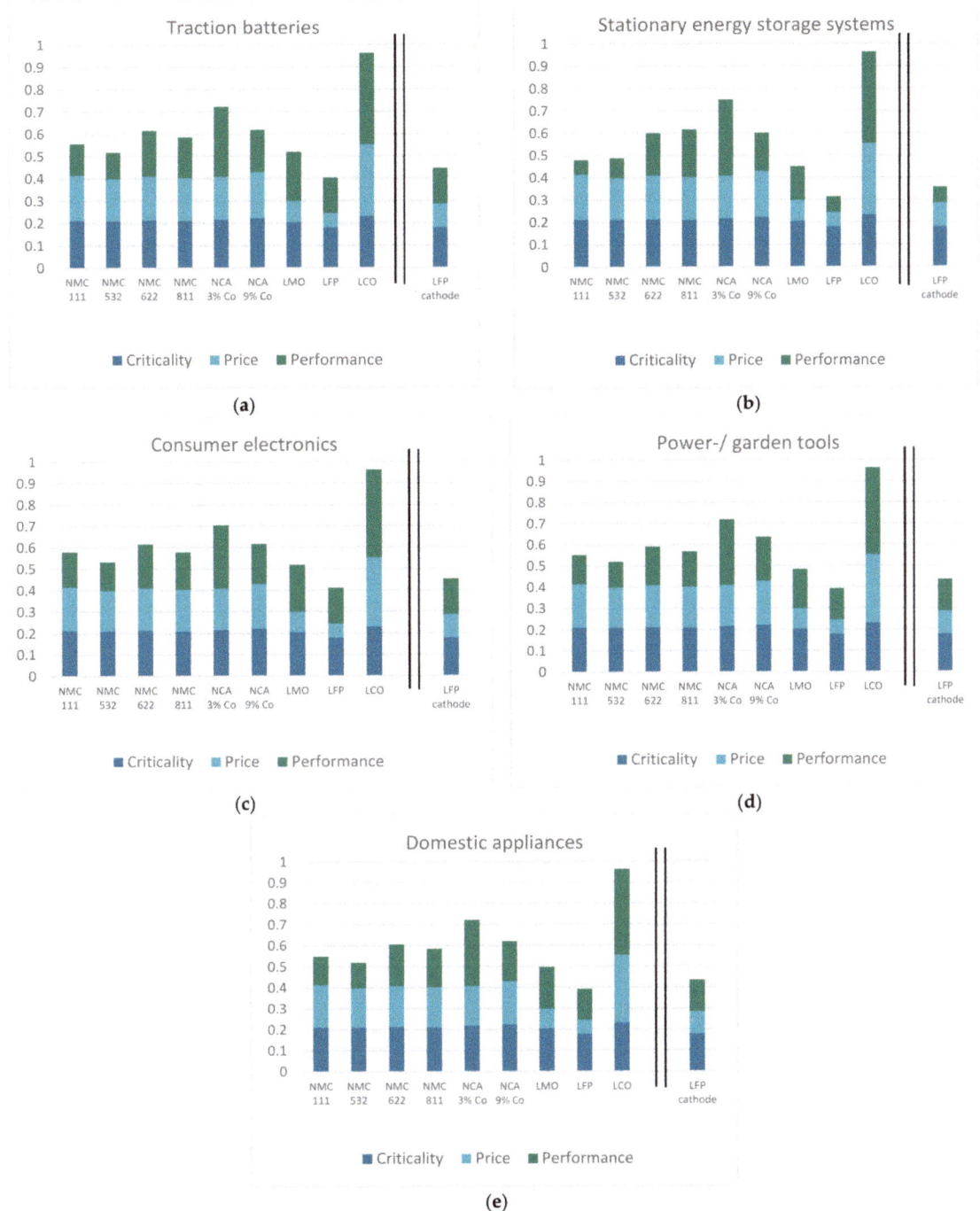

Figure 8. Results for the substitutability of cathode chemistries for the considered fields of application: (**a**) traction batteries; (**b**) stationary energy storage systems; (**c**) consumer electronics; (**d**) power-/garden tools; and (**e**) domestic appliances.

In contrast, LCO is not suggested in any of the considered fields of application as a potential substitute. Its high cobalt content, while still achieving the worst scores in the performance dimension for every field of application, disqualifies LCO as a suitable cathode chemistry for LIBs.

According to the presented evaluation, the second-best option for traction batteries would be the NMC 532 cathode chemistry. Although performing nowhere near NMC, LMO achieves almost the same substitutability level as NMC 532. This is due to the significantly lower costs of the raw materials, mainly owing to the cobalt-free composition. By taking the weighting obtained from the expert interviews into account, NMC 111 is preferable for utilization in traction batteries although it has higher cobalt content compared with NMC 622 and 811. This is contrary to the current trend. Both NCA cathode chemistries achieve high and thus poor substitutability ratings, only surpassed by LCO.

As LFP achieved the second-best performance rating for stationary energy storage systems (besides the best price and criticality rating), it is extremely well suited for this kind of application. With the underlying weighting, LMO is ranked in second place. However, it is to be expected that by individually increasing the weighting of the performance category, NMC 111 and NMC 532 will outperform LMO as the preferable cathode chemistry. From the assessed NMC derivates, NMC 811 receives the worst substitutability rating. While NCA 9% scores slightly better than NMC 811, the low-cobalt-containing NCA (NCA 3%) is placed a distant second.

Both the low-cobalt-containing NCA and NMC improved their performance score for consumer electronics compared with stationary energy storage systems. Although all other cathode chemistries were downgraded concerning the fulfillment of the performance parameters, not much changed in the general ranking of substitutability. NMC 811 is to be preferred to NMC 111 and NMC 622 but in total only ranks the fourth-best option for consumer electronics. Once again, LCO is ranked last by far. This is interesting, as LCO cathodes are currently commonly used for applications in the field of consumer electronics [8].

The ranking for power and garden tools, as well as domestic appliances, is the same as for consumer electronics, although all cathode chemistries fulfill the performance requirements a little better than for consumer electronics (except for the NCA cathodes).

4. Discussion

Considering the three categories (criticality, price, and performance of the substitute), the results prove the LFP cathode chemistry's dominance in every field of application. Even the additional evaluation with the adjusted price of the cathode materials shows the same outcome. From a material perspective, LFP remains the cheapest cathode chemistry (even when taking the price adjustment into account). Nevertheless, it has to be stated that other publications assess either NMC or NCA as the cheapest option [62,76]. As this is not justifiable considering the material prices, it is to be expected that optimized and efficient production processes contribute significantly to the overall manufacturing costs of battery cells (both NMC and NCA were scored with the highest "maturity score" [62]). Differing statements about the performance and costs of cathode chemistries result in uncertainties concerning the substitutability assessment and also demonstrate the complexity of the research field. This is identified as one of the reasons for the contradicting results of this article with publications concerned with the future market shares of the discussed cathode chemistries. Until recently, a significant decrease in the market share of LFP was predicted [77,78]. Nevertheless, as Tesla announced the use of LFP batteries in at least a few of their models [79], other car manufacturers might reconsider their current choice of cathode material. This might eventuate in a renaissance of the LFP battery chemistry. This article reveals the potential benefits of and argumentative enablers for this possible future trend.

In the literature, NMC is predicted to become the predominant cathode chemistry in the near future [8,77,78]. While this cannot directly be derived from the results presented in this article, at least one derivate of the considered NMC chemistries is in the top three alternatives for every considered field of application. While being assessed with very good scores in the performance category, the high price of the comprised materials prevents a top overall scoring. As already discussed, this assessment might change by extending the material perspective to the overall production costs of battery cells [62,76]. The NMC chemistries with high cobalt material shares (NMC 111 and 532) are the only cathode chemistries that could fare better than LFP by an alteration of the categories' weighting (a decrease/an increase in the importance of the price/performance, respectively). It is interesting to see that the expected progress to nickel-rich NMC cathodes [8] cannot be derived from the results of the present article. This is mainly due to their poorer performance concerning the safety indicator. Additionally, the durability and maturity of nickel-rich cathode chemistries are not yet on the level of battery generation 2a (NMC 111). The resulting score in the performance category cannot be compensated for by respective better scorings in the price category. Variances in the standardization of the category scores (scaling from minimum to maximum instead of zero to maximum), as well as the separate analysis of the NMC derivates (with neglection of the other cathode chemistries), result in small alterations in the conclusions. NMC 532 remains the preferable and NMC 622 the least preferable NMC cathode. However, compared with the initial assessment, NMC 811 is scored with better substitutability than NMC 111 in two more fields of application (traction batteries and power-/garden tools).

The market share of NCA batteries remains constant according to the literature [77,78]. However, this prediction is based on the assumption that Tesla will continue to use NCA batteries to the extent they did at the time the respective literature was published [78]. As stated before, Tesla introduced LFP batteries into its product portfolio in 2020. Hence, it is to be expected that the market shares of NCA batteries will decrease in the future. This agrees with the findings of the presented work, as the results suggest the substitution of NCA in every field of application.

The unfavorable substitutability score of LCO is validated by the significant decrease in the respective predicted market shares [77,78]. LCO cathodes are advised to be substituted in every field of application due to the significant amounts of cobalt in the material composition of the cathode. Concerning the material shares, differing statements can be found in the literature. However, by replacing the applied share (compare the results on the elemental level), the supply risk, environmental impact, are social implications are aggregated to the cathode-specific technology level. This is done by weighing the obtained values by mass shares of the respective raw material (or precursor material) within the considered cathode chemistries. This approach to calculating the criticality of material compounds or products is accepted in the scientific community [30–32]. The material compositions of the individual cathode chemistries can be obtained from Table 2. Thereby, lithium and manganese are applied as different derivates for some cathode chemistries. While lithium carbonate is used for NMC 111, NMC 532, NMC 622, LMO, and LCO [56,57], lithium hydroxide is required for the nickel-rich cathode NMC 811, LFP, and the NCA technologies [57–59]. The NMC active material is comprised of manganese sulfate [57], while for LMO manganese dioxide is applied [60]. The respective information is involved for the environmental dimension. The indicators of the considered LCIA endpoint categories are assessed in both a raw-material-specific and derivate-specific manner. As no dataset exists for cobalt sulfate hexahydrate in the ecoinvent database [47], a mixed calculation from sulfuric acid and cobalt based on stoichiometric calculations was conducted.

Table 2 contains more progressive values (1/3 Li_2CO_3 and 2/3 $CoSO_4$); however, the ranking concerning the overall substitutability score is not altered. This is due to the fact that even then the cobalt share is still very high compared with the other cathode chemistries.

A summary of the factors that increase the uncertainty in the results is presented in the following. In addition, outlooks for potential further research are mentioned.

The evaluation of the performance indicators is based on the acknowledged literature. Nevertheless, it has to be stated that the respective information is not consistent. An example is the scoring of the safety indicator. While Zubi et al. (who we use as the respective reference in this article) assess the NCA cathode as rather safe (1.5 points with a scale ranging from 0 to 2) [62], Meeus estimates the safety of NCA as being the lowest [76]. However, the estimate of Zubi et al. is supported by other references [80]. Another example of differing assessments can be observed for LMO cathodes. The costs, the energy, and the durability are scored rather differently. Expert interviews could reduce uncertainties concerning these aspects and are planned for further research. Additionally, the completeness of the considered performance indicators could be verified on this occasion (for example: is the temperature stability, which might be an important factor, especially for traction batteries, sufficiently covered by the durability indicator?)

The scoring for many indicators is fundamentally dependent on the material shares. As already discussed for LCO, differing statements concerning this aspect can be found in the literature. Hence, the results presented in this article should be seen as approximations. Detailed assessments must be conducted with the actual material shares of cathode chemistries utilized by the company that considers a substitution.

The price category has a relevant impact on the overall substitutability scoring of the considered cathode chemistries in all five fields of application. This article focuses only on the material prices and does not consider costs for the production of the battery cells and the respective manufacturing infrastructure. This is identified as a potential reason for partially significant differences between the price assessment in this article and the consulted literature [62,76]. The utilization of phosphate rock as the underlying material for the LFP price assessment is an inaccuracy worth mentioning. Due to this fact, an additional scenario with higher cathode prices for LFP derived from the literature was analyzed. Furthermore, the quality of historic data for the different cathode materials varies, which constitutes another potential source of uncertainty.

Criticality assessments are highly dependent on volatile data (e.g., the enormously increasing resources of lithium identified during the past few years). Consequently, criticality scores vary depending on the underlying base year. Hence, it is advisable to conduct such evaluations periodically. This is one of the reasons why the criticality of cathode chemistries was assessed in this article even though there are publications concerned with similar topics (compare [30,31]). In total, the criticality category has a relatively small impact on the selection of substitutes, as the scorings of the individual cathode chemistries do not differ significantly (with the biggest gap between LFP and LCO). This finding is in agreement with the work of Helbig et al. [30]. The integration of social and environmental aspects in addition to the dimension of supply risk does not change the result of small differences in the overall criticality scoring between the considered cathode chemistries.

In total, the assessment model presented in this article generates reasonable results that are consistent with the literature. Where this is not the case, the underlying reasons were identified and discussed. The introduced approach to evaluating the substitutability of LIB cathode chemistries is a basis for further research and sophistication. In particular, the company-specific adjustment of underlying data (for the price and performance categories) and the individual weighting of indicators and categories bear potential for increasing the applicability of the approach in the industry. Until then, the presented results, based on the described assumptions, provide a profound overview of the substitutability of cathode chemistries. The prospects for further research include the revision of the selected assumptions and the respective validation employing a broad expert survey. Particular focus in this regard should be put on the price category. The present article focuses on the material perspective. Hence, the applied indicators take the costs for the utilized raw materials into account and neglect the accompanying costs for producing the final composite material of the cathodes. The impact of a change from a raw material to a composite perspective in the price category should be analyzed in detail in further research.

This study did not involve a detailed sensitivity analysis but investigates the impact of an alteration in the selected indicators. Monte Carlo simulations are widely applied in the literature in order to conduct sensitivity analyses in the context of criticality assessments. The application of this method to the presented approach to assessing the substitutability of cathode chemistries should also be part of further research.

Finally, other cathode chemistries could be added as potential substitutes. Examples include blends (e.g., LMO/NMC) that are likely to obtain a considerable market share in the foreseeable future [63].

5. Conclusions

This article provides a decision support model for increasing the sustainability of product portfolios. For this purpose, the substitutability of products and material compounds is assessed based on the material criticality, performance, and price of the substitute. Thereby, the state of the art is compared with potential substitutes. The approach involves a variety of selected and relevant indicators to enable the derivation of the individual benefits and disadvantages of each evaluated option.

For validation, the developed methodology was applied to state-of-the-art LIB cathode chemistries (NMC 111, NMC 532, NMC 622, NMC 811, NCA 3%, NCA 9%, LMO, LFP, and LCO) as LIBs are some of the most relevant emerging technologies in the debate on future material availability. The assessment was conducted in a use-case-specific manner (traction batteries, stationary energy storage systems, consumer electronics, power-/garden tools, and domestic appliances). The outcomes show that LFP is the preferable cathode chemistry while LCO obtains the worst rating for all fields of application. The ranking based on the substitutability score of the other cathode chemistries varies per field of application. NMC 532, NMC 811, NMC 111, and LMO are to be regarded as recommendable types of cathodes.

The obtained results agree partially with predictions concerning future market shares of LIB cathode chemistries. Deviations are to be examined in further research. Potential causes, as well as options for improving the introduced methodology, are identified in this article. The main reasons for potential uncertainties are the applied assumptions concerning the price and performance categories. The price indicators are material-specific and based partly on historical data. It remains to be determined whether the applicability of the considered data is the same for each assessed material as well as whether a switch to cathode-chemistry-specific (instead of material-specific) prices would be feasible. The results concerning the category "performance of the substitute" could be refined and qualified by a further expert survey.

Supplementary Materials: The following are available online at https://www.mdpi.com/article/10.3390/resources10090087/s1, Figure S1: Questionnaire for the AHP, Table S1: Raw material criticality of LIB cathode chemistries, Table S2: Raw material criticality of LIB cathode chemistries, normalized on a scale from 0 to the maximum value, Table S3: Performance of cathode chemistries, Table S4: Importance of technical criteria/indicators per field of application, Table S5: Weighted performance of cathode chemistries for selected fields of application, normalized on a scale from 0 to the maximum value, Table S6: Price information—raw material level, Table S7: Price information—cathode chemistry level, Table S8: Price of cathode chemistries, normalized on a scale from 0 to the maximum value, Table S9: Consulted experts for the AHP, Table S10. Consulted experts for the AHP-process.

Author Contributions: Conceptualization, S.K.; methodology, S.K. and S.G.-C.; validation, S.K., S.G.-C. and R.M.; data curation, S.K. and M.S.; writing—original draft preparation, S.K. and L.W.; writing—review and editing, S.K., S.G.-C., L.W., M.S., A.S., and R.M.; visualization, S.K.; supervision, A.S. and R.M. All authors have read and agreed to the published version of the manuscript.

Funding: This research was funded by the Ministry of Economic Affairs, Labour and Tourism of Baden-Württemberg, grant number 017-105082/B7-RI.

Data Availability Statement: All data were obtained from publicly available literature sources and databases. The respective references are mentioned in the article. The only exceptions are selected indicators from the social dimension of the raw material criticality assessment. These were obtained from the fee-based "social hotspots database".

Acknowledgments: The authors would like to thank the external experts who returned a completed questionnaire for the AHP. These experts are, in alphabetical order, David Ensling, Johannes Full, Christoph Helbig, Sandra Huster, Gernot Kraberger, Sonja Rosenberg, Klaus Steinmüller, and Lars Wietschel.

Conflicts of Interest: The authors declare no conflict of interest. The funders had no role in the design of the study; in the collection, analyses, or interpretation of data; in the writing of the manuscript; or in the decision to publish the results.

References

1. United Nations. *Paris Agreement*; UN: New York, NY, USA, 2015; Available online: https://unfccc.int/process-and-meetings/the-paris-agreement/the-paris-agreement (accessed on 30 July 2021).
2. European Commission. *The European Green Deal*. COM 640; European Commission: Brussels, Belgium, 2019; Available online: https://eur-lex.europa.eu/legal-content/EN/TXT/?qid=1576150542719&uri=COM%3A2019%3A640%3AFIN (accessed on 30 July 2021).
3. European Commission. *'Fit for 55': Delivering the EU's 2030 Climate Target on the Way to Climate Neutrality*; COM(2021) 550 Final; European Commission: Brussels, Belgium, 2021; Available online: https://eur-lex.europa.eu/legal-content/EN/TXT/?uri=CELEX:52021DC0550 (accessed on 30 July 2021).
4. Ritchie, H. Cars, Planes, Trains: Where Do CO_2 Emissions from Transport Come from? 2020. Available online: https://ourworldindata.org/co2-emissions-from-transport (accessed on 30 July 2021).
5. Zhang, R.; Fujimori, S. The role of transport electrification in global climate change mitigation scenarios. *Environ. Res. Lett.* **2020**, *15*. [CrossRef]
6. IEA. *Energy Technology Perspectives 2020*; International Energy Agency: Paris, France, 2020; Available online: https://www.iea.org/reports/energy-technology-perspectives-2020 (accessed on 30 July 2021).
7. Köse, E.; Sauer, A.; Pelzel, C. Energieflexibel durch bivalente Produktionsanlagen/Energy flexibility through bivalent production facilities—Using bivalent production processes to reduce energy costs and stabilize the electricity grid. *Wt Werkstattstech. Online* **2017**, *107*, 366–372. [CrossRef]
8. Merriman, D. The EV Revolution: Impacts on Critical Raw Material Supply Chains. 2019. Available online: https://www.minersoc.org/wp-content/uploads/2019/05/3ICM-Merriman.pdf (accessed on 30 July 2021).
9. Karabelli, D.; Kiemel, S.; Singh, S.; Koller, J.; Ehrenberger, S.; Miehe, R.; Weeber, M.; Birke, K.P. Tackling xEV battery chemistry in view of raw material supply shortfalls. *Front. Energy Res.* **2020**, *8*. [CrossRef]
10. Olivetti, E.A.; Ceder, G.; Gaustad, G.G.; Fu, X. Lithium-ion battery supply chain considerations: Analysis of potential bottlenecks in critical metals. *Joule* **2017**, *1*, 229–243. [CrossRef]
11. Haglund, D.G. Strategic minerals: A conceptual analysis. *Resour. Policy* **1984**, *10*, 146–152. [CrossRef]
12. Erdmann, L.; Graedel, T.E. Criticality of non-fuel minerals: A review of major approaches and analyses. *Environ. Sci. Technol.* **2011**, *45*, 7620–7630. [CrossRef]
13. Graedel, T.E.; Barr, R.; Chandler, C.; Chase, T.; Choi, J.; Christoffersen, L.; Friedlander, E.; Henly, C.; Jun, C.; Nassar, N.T.; et al. Methodology of metal criticality determination. *Environ. Sci. Technol.* **2012**, *46*, 1063–1070. [CrossRef]
14. Kolotzek, C.; Helbig, C.; Thorenz, A.; Reller, A.; Tuma, A. A company-oriented model for the assessment of raw material supply risks, environmental impact and social implications. *J. Clean. Prod.* **2018**, *176*, 566–580. [CrossRef]
15. Miehe, R.; Schneider, R.; Baaij, F.; Bauernhansl, T. Criticality of material resources in industrial enterprises—Structural basics of an operational model. *Procedia CIRP* **2016**, *48*, 1–9. [CrossRef]
16. Schrijvers, D.; Hool, A.; Blengini, G.A.; Chen, W.-Q.; Dewulf, J.; Eggert, R.; van Ellen, L.; Gauss, R.; Goddin, J.; Habib, K.; et al. A review of methods and data to determine raw material criticality. *Resour. Conserv. Recycl.* **2020**, *155*, 104617. [CrossRef]
17. Panousi, S.; Harper, E.M.; Nuss, P.; Eckelman, M.J.; Hakimian, A.; Graedel, T.E. Criticality of Seven Specialty Metals. *J. Ind. Ecol.* **2016**, *20*, 837–853. [CrossRef]
18. Graedel, T.E.; Harper, E.M.; Nassar, N.T.; Nuss, P.; Reck, B.K. Criticality of metals and metalloids. *Proc. Natl. Acad. Sci. USA* **2015**, *112*, 4257–4262. [CrossRef]
19. U.S. Department of Energy. *Critical Materials Strategy*; U.S. Department of Energy: Washington, DC, USA, 2011.
20. European Commission. *Study on the EU's List of Critical Raw Materials*; Final Report; European Commission: Brussels, Belgium, 2020.
21. Nassar, N.T.; Barr, R.; Browning, M.; Diao, Z.; Friedlander, E.; Harper, E.M.; Henly, C.; Kavlak, G.; Kwatra, S.; Jun, C.; et al. Criticality of the geological copper family. *Environ. Sci. Technol.* **2012**, *46*, 1071–1078. [CrossRef]
22. Nuss, P.; Harper, E.M.; Nassar, N.T.; Reck, B.K.; Graedel, T.E. Criticality of iron and its principal alloying elements. *Environ. Sci. Technol.* **2014**, *48*, 4171–4177. [CrossRef]

23. Nassar, N.T.; Du, X.; Graedel, T.E. Criticality of the Rare Earth Elements. *J. Ind. Ecol.* **2015**, *19*, 1044–1054. [CrossRef]
24. USGS. *Mineral Commodity Summaries 2019*; US Geological Survey: Reston, VA, USA, 2019. Available online: https://pubs.er.usgs.gov/publication/70202434 (accessed on 30 July 2021).
25. USGS. *Mineral Commodity Summaries 2015*; US Geological Survey: Reston, VA, USA, 2015. Available online: https://www.usgs.gov/centers/nmic/mineral-commodity-summaries (accessed on 30 July 2021).
26. Viebahn, P.; Soukup, O.; Samadi, S.; Teubler, J.; Wiesen, K.; Ritthoff, M. Assessing the need for critical minerals to shift the German energy system towards a high proportion of renewables. *Renew. Sustain. Energy Rev.* **2015**, *49*, 655–671. [CrossRef]
27. Helbig, C.; Bradshaw, A.M.; Kolotzek, C.; Thorenz, A.; Tuma, A. Supply risks associated with CdTe and CIGS thin-film photovoltaics. *Appl. Energy* **2016**, *178*, 422–433. [CrossRef]
28. Kiemel, S.; Smolinka, T.; Lehner, F.; Full, J.; Sauer, A.; Miehe, R. Critical materials for water electrolysers at the example of the energy transition in Germany. *Int. J. Energy Res.* **2021**, *45*, 9914–9935. [CrossRef]
29. Yu, Y. Assessing the criticality of minerals used in emerging technologies in China. *PAN* **2020**, *36*, 5–20. [CrossRef]
30. Helbig, C.; Bradshaw, A.M.; Wietschel, L.; Thorenz, A.; Tuma, A. Supply risks associated with lithium-ion battery materials. *J. Clean. Prod.* **2018**, *172*, 274–286. [CrossRef]
31. Wentker, M.; Greenwood, M.; Asaba, M.C.; Leker, J. A raw material criticality and environmental impact assessment of state-of-the-art and post-lithium-ion cathode technologies. *J. Energy Storage* **2019**, *26*, 101022. [CrossRef]
32. Simon, B.; Ziemann, S.; Weil, M. Criticality of metals for electrochemical energy storage systems—Development towards a technology specific indicator. *Metall. Res. Technol.* **2014**, *111*, 191–200. [CrossRef]
33. Bach, V.; Finogenova, N.; Berger, M.; Winter, L.; Finkbeiner, M. Enhancing the assessment of critical resource use at the country level with the SCARCE method—Case study of Germany. *Resour. Policy* **2017**, *53*, 283–299. [CrossRef]
34. Pavel, C.; Marmier, A.; Alves Dias, P.; Blagoeva, D.; Tzimas, E.; Schüler, D.; Schleicher, T.; Jenseit, W.; Degreif, S.; Buchert, M. *Substitution of Critical Raw Materials in Low-Carbon Technologies: Lighting, Wind Turbines and Electric Vehicles*; Publications Office of the European Union: Luxembourg, 2016.
35. Halme, K.; Piirainen, K.; Vekinis, G.; Sievers, U.; Viljamaa, K. *Substitutionability of Critical Raw Materials*; European Parliament's Committee on Industry: Brussels, Belgium, 2012; Available online: https://op.europa.eu/de/publication-detail/-/publication/36145a5a-2da8-4730-9b69-4b5eb7444b25 (accessed on 30 July 2021).
36. Full, J.; Wanner, J.; Kiemel, S.; Miehe, R.; Weeber, M.; Sauer, A. Comparing technical criteria of various lithium-ion battery cell formats for deriving respective market potentials. In Proceedings of the 2020 IEEE Electric Power and Energy Conference (EPEC), Edmonton, AB, Canada, 9 November 2020; pp. 1–6. [CrossRef]
37. Achzet, B.; Helbig, C. How to evaluate raw material supply risks—An overview. *Resour. Policy* **2013**, *38*, 435–437. [CrossRef]
38. Goedkoop, M.; Heijungs, R.; Huijbregts, M.; de Schreyver, A.; Struijs, J.; van Zelm, R. *ReCiPe 2008 A life Cycle Impact Assessment Method which Comprises Harmonised Category Indicators at the Midpoint and the Endpoint Level. Report I: Characterisation*, 1st ed.; Ministerie van VROM: Den Haag, The Netherlands, 2009; Available online: https://www.leidenuniv.nl/cml/ssp/publications/recipe_characterisation.pdf (accessed on 30 July 2021).
39. Buchholz, P. *Angebotskonzentration bei Mineralischen Rohstoffen und Zwischenprodukten—Potenzielle Preis- und Lieferrisiken: DERA-Rohstoffliste 2012*; DERA Rohstoffinformationen 24; DERA: Hannover, Germany, 2014.
40. Stedman, A.; Green, K.P. *Fraser Institute Annual. Survey of Mining Companies 2018*; Fraser Institute: Vancouver, BC, Canada, 2019.
41. Kaufmann, D.; Kraay, A.; Mastruzzi, M. *The Worldwide Governance Indicators Project: Answering the Critics*; Policy Research Working Paper No. 4149; World Bank: Washington, DC, USA, 2007; Available online: https://openknowledge.worldbank.org/handle/10986/7203 (accessed on 30 July 2021).
42. UNDP. Human Development Index. 2019. Available online: http://hdr.undp.org/en/content/human-development-index-hdi (accessed on 30 July 2021).
43. Nassar, N.T.; Graedel, T.E.; Harper, E.M. By-product metals are technologically essential but have problematic supply. *Sci. Adv.* **2015**, *1*, e1400180. [CrossRef] [PubMed]
44. Marscheider-Weidemann, F.; Langkau, S.; Hummen, T.; Erdmann, L.; Tercero Espinoza, L.A.; Angerer, G.; Marwede, M.; Benecke, S. *Rohstoffe für Zukunftstechnologien 2016: Auftragsstudie*; DERA Rohstoffinformationen 28; DERA: Hannover, Germany, 2016.
45. Graedel, T.E.; Harper, E.M.; Nassar, N.T.; Reck, B.K. On the materials basis of modern society. *Proc. Natl. Acad. Sci. USA* **2015**, *112*, 6295–6300. [CrossRef] [PubMed]
46. Graedel, T.E.; Allwood, J.; Birat, J.-P.; Buchert, M.; Hagelüken, C.; Reck, B.K.; Sibley, S.F.; Sonnemann, G. What do we know about metal recycling rates? *J. Ind. Ecol.* **2011**, *15*, 355–366. [CrossRef]
47. Wernet, G.; Bauer, C.; Steubing, B.; Reinhard, J.; Moreno-Ruiz, E.; Weidema, B. The ecoinvent database version 3 (part I): Overview and methodology. *Int. J. Life Cycle Assess* **2016**, *21*, 1218–1230. [CrossRef]
48. Kaufmann, D.; Kraay, A.; Mastruzzi, M. *The Worldwide Governance Indicators, 2019 Update. Aggregate Governance Indicators 1996–2018*; World Bank: Washington, DC, USA, 2019; Available online: https://info.worldbank.org/governance/wgi/ (accessed on 30 July 2021).
49. Schwab, K. *The Global Competitiveness Report. Insight Report*; World Economic Forum: Geneva, Switzerland, 2019; Available online: https://www.weforum.org/reports/how-to-end-a-decade-of-lost-productivity-growth (accessed on 30 July 2021).

50. Wendling, Z.; Emerson, J.; Esty, D.C.; Levy, M.A.; Sherbinin, A. *Environmental Performance Index*; Global Metrics for the Environment: Ranking Country Performance on High-Priority Environmental Issues; Yale Center for Environmental Law & Policy: New Haven, CT, USA, 2018. [CrossRef]
51. The Fund for Peace. Fragile State Index 2019. 2019. Available online: https://fragilestatesindex.org/excel/ (accessed on 30 July 2021).
52. New Earth/Social Hotspots Database Project. 2021. Available online: http://www.socialhotspot.org/ (accessed on 30 July 2021).
53. World Health Organization. *Global Health Estimates 2016: Disease Burden by Cause, Age, Sex, by Country and by Region, 2000–2016*; WHO: Geneva, Switzerland, 2018; Available online: http://www.who.int/healthinfo/global_burden_disease/en/ (accessed on 30 July 2021).
54. Heidelberg Institute for International Conflict Research. Conflict Barometer 2018. 2019. Available online: https://hiik.de/2019/02/26/konfliktbarometer-2018/ (accessed on 30 July 2021).
55. UNDP. Gender Inequality Index. 2018. Available online: http://hdr.undp.org/en/content/gender-inequality-index-gii (accessed on 30 July 2021).
56. Dunn, J.B.; James, C.; Gaines, L.; Gallagher, K.; Dai, Q.; Kelly, J.C. *Material and Energy Flows in the Production of Cathode and Anode Materials for Lithium Ion Batteries*; Argonne National Laboratory: Argonne, IL, USA, 2015. Available online: https://greet.es.anl.gov/publication-anode-cathode-liion (accessed on 30 July 2021).
57. Dai, Q.; Kelly, J.C.; Dunn, J.B.; Benavides, P.T. *Update of Bill-of-Materials and Cathode Materials Production for Lithium-ion Batteries in the GREET Model*; Argonne National Laboratory: Argonne, IL, USA, 2018. Available online: https://greet.es.anl.gov/publication-update_bom_cm (accessed on 30 July 2021).
58. Benavides, P.T.; Dai, Q.; Kelly, J.; Dunn, J.B. *Addition of Nickel Cobalt Aluminum (NCA) Cathode Material to GREET2*; Argonne National Laboratory: Argonne, IL, USA, 2016. Available online: https://greet.es.anl.gov/publication-NCA-Cathode-2016 (accessed on 30 July 2021).
59. Xie, J.; Gao, F.; Gong, X.; Wang, Z.; Liu, Y.; Sun, B. Life cycle assessment of LFP cathode material production for power lithium-ion batteries. In *Advances in Energy and Environmental Materials. In Advances in Energy and Environmental Materials*; Han, Y., Ed.; Springer: Singapore, 2017; pp. 513–522. [CrossRef]
60. Notter, D.A.; Gauch, M.; Widmer, R.; Wäger, P.; Stamp, A.; Zah, R.; Althaus, H.-J. Contribution of Li-ion batteries to the environmental impact of electric vehicles. *Environ. Sci. Technol.* **2010**, *44*. [CrossRef]
61. Lu, Q.; Wu, P.F.; Shen, W.X.; Wang, X.C.; Zhang, B.; Wang, C. Life cycle assessment of electric vehicle power battery. *Mater. Sci. Forum* **2016**, *847*, 403–410. [CrossRef]
62. Zubi, G.; Dufo-López, R.; Carvalho, M.; Pasaoglu, G. The lithium-ion battery: State of the art and future perspectives. *Renew. Sustain. Energy Rev.* **2018**, *89*, 292–308. [CrossRef]
63. Ding, Y.; Cano, Z.P.; Yu, A.; Lu, J.; Chen, Z. Automotive Li-Ion batteries: Current status and future perspectives. *Electrochem. Energ. Rev.* **2019**, *2*, 1–28. [CrossRef]
64. Li, M. and Lu, J. Cobalt in lithium-ion batteries. *Science* **2020**, *367*, 979–980. [CrossRef] [PubMed]
65. Croy, J.R.; Long, B.R.; Balasubramanian, M. A path toward cobalt-free lithium-ion cathodes. *J. Power Sources* **2019**, *440*, 227113. [CrossRef]
66. Wood, M.; Li, J.; Ruther, R.E.; Du, Z.; Self, E.C.; Meyer, H.M.; Daniel, C.; Belharouak, I.; Wood, D.L. Chemical stability and long-term cell performance of low-cobalt, Ni-Rich cathodes prepared by aqueous processing for high-energy Li-Ion batteries. *Energy Storage Mater.* **2020**, *24*, 188–197. [CrossRef]
67. Zhang, S.S. Problems and their origins of Ni-rich layered oxide cathode materials. *Energy Storage Mater.* **2020**, *24*, 247–254. [CrossRef]
68. DERA; BGR. *Preismonitor. Dezember 2019*; Deutsche Rohstoffagentur—Bundesanstalt für Geowissenschaften und Rohstoffe: Berlin, Germany, 2019; Available online: https://www.deutsche-rohstoffagentur.de/DE/Themen/Min_rohstoffe/Produkte/Preisliste/pm_19_12.pdf?__blob=publicationFile&v=5 (accessed on 30 July 2021).
69. Shulga, M. US Pig Iron: Prices Go Down on Weakening Scrap Market. 2018. Available online: https://www.metalbulletin.com/Article/3825289/US-PIG-IRON-Prices-go-down-on-weakening-scrap-market.html (accessed on 30 July 2021).
70. Kelly, T.D.; Matos, G.R.; Buckingham, D.A.; DiFrancesco, C.A.; Porter, K.E. Historical Statistics for Mineral and Material Commodities in the United States. 2017–2018. Available online: https://www.usgs.gov/centers/nmic/historical-statistics-mineral-and-material-commodities-united-states (accessed on 30 July 2021).
71. Connors, L.A.; Hayward, B.E. *Trading für Profis. Mit Welchen Börsentechniken Sie von der Dummheit Vieler Anleger Profitieren*; Börsenverl: Rosenheim, Germany, 1998.
72. Saaty, T.L.; Vargas, L.G. *Decision Making in Economic, Political, Social and Technological Environments. With the Analytic Hierarchy Process*; The Analytic Hierarchy Process Series 7; RWS Publication: Pittsburgh, PA, USA, 1994.
73. Meixner, O.; Haas, R. *Wissensmanagement und Entscheidungstheorie: Mit 35 Tabellen*; Facultas: Vienna, Austria, 2010.
74. Velázquez-Martínez, V.; Santasalo-Aarnio, R.; Serna-Guerrero, R. A critical review of lithium-ion battery recycling processes from a circular economy perspective. *Batteries* **2019**, *5*, 68. [CrossRef]
75. Rudisuela, K. Battle of the batteries—Cost versus Performance. 2020. Available online: https://nickelinstitute.org/blog/2020/june/battle-of-the-batteries-cost-versus-performance/ (accessed on 17 July 2021).

76. Meeus, M. Overview of battery cell technologies. In Proceedings of the European Battery Cell R&I Workshop, Brussels, Belgium, 11–12 January 2018.
77. Pillot, C. Impact of the xEV market growth on lithium-ion batteries and raw materials supply 2018–2030. In Proceedings of the Automotive Battery Conference, Strasbourg, France, 27–31 January 2019; Available online: https://www.emove360.com/wp-content/uploads/2019/10/Impact-of-the-xEV-Market-growth-on-Lithium-Ion-batteries-and-raw-matterials-supply-2018-2030.pdf (accessed on 30 July 2021).
78. Or, T.; Gourley, S.W.D.; Kaliyappan, K.; Yu, A.; Chen, Z. Recycling of mixed cathode lithium-ion batteries for electric vehicles: Current status and future outlook. *Carbon Energy* **2020**, *2*, 6–43. [CrossRef]
79. Anderson, M. EV batteries shift into high gear: Advances in anodes, cathodes, and electrolytes are poised to appear in future electric vehicles. *IEEE Spectr.* **2020**, *57*, 8–9. [CrossRef]
80. Yoshio, M.; Brodd, R.J.; Kozawa, A. *Lithium-Ion Batteries. Science and Technologies*; Springer: New York, NY, USA, 2009; Available online: https://www.researchgate.net/publication/253140965_A_Review_of_Positive_Electrode_Materials_for_Lithium-Ion_Batteries (accessed on 30 July 2021).

Article

The Potential and Limitations of Critical Raw Material Recycling: The Case of LED Lamps

Julia S. Nikulski, Michael Ritthoff * and Nadja von Gries

Wuppertal Institute for Climate, Environment and Energy, Doeppersberg 19, 42103 Wuppertal, Germany; julia.nikulski@protonmail.com (J.S.N.); contact@nadjavongries.de (N.v.G.)
* Correspondence: michael.ritthoff@wupperinst.org

Abstract: Supply risks and environmental concerns drive the interest in critical raw material recycling in the European Union. Globally, waste electrical and electronic equipment (WEEE) is projected to increase by almost 40% until 2030. This waste stream can be a source of secondary raw materials. The determination of the economic feasibility of recycling and recovering specific materials is a data-intensive, time-consuming, and case-specific task. This study introduced a two-part evaluation scheme consisting of upper continental crust concentrations and raw material prices as a simple tool to indicate the potential and limitations of critical raw material recycling. It was applied to the case of light-emitting diodes (LED) lamps in the EU. A material flow analysis was conducted, and the projected waste amounts were analyzed using the new scheme. Indium, gallium, and the rare earth elements appeared in low concentrations and low absolute masses and showed only a small revenue potential. Precious metals represented the largest revenue share. Future research should confirm the validity and usefulness of the evaluation scheme.

Keywords: recycling economics; urban mining; LED lamps; material flow analysis

Citation: Nikulski, J.S.; Ritthoff, M.; von Gries, N. The Potential and Limitations of Critical Raw Material Recycling: The Case of LED Lamps. *Resources* **2021**, *10*, 37. https://doi.org/10.3390/resources10040037

Academic Editors: Andrea Thorenz and Armin Reller

Received: 2 March 2021
Accepted: 12 April 2021
Published: 16 April 2021

Publisher's Note: MDPI stays neutral with regard to jurisdictional claims in published maps and institutional affiliations.

Copyright: © 2021 by the authors. Licensee MDPI, Basel, Switzerland. This article is an open access article distributed under the terms and conditions of the Creative Commons Attribution (CC BY) license (https://creativecommons.org/licenses/by/4.0/).

1. Introduction

The annual global extraction of primary resources has grown almost fourfold between 1970 and 2010 and is significantly contributing to the loss of biodiversity, water stress, and climate change [1]. Wiedenhofer et al. [2] found that 53% of the global materials processed in 2014 entered the anthropogenic stock as part of buildings, technical infrastructure, durable consumer goods, or other long-lasting products. This ratio increased from 16% in 1900. Forecasts show that even considering efforts to stabilize the use of global primary stock-building materials total in-use material stocks will more than double between 2014 and 2050 worldwide [2]. These high raw material inputs are caused by current production and consumption patterns in linear economies [3]. The EU economy's growth is increasingly dependent on non-energy raw materials—such as metals and minerals—whose criticality was previously paid less attention to than that of oil and gas [4]. The European Commission estimated that roughly 30 million European jobs are contingent on the availability of raw materials [5]. According to a 2014 report by the European Commission [4] on critical raw materials, not only is the economic importance of some of these materials high, but also their supply risk. Around 91% of the overall non-energy raw materials used in the EU28 are imported from outside the member states. This means that procurement dependencies from countries with unstable governance systems (e.g., a weak rule of law, high levels of corruption, and political instability) can increase the uncertainty of material availability and jeopardize growth and jobs in Europe [4]. This supply risk could be lowered if critical materials were substituted or if materials were recycled from End-of-Life (EoL) products [4,6]. These economic as well as ecological concerns therefore lead to a growing interest in the resource potential of anthropogenic stockpiles and the recovery of secondary raw materials.

105

Within these stockpiles, electrical and electronic equipment (EEE)—such as monitors, lamps, and large and small household appliances—is a fast-growing portion. When this equipment reaches the end of its lifetime, it moves from stock to waste [7]. Forti et al. [7] estimated that in 2019 roughly 53.6 Mt of e-waste was generated globally. This amount is projected to increase to 74.7 Mt in 2030. Yet, only a fraction of this waste is collected and recycled, leaving valuable materials unrecovered in municipal solid waste or landfills [7]. The 2012 EU directive on waste electrical and electronic equipment (WEEE) addressed this issue by setting minimum standards and ratios for the collection, recycling, and recovery rates of e-waste [8]. Collection systems need to be expanded, and recycling technologies need to be enhanced or newly developed to achieve these targets.

Whether material recycling from the anthropogenic stock is ecologically and economically feasible compared with the extraction from primary raw materials is usually answered based on specific case studies. For this purpose, the case-specific processes for the extraction of the primary and secondary materials are then evaluated and compared with each other. Such a procedure, however, is time-consuming and linked to one specific technology. In many cases, no processes have yet been established for the recycling of materials and the production of secondary raw materials.

Light-emitting diodes (LED) lamps are an example of EEE with a relatively long lifespan and growing production, consumption, and EoL flows [9]. Yet, there is currently no established recycling technology available for LED lamps [10]. Only a few studies have investigated the recycling and material recovery of LED lighting. While some focused on technological development [11–13], others assessed the environmental impacts of LED lamps and their EoL phase [14] or discussed the economic potential of material recovery [9,15,16]. Studies analyzing the economic viability of LED lighting and other WEEE streams often focused on the absolute raw material amounts in the EoL products combined with their prices to derive recommendations for actions [16–18], or they focused on a cost–benefit analysis [19,20]. However, cost–benefit calculations require a significant amount of data input and often refer to specific recycling technologies. Meanwhile, a sole focus on raw material prices allows selecting the most profitable materials out of the ones considered. Still, it neglects the larger context and question of whether the amounts contained in the waste stream warrant to be recycled given their concentration in the overall amount of waste.

The question can be raised whether an initial assessment of recycling feasibility is possible based on simple and generally accessible information. Therefore, this paper introduces a two-part evaluation scheme to conduct such an initial assessment of economic feasibility for the material recycling of any WEEE stream. This scheme can be considered a precursor to cost–benefit analyses. It allows assessing the viability of recycling independent of a specific technology by evaluating two areas: First, material concentrations in the total LED lamp waste were compared to average material concentrations in the earth's crust, specifically the upper continental crust concentrations reported by Rudnick and Gao [21]. Second, raw material prices combined with the total amounts of materials embedded in the waste streams indicated which materials would generate the highest potential revenues. The main objective of introducing this evaluation scheme is to provide a method that can estimate the economic feasibility of recycling in a relatively fast and easy way by leveraging only easily accessible data. The results from this evaluation could indicate whether further investigations into new recycling methods are warranted and on which materials to focus.

LED lamp recycling was investigated in the European Union to illustrate the application of this evaluation scheme. This paper is structured as follows: Section 2 provides an overview of the methodology used to forecast the LED lamp waste generation between 2017 and 2030 in the EU28 member states. The data and the Weibull distribution used to model the future waste streams are described, and the proposed evaluation scheme is presented. Section 3 shows the results of the LED lamp waste forecast and the amounts of materials embedded in this waste stream. Using these results, the total potential revenue per material is calculated. The material concentration of the entire LED lamp waste is compared with the upper crust concentrations for each raw material. Section 4 discusses the results to determine whether the introduction of a new recycling technology would be feasible. Section 5 summarizes the findings of this study.

2. Materials and Methods

2.1. Data Collection

The following data were gathered to model the projected waste flows of LED lamps for the EU28 member states between 2017 and 2030. The material composition of white LEDs per average die area and the LED die area per LED lamp were used to calculate the mass of the specific material per lamp. The put on the market data, the average lifespan, and the average weight of LED lamps were combined to calculate the generated LED lamp waste. Applying the specific material per LED lamp to the total LED lamp waste yielded the total material weight contained in the waste. The consideration of collection, recycling, and material recovery rates for LED lamps allowed determining the recycling feasibility based on system and thermodynamic restrictions limiting the amount available for recovery. Because supply risk is a significant factor motivating the recovery of secondary raw materials, this case study focused on the materials included in LEDs, which are categorized as "critical" by the EU: cerium, europium, gadolinium, gallium, indium, palladium, terbium, yttrium [6]. Gold and silver were also included, given their high total material requirements (TMR) [22]. The relevant parts of an LED lamp that contain these materials are the chip, the interconnection technology, the phosphorus, and the printed circuit board (PCB) of white LEDs [15,23]. Different studies investigated the material composition of LED lamps with considerable differences in the reported amounts (e.g., [15,23–26]). The bill of materials used for this analysis was primarily taken from Deubzer et al. [23] and Buchert et al. [15]. They provided the most comprehensive list of critical raw materials included in LEDs. The weights for the materials included in the PCB were derived using material ratios published by Huisman et al. [27]. Those were applied to the weight of a PCB per one unit of LED reported by Scholand and Dillon [28]. The weights of the rare earth elements related to different types of phosphorus: YAG:Ce, TAG:Ce, ortho-silicate, or GAG:Ce [15,23]. The exact share of white LED lamps per phosphorus type was unknown. Therefore, the calculations in this study considered all rare earth elements that could potentially—with the given concentrations—be contained in a white LED. This was taken into account during the interpretation of the results. Bond wiring combined with gluing was assumed to be the most common interconnection technology and included in the bill of materials [23]. An overview of all materials considered is shown in Table 1.

Table 1. Material demand for selected LED lamp components and critical raw materials relating to 1 mm^2 die area of white LED.

Component	Material	Weight (mg)
Chip	Gallium	0.007 [23]
Chip	Indium	0.009 [23]
Phosphorus	Cerium	0.003 [23]
Phosphorus	Europium	0.003 [23]
Phosphorus	Gadolinium	0.015 [15]
Phosphorus	Terbium	0.165 [23]
Phosphorus	Yttrium	0.089 [23]
Printed circuit board	Gold	0.155 [27,28]
Printed circuit board	Silver	1.703 [27,28]
Printed circuit board	Palladium	0.093 [27,28]
Interconnection technology	Gold	0.019 [23]
Interconnection technology	Silver	0.276 [23]

The number of LED lamps put on the market (POM) in EU28 member states until 2030 was calculated using data from Marwede et al. [9], who estimated the development of POM amounts between 2008 and 2020. Buchert et al. [26] approximated that white LED lamps would have a market share of 95% in 2025 and be partially displaced by white OLED lamps until they would reach 75% market share in 2050. Based on this, we assumed that the growth of market share would slow until 100% is reached in 2030, and afterward decline. Figure 1 shows the overall EU-trend of POM amounts for white LED lamps. Table A1 shows the POM data used to derive this figure.

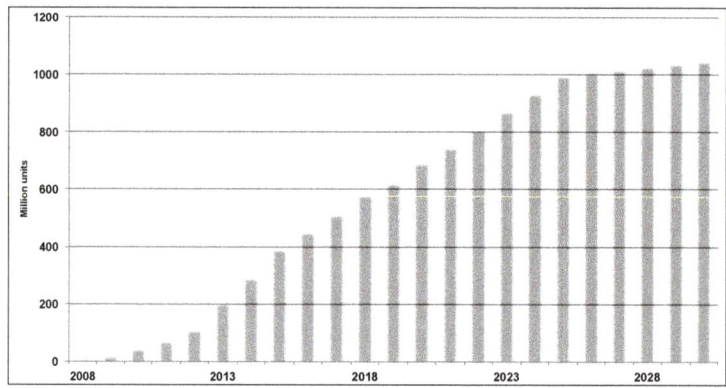

Figure 1. Development of put on the market amounts of LED lamps in EU28 countries until 2030.

The average lifespan, the average weight, and the average LED die area depend on different application types of lamps. Besides the application of LEDs in general lighting, there are many other applications of LEDs, such as in backlights of displays of electrical appliances. However, in this study, we focused on the application of LEDs in general lighting, which dominates the LED market [23]. Marwede et al. [9] differentiated between residential lamps and retrofits, commercial lamps and retrofits, industrial, outdoor, and architectural lamps based on common market segments. Data on average lifespans, weight, and die area for LED lamps is sparse. The length of the domestic service lifespan—describing the time from shipment of a product to its first user until the time of discard by its last user [29]—is difficult to determine for LED lamps. Therefore, technical lifespans combined with typical operating hours reported by Marwede et al. [9] were used. The average weight of LED lamps differed by application and was based on individual case studies. Aside from the case of residential retrofits [30], no studies were found which investigated the weight of several different lamps of the same application type. The average size of the LED die area per LED lamp determines the total amounts of critical raw materials included in one lamp. This size varies depending on different applications, and very few studies previously dealt with the determination of this size [15,23]. Deubzer et al. [23] used differentiated die areas for each application type with a lower and upper limit. For this study, the weighted average of these die areas was used. Table 2 shows the average lifespan, weight, and die area for each application type used in this study.

Collection rates of LED lamps per EU28 member state were approximated using Eurostat data on the collection rate of lighting equipment because there are no data available on the collection amounts of LED lamps. These data are shown in Table A2. The latest available data for the EU28 were used to extrapolate the development of the rate until 2030. Scenario 1 assumed that the 2017 collection rate across all EU28 member states would remain the same until 2030 at 14%. Scenario 3 assumed that the EU target of a collection rate of 85% would be achieved [7], while scenario 2 represented achieving half of the 85% target by 2030. The EU collection rate target of 85% relates to the amount collected compared to the total waste generated in a member state in a given year [7]. No recycling rate for LED lamps is currently captured because there does not exist an LED recycling technology. Therefore, the recycling rate baseline scenario (scenario 1) was linked to the application of current recycling technologies to the recycling of LED lamps.

Table 2. The average lifespan, average weight, and average die area for different LED lamp types.

Application Type	Average Lifespan (Years)	Average Weight (kg)	Average Die Area (mm^2)
Residential lamps	18.8 [9]	0.520 [31]	9 ± 2 [23]
Residential retrofits	12.5 [9]	0.2452 [30]	9 ± 2 [2]
Commercial lamps	5.9 [9]	1.75 [14]	11 ± 3 [23]
Commercial retrofits	2.9 [9]	0.2452 [1]	11 ± 3 [2]
Outdoor	10.0 [9]	15.0 [32]	17 ± 4 [23]
Industrial	8.3 [9]	3.5 [23]	40 ± 9 [23]
Architectural	10.0 [9]	4.5 [23]	79 ± 42 [23]

[1] Assumption that commercial retrofits weigh the same as residential retrofits. [2] Assumption that residential and commercial retrofits have the same average die area as lamps.

Reuter and van Schaik [33] simulated how much of the different materials included in LED lamps could be recycled in pre-processing steps depending on the LED lamp design to achieve a metal-rich fraction. While they did not report an overall recycling rate for the entire lamp, they did publish material-specific rates. For aluminum, for example, they predicted a recycling rate of around 75%, while the rate for copper was between 40% and 45%. These values were derived from an optimized recycling process aimed at increasing the recovery of the metal fraction contained in LED lamps [33]. Thus, the assumed baseline scenario with an overall recycling rate of 50% was lower than the reported values by Reuter and van Schaik to account for a non-specific technology applied. This rate assumed that 50% of the weight of the LED lamp could be recycled using current technologies. Scenario 3 reflected the achievement of the 80% target for recycling of WEEE [7]. In scenario 2, the recycling rate reached 65%, assuming that the newly developed technology could only reach the target halfway. The yearly recycling and collection rates between 2017 and 2030 for scenarios 2 and 3 resulted from interpolating the values in scenario 1 as starting values and the scenario values of 2 and 3 as end values. This approach represents continuous progress in expanding collection systems and optimizing recycling technologies. The different scenarios are shown in Table 3. The interpolated collection and recycling rates for each year are shown in Table A3.

Table 3. Collection and recycling scenarios of LED lamp waste for all EU28 member states in 2030.

	Scenario 1	Scenario 2	Scenario 3
Collection rate (%)	14	50	85
Recycling rate (%)	50	65	80

After the collection and recycling steps, the retained materials need to be recovered. How much of each material is recovered depends on whether the metallurgical processes applied are compatible with each other, as well as the chemistry and the concentration of the materials [33]. As shown in [33], the Metal Wheel illustrates the different recovery paths for materials included in EoL products. The recycling steps used to separate elements and components of products into different fractions determine which metallurgical processes would be applied and which materials would be lost or recovered. The different metal routes shown in the Metal Wheel illustrate the incompatibility of the recovery of gallium and indium with rare earth elements, as well as the limited possibilities to recover precious metals in the same process as either rare earth elements or indium/gallium [33]. Information on specific material recovery rates for LED lamps is rare. However, the limited available research was used to derive different groups for the recovery of different critical raw materials from LED lamp waste. Each group focused on a different combination of materials that were compatible with each other and could be extracted using a particular process. The compatibility of the materials and the potential recovery rates were derived from studies investigating the recovery of these materials from LED lamps or compact fluorescent lamps (CFLs). Table 4 provides an overview of the recovery rate groups.

Table 4. Material recovery groups are derived for different critical raw materials based on values reported in cited studies.

	Group A—Indium and Gallium Recovery	Group B—Rare Earth Elements Recovery	Group C—Precious Metals Recovery
Gallium	90–99% [11–13]	–	–
Indium	95% [11,34]	–	–
Cerium	–	60–100% [35–37]	–
Europium	–	90–100% [35–37]	–
Gadolinium	–	50% [37]	–
Terbium	–	77% [37]	–
Yttrium	–	76–100% [35–37]	–
Gold	–	38–68% [36]	50–100% [33,38]
Silver	–	–	50–81% [33,38]
Palladium	–	–	13–60% [33,38]

2.2. Data Modeling

According to Oguchi et al., the amount of LED waste generated G_t at the end of year t could be calculated by the product sum of the amounts of products put on the market (POM) between years i and $t - i$, multiplied by the average weight w of an LED lamp and the percentage of LED lamps discarded in year t, expressed by $f_t(i)$ [39]. It was assumed that the average weight of an LED lamp is constant over time. The corresponding equation proposed by Oguchi et al. [39] is

$$G_t = \sum_i POM_{t-1} \cdot w \cdot f_t(i), \qquad (1)$$

where $f_t(i) = W_t(i + 0.5) - W_t(i - 0.5)$, \qquad (2)

and $W_t(y) = 1 - \exp\left[-\left\{\frac{y}{y_{av}}\right\}^k \cdot \left\{\Gamma\left(1 + \frac{1}{k}\right)\right\}^k\right].$ \qquad (3)

To calculate the share of lamps leaving the use phase, the lifetime distribution of LED lamps had to be modeled, which was carried out with the help of the cumulative Weibull distribution function W_t [40]. This function, given by Equation (3), is defined by the average lifespan of LED lamps y_{av}, the shape factor k of the distribution, the lifespan y, and the gamma function Γ [39]. The Weibull function is a common distribution function to model data on product survival [40]. Factor k determines when the majority of the LED lamps are discarded: a small value indicates early disposal, while a large value signifies that products remain longer in use [41]. No previous studies were found which reported k values for LED lamps. Therefore, studies investigating other EEE were reviewed to choose an approximate value for k. Kalmykova et al. [42] collected lifespan data from discarded LED TVs and monitors. The reported k for LED TVs of 3.75014 was used in this study, implying that LED lamps are more likely to be used until the end of their technical lifespan compared to being disposed of early due to consumer preferences. The materials included in the calculated LED waste G_t were determined based on the units of LED lamps discarded and the materials contained in one unit of LED lamp.

2.3. Data Evaluation

The economic feasibility criteria introduced in this study are comprised of the upper crust material concentrations and the raw material prices. The natural accumulation of usable minerals and rocks is called a deposit if the exploitation of this accumulation can be realized economically depending on its size and contents [43]. There is no comparable definition for the materials contained in the anthropogenic stock. Therefore, all amounts of all materials in this stock are usually considered to be anthropogenic deposits (e.g., [44]). Yet, similar to natural deposits, anthropogenic deposits need to be judged based on their size and contents to determine whether the contained resources are mineable and exploitation is economically feasible. Moreover, this assessment of anthropogenic deposits needs to be compared to natural deposits to decide if secondary raw materials are ecologically and economically favorable over primary resources. In this context, it can be worthwhile to draw on information introduced in geosciences. Various researchers investigated and determined the concentrations of elements in the upper continental crust. The best-known results came from Clarke and Washington [45] as well as Goldschmidt [46]. However, these publications do not cover all elements, and specifically, they lack values for the trace elements important to our study. Further investigations and calculations followed, and data on trace elements were gathered. In this study, reference is therefore made to the relatively recent work of Rudnick and Gao [21].

In order to form an orebody, the element under consideration must be enriched many times above the normal abundance in the earth's crust. The minimum content of a mineable deposit and the degree of enrichment—called enrichment factor—differ between various elements [47]. The limit of the economic feasibility for mining would only be below the average content of the earth's crust under exceptional circumstances, e.g., when particularly simple and cost-effective processing is possible. This applies, for example, to the extraction of titanium raw materials from marine soaps [43], which is a rare exception. Thus, the concentration of elements in the upper crust of the earth can usually be considered as the lowest limit for the extraction of raw materials. Furthermore, the various elements in the anthroposphere are usually not easily accessible and separable, but rather, they are often present in complex products and material compounds. Therefore, it can be assumed that the material recovery from the anthropogenic stockpile is significantly less ecologically and economically advantageous compared to the extraction from enriched natural deposits with higher concentrations.

In addition to the upper crust concentrations, raw material prices were used for the evaluation to determine the overall revenue potential. The data availability for raw material prices is scattered. Two different sources relating to slightly different time horizons needed to be considered to calculate revenue estimations for all critical raw materials included in this study. Data from the United States Geological Survey (USGS) for the year 2019 was used for gold, silver, palladium, indium, gallium, and yttrium [48]. For cerium, europium, gadolinium, and terbium, data from the Institute for Rare Earths and Strategic Metals (ISE) were collected to calculate yearly averages between October 2019 and September 2020 [49–60]. An overview of the considered prices is given in Table A4.

3. Results

3.1. Total Mass of Critical Raw Materials Included in LED Lamp Waste between 2017 and 2030

According to our calculations, more than 2.6 million tons of LED lamp waste would be generated between 2017 and 2030 in the EU28 member states. That corresponds to 2.4 million LED lamps that would be discarded. Considering the average weight and material composition per lamp, Table 5 shows the total mass of critical raw materials included in the generated waste. Silver was the material with the highest mass contained in this waste with more than 60 t. At the same time, cerium and europium had the lowest shares with only 91.02 kg. Precious metals accounted for the largest share of critical raw material mass. Indium and gallium, on the other hand, were contained in lower masses than rare earth elements. These amounts represented the theoretical potential for the recycling and recovery of critical raw materials included in the LED lamp waste. However, collection system limitations, recycling inefficiencies, and recovery process constraints reduce the actual amount that can be extracted from the waste. Therefore, how much of the total mass of materials can be recovered and used as secondary raw materials depends on the change of the collection rate, recycling rate, and the assumed material recovery rate.

Table 5. The theoretical potential of materials available for recovery in LED lamp waste generated between 2017 and 2030 in EU28 member states.

Element	Total Mass in LED Lamp Waste (kg)
Gallium	212.38
Indium	273.05
Cerium	91.02
Europium	91.02
Gadolinium	455.09
Terbium	5006.00
Yttrium	2700.21
Gold	5274.15
Silver	60,048.40
Palladium	2818.62

3.2. Material Recovery Potential after Collection and Recycling Steps between 2017 and 2030

Table 6 gives an overview of the potential for the material recovery from the waste amounts depending on the different collection and recycling rate scenarios. For example, the total amount of cerium included in the LED lamp waste between 2017 and 2030 was 91.02 kg. The maximum amount that could be extracted from the lamp waste only ranged between 6.42 kg and 43.41 kg, depending on the collection and recycling rate scenario. Recovering cerium would also mean that materials such as indium, gallium, palladium, and silver would be lost during the recovery process. The amount of cerium was small compared to how much silver could be extracted from the lamp waste. Of the more than 60 t of silver stored in this waste, only a maximum between 3.4 t and 23 t could be recovered. To determine which critical raw materials should be treated as a priority in the material recovery process and whether introducing a new recycling technology would be economically feasible, the upper continental crust concentrations and raw material prices were introduced as scenario evaluation metrics.

Table 6. Weights of materials in LED lamp waste that could be recovered between 2017 and 2030 in EU28 member states [1].

		Scenario 1 (kg)	Scenario 2 (kg)	Scenario 3 (kg)
Group A	Gallium	13.49–14.84	46.68–51.34	90.59–99.65
	Indium	18.31	63.35	122.95
Group B	Cerium	3.85–6.42	13.34–22.23	25.88–43.14
	Europium	5.78–6.42	20.00–22.23	38.83–43.14
	Gadolinium	16.06	55.57	107.85
	Terbium	272.04	941.32	1826.95
	Yttrium	144.83–190.57	501.15–659.40	972.65–1279.80
	Gold	141.45–253.11	489.43–875.82	949.91–1699.83
Group C	Gold	186.11–372.22	643.99–1287.97	1249.88–2499.76
	Silver	2118.97–3432.73	7332.06–11,877.93	14,230.38–23,053.21
	Palladium	25.86–119.36	89.48–412.99	173.67–801.55

[1] Scenario 1: a 14 % collection rate and a 50% recycling rate for all years. Scenario 2: increasing collection rate from 14% to 50% and increasing recycling rate from 50% to 65% between 2017 and 2030. Scenario 3: increasing collection rate from 14% to 85% and increasing recycling rate from 50% to 80% between 2017 and 2030. Exact values for each year can be found in Table A3 in Appendix A.

3.3. Evaluation of Economic Feasibility

The previously estimated amounts of critical raw materials contained in the total LED lamp waste were compared to the material concentrations in the upper continental crust. Table 7 displays the ratio of these concentrations. They indicated which raw materials occur in higher concentrations in the waste than in the upper continental crust. Indium and terbium appeared only slightly more frequently in the LED lamp waste. Silver, gold, and palladium, on the other hand, were significantly more highly concentrated in the waste. Cerium showed the lowest concentration in the lamp waste. In addition, gallium, europium, gadolinium, and yttrium all appeared less frequently compared to the upper continental crust. This implies that LED lamp waste is not an adequate urban mine for these types of materials. It would likely be more cumbersome and expensive to extract these low concentrations from the waste than from natural deposits or other waste products containing higher critical raw material volumes. For the materials with only slightly higher concentrations—indium and terbium have a ratio between one and three—it is questionable whether it would be lucrative to extract them. As previously mentioned, not all of these materials could be recovered in the same metallurgical processes. The ratios of concentrations can put into perspective how difficult it will be to extract certain materials from the waste, given that lower concentrations can increase the likelihood that materials will be lost during recycling processes [33]. However, to determine the economic feasibility of the recovery of certain elements, it is necessary to consider the total mass contained in the waste in combination with raw material prices.

Table 7. Upper crust concentrations, as reported by Rudnick and Gao [21], compared to critical raw material concentrations in the LED lamp waste in EU28 member states.

Raw Materials	Upper Crust Concentration (ppm)	Concentration in LED Lamp Waste (ppm)	Ratio of Waste to Upper Crust Concentration
Gallium	17.5	0.081	0.005
Indium	0.056	0.104	1.856
Cerium	63.0	0.035	0.001
Europium	1.0	0.035	0.035
Gadolinium	4.0	0.173	0.043
Terbium	0.7	1.906	2.723
Yttrium	21.0	1.028	0.049
Gold	0.0015	2.008	1338.701
Silver	0.053	22.863	431.368
Palladium	0.00052	1.073	2063.743

The calculated amounts of critical raw materials were used to determine the potential revenues that could be generated from the recycling of LED lamp waste. Scenario 2 was chosen as the most likely future development, considering that thirteen years to achieve the EU targets in 2030—as assumed in scenario 3—is little time. Table 8 shows the estimated revenue that could be generated if the LED lamp waste between 2017 and 2030 were collected and recycled according to scenario 2. Recovery group A—consisting of indium and gallium—yielded the lowest revenue with a maximum of USD 53,972. Group B generated between USD 22.8 MM and USD 40.2 MM. The largest proportion of this revenue came from gold (USD 22 MM–USD 39.4 MM). As previously mentioned, the masses of the different rare earth elements would not occur at the same time in the total LED lamp waste because they depend on the specific phosphorus used in the white LEDs. The recovery of terbium yielded USD 699,361. The remaining rare earth elements—cerium, europium, gadolinium, and yttrium—only amounted to between USD 24,380 and USD 30,438 in total. This was due to a combination of low masses and material prices. Cerium, gadolinium, and yttrium had the lowest prices per kg—less than USD 35. Cerium, europium, and gadolinium also had very low masses—all below 56 kg. Therefore, a lower share of TAG:Ce phosphorus would lower the revenue more significantly compared to a lower share of YAG:Ce. Irrespective of the different phosphorus applied, the rare earth elements contributed a maximum of 1.8% to the overall potential revenue of group B. The precious metals in group C provided the highest revenue with a range between USD 37.1 MM and USD 84.1 MM. Due to its lower price per kg compared to gold and palladium, the recovery of silver only accounted for 7% of the total group C revenue, even though silver had the highest share of mass.

Table 8. Estimated revenue that could be generated from LED lamp waste recycling in the EU28 between 2017 and 2030.

Recovery Group	Raw Materials	Scenario 2 (kg)	Raw Material Price [1] (USD/kg)	Estimated Revenue (MM USD)
Group A	Gallium	46.68–51.34	570.00	0.051–0.054
	Indium	63.35	390.00	
Group B	Cerium	13.34–22.23	4.58	22.8–40.2
	Europium	20.00–22.23	286.33	
	Gadolinium	55.57	27.94	
	Terbium	941.32	742.96	
	Yttrium	501.15–659.40	34.00	
	Gold	489.43–875.82	45,010.98	
Group C	Gold	643.99–1287.97	45,010.98	37.1–84.1
	Silver	7332.06–11,877.93	520.84	
	Palladium	89.48–412.99	48,226.05	

[1] Raw material prices for gallium, indium, yttrium, gold, silver, and palladium refer to the average annual prices in 2019. Prices for cerium, europium, gadolinium, and terbium are averages of the monthly prices between October 2019 and September 2020. Values are also shown in Table A4 in the Appendix A.

Considering the potential revenue and ratios of concentration and comparing these among the different materials, all rare earth elements aside from terbium generated negligible amounts of revenue while appearing in lower concentrations in the waste than in the upper continental crust. Although indium appeared in the LED lamp waste almost twice as frequently compared to the concentration in the upper crust, its potential revenue contribution of USD 24.705 was very small. Terbium had the second-highest ratio of concentrations. However, with a maximum revenue of less than USD 700,000, terbium appeared to be less significant than any of the precious metals. Even silver, which had a lower price per kg than terbium, could earn between USD 3.8 MM and USD 6.2 MM because its mass was between 7 to 12 times higher than that of terbium. Precious metals generated the highest revenue, appeared in significantly higher concentrations in the waste compared to the upper continental crust, and had some of the highest total masses in the LED lamp waste of all considered critical raw materials. Therefore, the focus of recycling and recovery should lie on precious metals if a new recycling technology is developed.

4. Discussion

The differences in recovered material mass across the scenarios were significant. They showed the effect of losses that could occur due to inefficiencies in collection systems and recycling processes. The results also illustrated that the recovery of rare earth elements yielded negligibly small masses and revenue potentials. Several other studies reached similar conclusions on the economic viability of rare earth elements, indium, and gallium, and the contribution of precious metals to the overall revenue potential. Cenci et al. [16] found that gold was the most important material to recover in terms of economic value. Cucchilla et al. [17] investigated WEEE other than LED lamps—including LCD and LED monitors, smartphones, and notebooks—and discovered that gold contributed to more than half of the potential revenue that could be generated from all of these products. Reuter and van Schaik's [33] recycling simulation of LED lamps disregarded indium, gallium, and rare earth elements entirely, focusing instead on metal-rich fractions. In general, gallium is difficult to recycle and recover because it appears in material compounds that are challenging to untangle, and the amounts it appears in are very small [61]. Ylä-Mella and Pongrácz [62] mentioned in connection with indium that low material concentrations in products and the loss of quality during the recycling process pose economic barriers to recycling. Similar issues surrounding the recyclability of rare earth elements were discussed by Balaram [63], who highlighted their occurrence in low amounts and the difficulty of separating the rare earth elements individually. Moreover, the cost of recycling these elements from any EoL products exceeds the potential revenue that could be generated from them and is therefore not economically feasible [61–63].

As previously mentioned, one of the EU's objectives is to reduce the supply risks of critical raw materials contained in WEEE. Even if the potential revenue from rare earth elements, indium, and gallium is small, their mass could still be relevant to decrease import dependencies from other countries. However, comparing the amounts of some of these critical raw materials contained in the total mass with the yearly consumption of these elements in Europe shows that they contribute insignificantly to reducing supply risk. For example, Germany's total annual gallium demand was estimated at 30–40 t in 2015 [61]. The gallium in the cumulated LED lamp waste between 2017 and 2030 would only account for around 0.1% of Germany's yearly consumption. Little information is available on the demand for indium and rare earth elements. Global indium production was estimated to be 790 t in 2013 [64], demonstrating the small impact the recovered indium from LED lamps would have. According to a communication from the European Commission [6], the EU has a 0% import reliance on indium. However, all of the rare earth elements considered in this study pose a 100% import reliance for the EU [6]. No final determination can be made on the relation between the availability of recovered rare earth elements to their annual demand in the EU. Considering the absolute mass of these elements in the LED lamp waste, only yttrium and terbium seem to be of a relevant size to affect the supply risk.

The environmental perspective is another reason why LED lamps should be recycled. According to the review of LCA studies on LED lamps conducted by Franz and Wenzel [65], the disposal phase accounts for up to 27% of the total environmental impact of an LED lamp. However, recycling or energy recovery of the lamp can also create an environmental benefit. Most LCA studies show the highest environmental burden during the use phase [65]. In some cases, recycling can be very energy-intensive and more environmentally harmful than natural resource mining because of complex recycling processes required to untangle material compounds in complex products, as suggested in the case of indium recycling from LCD screens [66,67].

Whether the main incentive for recycling is economic or environmental, during the product design process, producers should already consider the EoL phase to make disassembly of LED lamps as easy as possible and to enable the recovery of as much material mass as possible [30,33]. Such eco-design strategies can increase the efficiency of the recycling process, as well as the environmental benefit [16]. While Dzombak et al. [30] saw some improvements in LED lamp design over a period of seven years, for example, a lower overall material mass, the majority of the examined lamps were still not easy to disassemble and contained elements and materials hindering high levels of material recovery. Another option to reduce the environmental impact of products is to focus on material efficiency and use less new material through light-weight design [68].

The results of this study have several limitations. First, the data used for the calculations were based on scarce information available in the literature. Various assumptions had to be made about the average weight and lifespans of different LED lamp applications and the development of LED lamp sales numbers in the EU. The effect of different recycling and collection rates on the results was considered by applying different scenarios. Second, the limited availability of studies investigating the material recovery of LED lamps led to the derivation of exemplary recovery groups that cannot fully represent all possible material combinations. For example, it is likely that gold could be recovered in the same process with indium and gallium, as suggested by the Metal Wheel [33]. Third, this study only considers critical raw materials and disregards the revenue potential of non-critical materials such as aluminum, copper, tin, and plastics. Integrating these into the analysis might change what courses of action are derived and which material groups are most profitable. Fourth, the raw material prices used are the average prices for primary resources. It is debatable whether the same prices can be achieved for secondary raw materials. However, because it is more difficult to gain information on secondary raw material prices, the primary resource prices were used as an approximation. Finally, this study only considered the LED package in the lamp, but not the lamp housing, PCB, or other electronic parts, disregarding further non-critical materials that could be recovered to be used as secondary materials and generate profits.

This study highlights several areas for future research. First, the proposed two-part evaluation scheme for economic feasibility should be applied to other WEEE streams and compared with results of earlier studies, which determined whether and in what way the recycling of certain WEEE streams is economically feasible. This could validate the usefulness of this evaluation scheme. Furthermore, the suggested evaluation dimensions—the upper crust concentrations and raw material prices—should be supplemented with additional dimensions and data to enhance the validity and expand the applicability of the scheme. Only data that are readily available and easy to collect should be integrated to preserve the main objective of the evaluation system: providing a fast and easy way to determine whether further investigations into the recyclability and feasibility are warranted and on which materials the focus should lie. Finally, this scheme could be applied to determine the economic feasibility of the recovery of materials other than elements. In this case, upper crust concentrations—which only relate to elements—could not be used. An alternative metric in addition to raw material prices would need to be applied.

5. Conclusions

The purpose of this study was to introduce a simple evaluation scheme that could be used to determine the potential and limitations of critical raw material recycling. The two-part evaluation system consisting of upper continental crust concentrations and raw material prices does not require much data collection effort. It represents a simple tool that can be applied to various WEEE streams and expanded to materials other than elements. The usefulness of the evaluation scheme was demonstrated in the case of LED lamps. In this context, this study also contributed to the LED literature. It addressed the research gap concerning the economic feasibility of LED lamp recycling, as mentioned by Cenci et al. [16] and Rahman et al. [10], with a focus on critical raw materials. Previous studies focused mainly on non-critical materials, disregarded collection and recycling rates when calculating the material mass available for recovery, and considered only raw material prices without comparing natural occurrence with material concentrations in LED lamp waste. Moreover, these investigations required high effort, a lot of time, and a lot of data input. These shortcomings were addressed in this study by examining the economic feasibility of recycling critical raw materials—specifically addressing the potential of indium, gallium, and rare earth elements—as well as accounting for losses during the collection and recycling steps.

The results of this study show that precious metals—particularly, gold—are the most economically viable materials contained in the LED part of an LED lamp. These materials are contained in higher concentrations in the lamp waste than the upper continental crust. They comprise high total masses, and they generate the most revenue out of the three different material groups investigated. Indium, gallium, and rare earth elements have low concentrations, low total masses, and generate low potential revenue. Therefore, new recycling technologies for LED lamps should focus on precious metals and be optimized to lose as little as possible of those elements in the process. Whether this amount of revenue would suffice to develop and implement an appropriate LED recycling technology needs to be investigated by a cost–benefit analysis considering the costs of the specific technology. The specific economic potential of the recycling of LED lamps depends on the recycling technology applied. Pre-treatment and pre-concentration steps that require manual labor will increase recycling costs. At the same time, not only recycling but also collection steps need to be considered. The currently low collection rate of 14% for waste lighting equipment in the EU shows that significant improvements are required to reach the EU collection target and to increase revenue through larger material quantities available for recycling.

Future research endeavors should include further studies on WEEE recycling, which leverage the herein proposed two-part evaluation scheme to validate its usefulness. Moreover, additional easily accessible metrics and data to estimate the economic feasibility of material recycling of other WEEE streams should be suggested.

Author Contributions: J.S.N. contributed to the conceptualization, methodology, validation, formal analysis, data curation, investigation, visualization, writing, review, and editing. M.R. added to the conceptualization, writing, review and editing, supervision, and project administration. N.v.G. made contributions to the conceptualization, methodology, investigation, review, and editing. All authors have read and agreed to the published version of the manuscript.

Funding: This research was funded by EIT RawMaterials as part of the project "REDLED: Recycling EnD-of-life LED" with the project number 18039. Financial support was provided by Wuppertal Institut für Klima, Umwelt, Energie gGmbH within the funding program Open Access Publishing.

Institutional Review Board Statement: Not applicable.

Informed Consent Statement: Not applicable.

Data Availability Statement: The data used to derive the results are freely available and contained in the article and Appendix A.

Conflicts of Interest: The authors declare no conflict of interest. The funders had no role in the design of the study; in the collection, analyses, or interpretation of data; in the writing of the manuscript, or in the decision to publish the results.

Appendix A

The values in Table A1 between 2008 and 2020 relating to the total POM amounts as well as the different LED lamp applications were taken from Marwede et al. [9]. The values between 2021 and 2025 were extrapolated based on the data until 2020. Given the assumption by Buchert et al. [26] that the LED demand will reach 95% of the market share, the value in 2030 was calculated to represent 100%. The data for the years between 2025 and 2030 were interpolated.

Table A1. Yearly put-on-the-market (POM) amounts in units of lamps for different LED lamp applications in EU28 member states.

Year	Residential (pcs)	Commercial (pcs)	Industrial (pcs)	Outdoor (pcs)	Architectural (pcs)	Residential Retrofits (pcs)	Commercial Retrofits (pcs)	Total POM (pcs)
2030	576,816,079	73,412,955	10,487,565	10,487,565	20,975,130	314,626,952	41,950,260	1,048,756,507
2029	571,047,918	72,678,826	10,382,689	10,382,689	20,765,379	311,480,682	41,530,758	1,038,268,942
2028	565,279,757	71,944,696	10,277,814	10,277,814	20,555,628	308,334,413	41,111,255	1,027,781,377
2027	559,511,596	71,210,567	10,172,938	10,172,938	20,345,876	305,188,143	40,691,752	1,017,293,811
2026	553,743,436	70,476,437	10,068,062	10,068,062	20,136,125	302,041,874	40,272,250	1,006,806,246
2025	547,975,275	69,742,308	9,963,187	9,963,187	19,926,374	298,895,604	39,852,747	996,318,681
2024	513,482,418	65,352,308	9,336,044	9,336,044	18,672,088	280,081,319	37,344,176	933,604,396
2023	478,989,560	60,962,308	8,708,901	8,708,901	17,417,802	261,267,033	34,835,604	870,890,110
2022	444,496,703	56,572,308	8,081,758	8,081,758	16,163,516	242,452,747	32,327,033	808,175,824
2021	410,003,846	52,182,308	7,454,615	7,454,615	14,909,231	223,638,462	29,818,462	745,461,538
2020	379,500,000	48,300,000	6,900,000	6,900,000	13,800,000	207,000,000	27,600,000	690,000,000
2019	341,000,000	43,400,000	6,200,000	6,200,000	12,400,000	186,000,000	24,800,000	620,000,000
2018	319,000,000	40,600,000	5,800,000	5,800,000	11,600,000	174,000,000	23,200,000	580,000,000
2017	280,500,000	35,700,000	5,100,000	5,100,000	10,200,000	153,000,000	20,400,000	510,000,000
2016	247,500,000	31,500,000	4,500,000	4,500,000	9,000,000	135,000,000	18,000,000	450,000,000
2015	214,500,000	27,300,000	3,900,000	3,900,000	7,800,000	117,000,000	15,600,000	390,000,000
2014	159,500,000	20,300,000	2,900,000	2,900,000	5,800,000	87,000,000	11,600,000	290,000,000
2013	110,000,000	14,000,000	2,000,000	2,000,000	4,000,000	60,000,000	8,000,000	200,000,000
2012	60,500,000	7,700,000	1,100,000	1,100,000	2,200,000	33,000,000	4,400,000	110,000,000
2011	38,500,000	4,900,000	700,000	700,000	1,400,000	21,000,000	2,800,000	70,000,000
2010	24,200,000	3,080,000	440,000	440,000	880,000	13,200,000	1,760,000	44,000,000
2009	11,000,000	1,400,000	200,000	200,000	400,000	6,000,000	800,000	20,000,000
2008	5,500,000	700,000	100,000	100,000	200,000	3,000,000	400,000	10,000,000

The data for the development of the collection scenarios were taken from Eurostat and are displayed in Table A2. The collection rate is calculated according to the EU Directive

2012/19/EU [8] by dividing the waste collected in year t by the average POM amount of the three previous years. The data were accessed under https://ec.europa.eu/eurostat/databrowser/view/env_waselee/default/table?lang=en (accessed on 28 August 2019).

Table A2. Data used to calculate collection rates for lighting equipment in EU28 member states.

Year	POM (t)	Waste Collected (t)	Collection Rate (%)
2017	518,852	68,940	14
2016	485,245	54,914	13
2015	560,470	36,713	10
2014	390,760	27,774	7
2013	350,599	24,955	–
2012	389,443	20,461	–
2011	379,300	18,185	–

Table A3. Yearly estimated collection and recycling rates for EU28 member states between 2017 and 2030.

Year	Scenario 1 Rates (%)		Scenario 2 Rates (%)		Scenario 3 Rates (%)	
	Collection	Recycling	Collection	Recycling	Collection	Recycling
2030	14	50	50	65	85	80
2029	14	50	47	64	80	78
2028	14	50	45	63	74	75
2027	14	50	42	62	69	73
2026	14	50	39	60	63	71
2025	14	50	36	59	58	68
2024	14	50	34	58	52	66
2023	14	50	31	57	47	64
2022	14	50	28	56	42	62
2021	14	50	25	55	36	59
2020	14	50	23	53	31	57
2019	14	50	20	52	25	55
2018	14	50	17	51	20	52
2017	14	50	14	50	14	50

Table A4. Prices for critical raw materials from the Institute for Rare Earths and Strategic Metals for cerium, europium, gadolinium, and terbium [49–60], and from United States Geological Survey for gallium, gold, indium, palladium, silver, and yttrium [48].

Month and Year	Cerium	Europium	Gadolinium	Gallium	Gold	Indium	Palladium	Silver	Terbium	Yttrium
				Prices in USD/kg						
October 2019	4.88	N.A.[1]	26.98	–[2]	–	–	–	–	716.82	–
November 2019	4.97	N.A.	26.10	–	–	–	–	–	660.70	–
December 2019	4.61	N.A.	26.98	–	–	–	–	–	628.28	–
January 2020	4.66	N.A.	28.70	–	–	–	–	–	645.00	–
February 2020	4.71	N.A.	28.58	–	–	–	–	–	645.81	–
March 2020	4.76	N.A.	30.27	–	–	–	–	–	763.91	–
April 2020	4.50	288.00	N.A.	–	–	–	–	–	715.00	–
May 2020	4.50	288.00	N.A.	–	–	–	–	–	712.00	–
June 2020	4.40	285.00	N.A.	–	–	–	–	–	820.00	–
July 2020	4.35	285.00	N.A.	–	–	–	–	–	835.00	–

Table A4. *Cont.*

Month and Year	Cerium	Europium	Gadolinium	Gallium	Gold	Indium	Palladium	Silver	Terbium	Yttrium
				Prices in USD/kg						
August 2020	4.35	286.00	N.A.	–	–	–	–	–	853.00	–
September 2020	4.30	286.00	N.A.	–	–	–	–	–	920.00	–
Yearly	4.58	286.33	27.94	570.00	45,010.98	390.00	48,226.05	520.84	742.96	34.00

[1] N.A. signifies that the values for these dates were not available through the Institute for Rare Earths and Strategic Metals. [2] The dash (–) signifies values for these dates are not applicable to these elements because only yearly data were used from U.S.G.S.

References

1. Oberle, B.; Bringezu, S.; Hatfield-Dodds, S.; Hellweg, S.; Schandl, H.; Clement, J.; Cabernard, L.; Che, N.; Chen, D.; Droz-Georget, H.; et al. *Global Resources Outlook 2019: Natural Resources for the Future We Want. IRP (2019)*; United Nations Environment Programme: Nairobi, Kenya, 2019. Available online: https://www.resourcepanel.org/reports/global-resources-outlook (accessed on 31 August 2020).
2. Wiedenhofer, D.; Fishman, T.; Lauk, C.; Haas, W.; Krausmann, F. Integrating material stock dynamics into economy-wide material flow accounting: Concepts, modelling, and global application for 1900–2050. *Ecol. Econ.* **2019**, *156*. [CrossRef]
3. Martins, F.F.; Castro, H. Raw material depletion and scenario assessment in European Union—A circular economy approach. *Energy Rep.* **2020**, *6*. [CrossRef]
4. European Commission. Report on Critical Raw Materials for the EU—Report of the Ad Hoc Working Group on Defining Critical Raw Materials. 2014. Available online: https://www.google.de/url?sa=t&rct=j&q=&esrc=s&source=web&cd=&ved=2ahUKEwje5d_GqvvvAhU3hf0HHVe2DPgQFjAAegQIDBAD&url=https%3A%2F%2Fec.europa.eu%2Fdocsroom%2Fdocuments%2F10010%2Fattachments%2F1%2Ftranslations%2Fen%2Frenditions%2Fpdf&usg=AOvVaw0wJWTi1phbJWSxhMT_1dnG (accessed on 31 August 2020).
5. European Commission. Communication from the Commission to the European Parliament and the Council, the Raw Materials Initiative—Meeting Our Critical Needs for Growth and Jobs in Europe. 2008. Available online: https://eur-lex.europa.eu/LexUriServ/LexUriServ.do?uri=COM:2008:0699:FIN:EN:PDF (accessed on 31 August 2020).
6. European Commission. Communication from the Commission to the European Parliament, the Council, the European Economic and Social Committee and the Committee of the Regions—Critical Raw Materials Resilience: Charting a Path towards Greater Security and Sustainability. 2020. Available online: https://eur-lex.europa.eu/legal-content/EN/TXT/?uri=CELEX%3A52020DC0474 (accessed on 10 February 2021).
7. Forti, V.; Balde, C.P.; Kuehr, R.; Bel, G. *The Global E-Waste Monitor 2020: Quantities, Flows and the Circular Economy Potential*; United Nations University/United Nations Institute for Training and Research, International Telecommunication Union, and International Solid Waste Association: Bonn, Germany; Geneva, Switzerland; Rotterdam, The Netherlands, 2020.
8. Directive 2012/19/EU of the European Parliament and of the Council of 4 July 2012 on Waste Electrical and Electronic Equipment (WEEE) Text with EEA Relevance. 2012. Available online: https://eur-lex.europa.eu/legal-content/EN/TXT/?uri=celex%3A32012L0019 (accessed on 31 August 2020).
9. Marwede, M.; Chancerel, P.; Deubzer, O.; Jordan, R.; Nissen, N.F.; Lang, K.-D. Mass Flows of Selected Target Materials in LED Products. In Proceedings of the 2012 Electronics Goes Green 2012+, Berlin, Germany, 9–12 September 2012. Available online: https://www.researchgate.net/publication/261243071_Mass_flows_of_selected_target_materials_in_LED_products (accessed on 31 August 2020).
10. Rahman, S.M.; Kim, J.; Lerondel, G.; Bouzidi, Y.; Nomenyo, K.; Clerget, L. Missing research focus in end-of-life management of light-emitting diode (led) lamps. *Resour. Conserv. Recycl.* **2017**, *127*, 256–258. [CrossRef]
11. Zhan, L.; Xia, F.; Ye, Q.; Xiang, X.; Xie, B. Novel recycle technology for recovering rare metals (Ga, In) from waste light-emitting diodes. *J. Hazard. Mater.* **2015**, *299*, 388–394. [CrossRef]
12. Nagy, S.; Bokányi, L.; Gombkötő, I.; Magyar, T. Recycling of gallium from end-of-life light emitting diodes. *Arch. Met. Mater.* **2017**, *62*, 1161–1166. [CrossRef]
13. Zhou, J.; Zhu, N.; Liu, H.; Wu, P.; Zhang, X.; Zhong, Z. Recovery of gallium from waste light emitting diodes by oxalic acidic leaching. *Resour. Conserv. Recycl.* **2019**, *146*, 366–372. [CrossRef]
14. Principi, P.; Fioretti, R. A comparative life cycle assessment of luminaires for general lighting for the office—compact fluorescent (cfl) vs light emitting diode (led)—A case study. *J. Clean. Prod.* **2014**, *83*, 96–107. [CrossRef]
15. Buchert, M.; Manhart, A.; Bleher, D.; Pingel, D. *Recycling Critical Raw Materials from Waste Electronic Equipment*; Öko-Institut eV: Freiburg, Germany, 2012; pp. 30–40.
16. Cenci, M.P.; Dal Berto, F.C.; Schneider, E.L.; Veit, H.M. Assessment of LED Lamps Components and Materials for a Recycling Perspective. *Waste Manag.* **2020**, *107*, 285–293. [CrossRef]
17. Cucchiella, F.; D'Adamo, I.; Koh, S.L.; Rosa, P. Recycling of WEEEs: An economic assessment of present and future e-waste streams. *Renew. Sustain. Energy Rev.* **2015**, *51*, 263–272. [CrossRef]

18. Vanegas, P.; Peeters, J.R.; Cattrysse, D.; Dewulf, W.; Duflou, J.R. Improvement potential of today's WEEE Recycling Performance: The case of LCD TVs in Belgium. *Front. Environ. Sci. Eng.* **2017**, *11*, 13. [CrossRef]
19. Choi, J.-K.; Fthenakis, V. Crystalline silicon photovoltaic recycling planning: Macro and micro perspectives. *J. Clean. Prod.* **2014**, *66*, 443–449. [CrossRef]
20. D'Adamo, I.; Miliacca, M.; Rosa, P. Economic feasibility for recycling of waste crystalline silicon photovoltaic modules. *Int. J. Photoenergy* **2017**. [CrossRef]
21. Rudnick, R.L.; Gao, S. Composition of the continental crust. *Crust* **2003**, *3*, 1–64.
22. Wuppertal Institut for Climate, Environment and Energy, Material Intensity of Materials, Fuels, Transport Services, Food. 2014. Available online: https://wupperinst.org/uploads/tx_wupperinst/MIT_2014.pdf (accessed on 31 August 2020).
23. Deubzer, O.; Jordan, R.F.I.; Marwede, M.; Chancerel, P. *CycLED—Cycling Resources Embedded in Systems Containing Light Emitting Diodes. Deliverable 2.1: Categorization of LED Products*; Fraunhofer Verlag: Stuttgart, Germany, 2012.
24. Angerer, G.; Marscheider-Weidemann, F.; Lüllmann, A.; Erdmann, L.; Scharp, M.; Handke, V.; Marwede, M. *Rohstoffe für Zukunftstechnologien: Einfluss des Branchenspezifischen Rohstoffbedarfs in Rohstoffintensiven Zukunftstechnologien auf die Zukünftige Rohstoffnachfrage*; Fraunhofen Insititut System- und Innovationforschung: Karlsruhe, Germany, 2009. (In Germany)
25. Zimmermann, T.; Gößling-Reisemann, S. Critical Materials and Dissipative Losses: A Screening Study. *Sci. Total Environ.* **2013**, *461–462*, 774–780. [CrossRef]
26. Buchert, M.; Degreif, S.; Schüler, D.; Prakash, S.; Möller, M.; Koehler, A.; Behrendt, S.; Nolte, R.; Röben, A. *Substitution Als Strategie Zur Minderung Der Kritikalität von Rohstoffen Für Umwelttechnologien-Potentialermittlung Für Second-Best-Lösungen. Arbeitsbericht 2: Abschätzung Des Materialbedarfs Der 40 Prioritären Umwelttechnologien in Den Szenarien Business-As-Usual Und Green-Economy*; Umweltbundesamt: Dessau-Roßlau, Germany, 2019.
27. Huisman, J.; Magalini, F.; Kuehr, R.; Maurer, C.; Ogilvie, S.; Poll, J.; Delgado, C.; Artim, E.; Szlezak, J.; Stevels, A. *Review of Directive 2002/96 on Waste Electrical and Electronic Equipment (WEEE)*; United Nations University: Bonn, Germany, 2007.
28. Scholand, M.; Dillon, H.E. *Life-Cycle Assessment of Energy and Environmental Impacts of LED Lighting Products Part 2: LED Manufacturing and Performance*; Pacific Northwest National Lab (PNNL): Richland, WA, USA, 2012.
29. Murakami, S.; Oguchi, M.; Tasaki, T.; Daigo, I.; Hashimoto, S. Lifespan of commodities, part I: The creation of a database and its review. *J. Ind. Ecol.* **2010**, *14*, 598–612. [CrossRef]
30. Dzombak, R.; Padon, J.; Salsbury, J.; Dillon, H. Assessment of end-of-life design in solid-state lighting. *Env. Res. Lett.* **2017**, *12*, 084013. [CrossRef]
31. Navigant Consulting Europe Ltd. *Life Cycle Assessment of Ultra-Efficient Lamps*; Department of the Environment and Energy: London, UK, 2009.
32. Tähkämö, L.; Halonen, L. Life cycle assessment of road lighting luminaires—Comparison of Light-Emitting Diode and High-Pressure Sodium Technologies. *J. Clean. Prod.* **2015**, *93*. [CrossRef]
33. Reuter, M.; van Schaik, A. Product-centric simulation-based design for recycling: Case of LED lamp recycling. *J. Sustain. Metall.* **2015**, *1*, 4–28. [CrossRef]
34. Zimmermann, Y.-S.; Niewersch, C.; Lenz, M.; Kül, Z.Z.; Corvini, P.F.-X.; Schäffer, A.; Wintgens, T. Recycling of indium from CIGS photovoltaic cells: Potential of combining acid-resistant nanofiltration with liquid–liquid extraction. *Environ. Sci. Technol.* **2014**, *48*, 13412–13418. [CrossRef]
35. Pavón, S.; Fortuny, A.; Coll, M.T.; Sastre, A.M. Rare earths separation from fluorescent lamp wastes using ionic liquids as extractant agents. *Waste Manag.* **2018**, *82*, 241–248. [CrossRef]
36. Marra, A.; Cesaro, A.; Belgiorno, V. Recovery opportunities of valuable and critical elements from WEEE treatment residues by hydrometallurgical processes. *Environ. Sci. Pollut. Res.* **2019**, *26*, 19897–19905. [CrossRef]
37. Innocenzi, V.; Ippolito, N.M.; Pietrelli, L.; Centofanti, M.; Piga, L.; Vegliò, F. Application of solvent extraction operation to recover rare earths from fluorescent lamps. *J. Clean. Prod.* **2018**, *172*, 2840–2852. [CrossRef]
38. Batnasan, A.; Haga, K.; Shibayama, A. recovery of precious and base metals from waste printed circuit boards using a sequential leaching procedure. *JOM* **2018**, *70*. [CrossRef]
39. Oguchi, M.; Kameya, T.; Yagi, S.; Urano, K. Product flow analysis of various consumer durables in Japan. *Resour. Conserv. Recycl.* **2008**, *52*, 463–480. [CrossRef]
40. Nordic Council of Ministers Method to Measure the Amount of WEEE Generated. Report to Nordic Council's Subgroup on EEE Waste. 2009. Available online: https://www.norden.org/en/publication/method-measure-amount-weee-generated (accessed on 31 August 2020).
41. Sander, K.; Gößling-Reisemann, S.; Zimmermann, M.; Marscheider-Weidemann, F.; Wilts, H.; Schebeck, L.; Wagner, J.; Heegn, H.; Pehlken, A. Recyclingpotenzial Strategischer Metalle (ReStra). *Uba-Texte* **2017**, *68*, 2017.
42. Kalmykova, Y.; Patrício, J.; Rosado, L.; Berg, P.E. Out with the old, out with the new—The effect of transitions in TVs and monitors technology on consumption and WEEE generation in Sweden 1996–2014. *Waste Manag.* **2015**, *46*, 511–522. [CrossRef] [PubMed]
43. Pohl, W. *Mineralische Und Energie-Rohstoffe: Eine Einführung zur Entstehung und Nachhaltigen Nutzung von Lagerstätten*, 5th ed.; Schweizerbartsche Verlagsbuchhandlung: Stuttgart, Germany, 2005.
44. Schiller, G.; Ortlepp, R.; Krauß, N.; Steger, S.; Schütz, H.; Acosta Fernandez, J.; Reichenbach, J.; Wagner, J.; Baumann, J. Kartierung Des Anthropogenen Lagers in Deutschland Zur Optimierung Der Sekundärrohstoffwirtschaft. 2015. Available

online: https://www.umweltbundesamt.de/publikationen/kartierung-des-anthropogenen-lagers-in-deutschland (accessed on 31 August 2020).
45. Clarke, F.W.; Washington, H.S. *The Composition of the Earth's Crust*; Government Printing Office: Washington, DC, USA, 1924.
46. Goldschmidt, V.M. Grundlagen der quantitativen Geochemie. In *Fortschritte der Mineralogie, Kristallographie und Petrographie*; Deutsche Mineralogische Gesellschaft: Berlin, Germany, 1933.
47. Evans, A.M. *Erzlagerstättenkunde*; Ferdinand Enke Verlag: Stuttgart, Germany, 1992. (In Germany)
48. U.S. Geological Survey. *Mineral Commodity Summaries 2020: U.S. Geological Survey*. 2020. Available online: https://doi.org/10.3133/ (accessed on 10 February 2021).
49. Institute of Rare Earths and Strategic Metals Rare Earth Prices in September 2020. Available online: https://en.institut-seltene-erden.de/preise-fuer-seltene-erden-im-september-2020/ (accessed on 7 September 2020).
50. Institute of Rare Earths and Strategic Metals Rare Earth Prices in August 2020. Available online: https://en.institut-seltene-erden.de/rare-earth-prices-in-august-2020/ (accessed on 7 September 2020).
51. Institute of Rare Earths and Strategic Metals Rare Earth Prices in July 2020. Available online: https://en.institut-seltene-erden.de/prices-for-rare-earths-in-july-2020/ (accessed on 7 September 2020).
52. Institute of Rare Earths and Strategic Metals Rare Earth Prices in June 2020. Available online: https://en.institut-seltene-erden.de/prices-for-rare-earths-in-june-2020/ (accessed on 7 September 2020).
53. Institute of Rare Earths and Strategic Metals Rare Earth Prices in May 2020. Available online: https://en.institut-seltene-erden.de/prices-for-rare-earths-in-may-2020/ (accessed on 7 September 2020).
54. Institute of Rare Earths and Strategic Metals Rare Earth Prices in April 2020. Available online: https://en.institut-seltene-erden.de/prices-for-rare-earths-in-april-2020/ (accessed on 7 September 2020).
55. Institute of Rare Earths and Strategic Metals Rare Earth Prices in March 2020. Available online: https://en.institut-seltene-erden.de/prices-for-rare-earths-in-march-2020/ (accessed on 7 September 2020).
56. Institute of Rare Earths and Strategic Metals Rare Earth Prices in February 2020. Available online: https://en.institut-seltene-erden.de/rare-earth-prices-in-february-2020/ (accessed on 7 September 2020).
57. Institute of Rare Earths and Strategic Metals Rare Earth Prices in January 2020. Available online: https://en.institut-seltene-erden.de/prices-for-rare-earths-in-january-2020/ (accessed on 7 September 2020).
58. Institute of Rare Earths and Strategic Metals Rare Earth Prices in December 2019. Available online: https://en.institut-seltene-erden.de/Rare-earth-prices-in-december-2019/ (accessed on 7 September 2020).
59. Institute of Rare Earths and Strategic Metals Rare Earth Prices in November 2019. Available online: https://en.institut-seltene-erden.de/Rare-earth-prices-in-november-2019/ (accessed on 7 September 2020).
60. Institute of Rare Earths and Strategic Metals Rare Earth Prices in October 2019. Available online: https://en.institut-seltene-erden.de/rare-earth-prices-in-october-2019/ (accessed on 7 September 2020).
61. Liedtke, M.; Huy, D. *Rohstoffbewertung—Gallium*; Deutschae Rohstoffagentur (DERA) in der Bundesanstalt für Geowissenschaften und Rohstoffe: Berlin, Germany, 2018. (In Germany)
62. Ylä-Mella, J.; Pongrácz, E. Drivers and constraints of critical materials recycling: The case of indium. *Resources* **2016**, *5*, 34. [CrossRef]
63. Balaram, V. Rare Earth Elements: A review of applications, occurrence, exploration, analysis, recycling, and environmental impact. *Geosci. Front.* **2019**, *10*. [CrossRef]
64. Marscheider-Weidemann, F.; Langkau, S.; Hummen, T.; Erdmann, L.; Espinoza, L.A.T.; Angerer, G.; Marwede, M.; Benecke, S.; Mikrointegration, F.-I. *Für Zukunftstechnologien 2016: Auftragsstudie*; Deutsche Rohstoffagentur (DERA) in der Bundesanstalt für Geowissenschaften und Rohstoffe: Berlin, Germany, 2016. (In Germany)
65. Franz, M.; Wenzl, F.P. Critical review on life cycle inventories and environmental assessments of LED-lamps. *Crit. Rev. Environ. Sci. Technol.* **2017**, *47*, 2017–2078. [CrossRef]
66. Schmidt, M.; Schäfer, P.; Rötzer, N. Primär- und Sekundärmetalle und Ihre Klimarelevanz. In *Recycling und Sekundärrohstoffe, Band 13*; Thomé-Kozmiensky Verlag: Nietwerder, Germany, 4 March 2020. (In Germany)
67. Tercero Espinoza, L.A.; Rostek, L.; Loibl, A.; Stijepic, D. *The Promise and Limits of Urban Mining*; Fraunhofer ISI: Karlsruhe, Germany, 2020.
68. Carruth, M.A.; Allwood, J.M.; Moynihan, M.C. The technical potential for reducing metal requirements through lightweight product design. *Resour. Conserv. Recycl.* **2011**, *57*, 48–60. [CrossRef]

Article

How to Identify Potentials and Barriers of Raw Materials Recovery from Tailings? Part I: A UNFC-Compliant Screening Approach for Site Selection

Rudolf Suppes [1,2,*] and Soraya Heuss-Aßbichler [3]

1 Institute of Mineral Resources Engineering (MRE), RWTH Aachen University, Wüllnerstr. 2, 52062 Aachen, Germany
2 CBM GmbH—Gesellschaft für Consulting, Business und Management mbH, Horngasse 3, 52064 Aachen, Germany
3 Department of Earth and Environmental Sciences, Ludwig-Maximilians-Universität München, Theresienstr. 41, 80333 Munich, Germany; soraya@min.uni-muenchen.de
* Correspondence: suppes@cbm-ac.de

Citation: Suppes, R.; Heuss-Aßbichler, S. How to Identify Potentials and Barriers of Raw Materials Recovery from Tailings? Part I: A UNFC-Compliant Screening Approach for Site Selection. *Resources* **2021**, *10*, 26. https://doi.org/10.3390/resources10030026

Academic Editors: Andrea Thorenz and Armin Reller

Received: 30 January 2021
Accepted: 5 March 2021
Published: 16 March 2021

Publisher's Note: MDPI stays neutral with regard to jurisdictional claims in published maps and institutional affiliations.

Copyright: © 2021 by the authors. Licensee MDPI, Basel, Switzerland. This article is an open access article distributed under the terms and conditions of the Creative Commons Attribution (CC BY) license (https://creativecommons.org/licenses/by/4.0/).

Abstract: Mapping the raw material (RM) potential of anthropogenic RMs, such as tailings, requires a comprehensive assessment and classification. However, a simple procedure to quickly screen for potentially viable RMs recovery projects similar to reconnaissance exploration of natural mineral RMs is missing. In this article, a quick and efficient approach to systematically screen tailings storage facilities (TSFs) is presented to evaluate if a particular TSF meets the criteria to be assessed in a more advanced study including costly on-site exploration. Based on aspects related to a TSF's contents, physical structure, surroundings, potential environmental and social impacts, and potentially affected stakeholders, it guides its user in compiling the information at local scale in a structured manner compliant with the United Nations Framework Classification for Resources (UNFC). The test application to the TSF Bollrich (Germany), situated in a complex environment close to various stakeholders, demonstrates that a quick and remote assessment with publicly accessible information is possible. Since an assessment of tailings under conventional classification codes from the primary mining industry neglects relevant aspects, it is concluded that tailings should be considered as anthropogenic RMs. The developed screening approach can help to create a TSF inventory which captures project potentials and barriers comprehensively.

Keywords: anthropogenic raw materials; Bollrich; critical raw materials; tailings; environmental and social risks; resource management; United Nations Framework Classification for Resources (UNFC)

1. Introduction

Humanity faces the challenge of supplying a growing world population with electric energy while transitioning to a decarbonised electric energy generation. The construction and operation of decarbonised electric energy generation will significantly increase the demand for industrial minerals, as well as for base and high-tech metals [1–4]. However, most of the required raw materials (RMs) for the energy transition are produced outside the European Union (EU) [5]. This induces a potential supply risk which is aggravated by political conflicts, speculations on stock markets and the fact that many mineral RMs are produced in a few countries only. For instance, China is the global main producer of 24 out of 53 mineral RMs assessed by the German Raw Materials Agency (DERA) and is amongst the top 3 producers of other 11 mineral RMs [6].

For import-dependent regions as the EU [7–9], one way to decrease the supply risk is to diversify the mineral RMs sourcing. In the last two decades, there has been a growing interest in RMs recovery from waste [10–13] and the potential is vast: taking mineralised waste as an example, 624 Mt were produced in the EU in 2016, which is equivalent to 28% of the total generated waste [14]. Part of the mineralised waste is produced by processing ores during which ore minerals are concentrated and unwanted minerals are rejected. The rejected minerals are called *tailings* and they consist of finely ground rocks and chemical additives which are often stored in tailings storage facilities (TSFs) [15]. Tailings can contain residual (non-)metalliferous minerals that can be valorised due to less efficient processing technologies of the past or because the contained minerals were not exploitable but are used as RMs in modern technologies [16–21]. Indeed, there are efforts to improve tailings-related safety by monitoring or the removal of contaminants for instance [22–24]. However, tailings still pose a risk to human health and life, the environment, and the economy; for instance by acid mine drainage (AMD) as a result of the oxidation of sulphide minerals in contact with air and water, heavy metal-laden dust emissions or structural collapse due to the often poor construction of TSFs in the past [15,25,26]. Hence, TSFs can be regarded as ecologically critical legacies with a RM potential.

Currently, the RM potential of tailings is not captured due to a general lack of data collection, and non-standardised practices in their exploration and classification [21,27–30]. In the primary mining industry, the classification of mineral RMs is a standardised practice to communicate economic viability [31]. For instance, the classification scheme by the Committee for Mineral Reserves International Reporting Standards (CRIRSCO) is globally accepted and its principles have also been applied in the exploration of tailings [32–34]. However, a systematic screening for potentially viable tailings is currently missing. Additionally, the primary mining industry is strongly driven by economic factors [35,36] so the classification standards mainly address the needs of investors [31]. Hence, the application of the CRIRSCO to tailings neglects the negative environmental impacts and social conflicts that are often associated with TSFs [37–41].

Since 1997, the United Nations Framework Classification of Resources (UNFC) has been developed to make the classification of natural mineral and energy RMs comparable. It has recently been placed in a larger context of resource management in order to support resource policies [42]; thus it contributes to coping with RMs supply risks. The advantage of the UNFC is that it considers environmental and social aspects as a project's potential key drivers beside economic ones [42]. Since 2018, a specification document is available to make the application of the UNFC to anthropogenic RMs possible [43]. The application of the UNFC to natural and anthropogenic RMs enables a consistent and comparable assessment of both RM types. This promotes a comprehensive overview of the available RMs. However, there is currently no standardised procedure for their assessment and classification [44]. A comparative case study applying CRIRSCO and UNFC principles to a metalliferous tailings deposit in Portugal demonstrates that the inclusion of environmental and social aspects can affect the classification result substantially [45].

In natural mineral RMs assessment, a mineral deposit must first be identified. A typical first step is reconnaissance exploration where an analysis at regional scale aims to identify areas of mineral occurrences that qualify for further investigation [46]. The following prospection and exploration aim to generate detailed geological knowledge [31]. In contrast, there is currently no standardised approach for project development of anthropogenic mineral RMs. The locations of TSFs are usually known but little information is available to evaluate a potentially viable project. For resource managers the question arises how to select tailings as a potentially viable RM? The exploration and inventory of TSFs to capture RMs availability requires in-depth research, stakeholder consultation, and on-site investigation, which is generally time-consuming and costly. This can be remedied with a pre-selection of potentially viable projects through screening comparable to reconnaissance exploration. This aspect has not yet been considered in the existing classification codes so there are no corresponding guidelines for a first TSF assessment and classification.

The goal of this article is to develop and test a systematic approach for a quick and efficient pre-selection of potentially viable tailings by screening in a structured UNFC-compliant manner. 5 steps are defined in order to systematically collect the necessary information. Assessment criteria are established in order to be able to carry out a first compilation and interpretation of the data on metalliferous tailings. This includes geological, technological, economic, environmental, social and legal aspects. Based on the assessment result, it can be decided whether the selected TSF fulfils the criteria for further assessment including on-site exploration or whether it is to be inventoried for a future re-assessment due to a lack of information. The approach builds on remote data collection from publicly accessible internet sources, satellite images, scientific databases and thematic geoscientific maps. It is the first attempt to screen TSFs in a systematic and comprehensive manner. The TSF Bollrich (Germany) is chosen for the test application since it contains economically highly important RMs, such as $BaSO_4$, Cu, In, Pb, and Zn, and because it is located in a complex environment so that environmental and social aspects gain essential importance.

The research questions are: (1) should tailings be considered as anthropogenic RMs, (2) which information is necessary for TSF screening to reveal the driving factors and barriers for project development, and (3) can remotely assessed TSFs be classified with the current UNFC concept?

The research is structured as follows:

- considerations necessary for the UNFC's application to anthropogenic RMs
- argumentation for the consideration of tailings as anthropogenic RMs
- development of a quick and efficient UNFC-compliant approach for a systematic TSF screening
- case study on the TSF Bollrich with recommendations for further assessment
- discussion of the limitations of the developed systematic approach due to data uncertainty
- discussion of the developed approach in the context of RMs classification

2. Considerations for Anthropogenic Raw Materials Assessment

In this section, (1) terminology used interchangeably in the literature is defined as used in this article, (2) gaps in the current application of the UNFC to anthropogenic RMs are outlined, (3) the features of tailings in the context of natural mineral and anthropogenic RMs are analysed to outline necessary aspects that need to be considered in the assessment of tailings.

2.1. Key Words and Definitions

Metalliferous tailings from industrial processes are focussed and other mineralised waste (e.g., overburden, slags) is excluded. *TSF* refers to a physical structure to store tailings in and *(tailings) deposit* refers to a potential RM source. Generally, every TSF is a mineral occurrence in exploration terms and can potentially become a mineral RM deposit [47] (p. 124). *Target minerals* are intended for valorisation in contrast to the remaining *other minerals*. The categorisation depends on the intended valorisation path. *Recovery* refers to the physical tailings extraction and *tailings mining* refers to the whole process from exploration, recovery and processing to reclamation. *Screening* is defined as the first remote *study/assessment* to evaluate project potentials and barriers to select potentially viable projects for further assessment. It is comparable to *reconnaissance exploration* of natural mineral RMs.

2.2. Brief Introduction of the UNFC and Considerations for Its Application to Anthropogenic Raw Materials

The following description is based on Reference [42] (p. 2): the UNFC is a 'principles-based system in which products of a resource project are classified on the basis of three fundamental criteria: environmental-socio-economic viability (E), technical feasibility (F), and degree of confidence in the estimate (G), using a numerical coding system'. In a three-dimensional system (cf., Figure 1), these criteria are combined to classes with different categories (e.g., E1, E2, E3) and, where appropriate, to subcategories (e.g., E1.1). For that matter:

- the E category 'designates the degree of favourability of environmental-socio-economic conditions in establishing the viability of the project, including consideration of market prices and relevant legal, regulatory, social, environmental and contractual conditions',

- the F category 'designates the maturity of technology, studies and commitments necessary to implement the project. These projects range from early conceptual studies through to a fully developed project that is producing, and reflect standard value chain management principles',
- and the G category 'designates the degree of confidence in the estimate of the quantities of products from the project'.

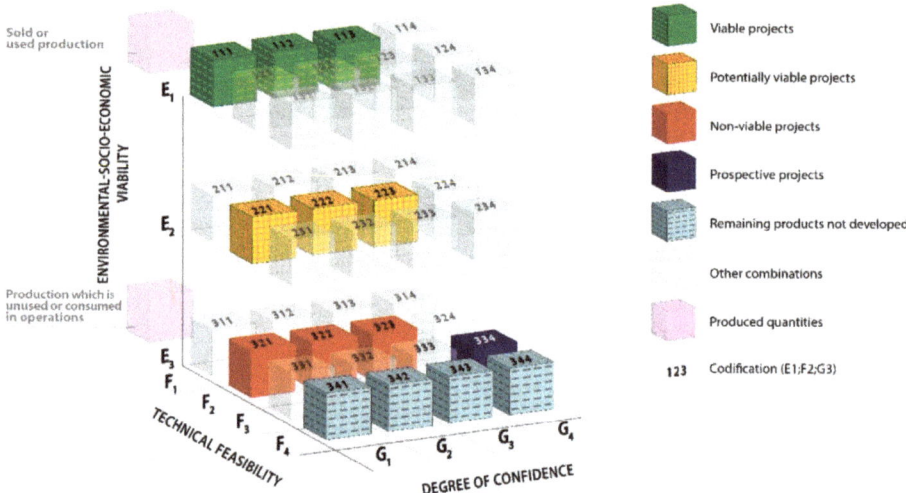

Figure 1. United Nations Framework Classification for Resources (UNFC) categories and examples of classes (from Update 2019 of the UNFC, by United Nations Economic Commission for Europe (UNECE) Expert Group on Resource Management (EGRM), ©2020 United Nations. Reprinted with the permission of the United Nations).

In pioneering case studies, the UNFC has been successfully applied to anthropogenic RMs, such as landfills [48], municipal waste incineration residues [49,50], electronic waste [48,51], and metalliferous tailings [45]. They deal with already identified RMs recovery projects, partially in advanced stages, showing that quality-assured data on recoverable anthropogenic RMs quantities can be evaluated with the UNFC [44]. However, the specifications document developed for its application to anthropogenic RMs merely defines relevant terminology and principles [43]. Hence, specifications are missing on how to develop a case study, which knowledge is required, and which factors and criteria should be considered for the rating of the G, F, and E categories.

An efficient resource management of anthropogenic RMs within the UNFC requires a systematic approach to identify potential projects in exploratory studies. Heuss-Aßbichler et al. [44] (pp. 7–11) make the following recommendations for the development of a sustainable resource management, which are considered in the developed approach:

- ESG issues must be addressed for the recycling of RMs,
- a broad spectrum of stakeholder perspectives must be included, and
- environmental and social impacts must be assessed and classified.

Furthermore, a first screening for a potential project should outline the requirements for further detailed investigations including the definition and characterisation phases [44] (p. 1). Moreover, it should give an overview of the potentials, barriers and relevant stakeholders.

2.3. Justification for the Assessment of Tailings as Anthropogenic Raw Materials

Similarities: compared to products of high purity, such as metals or complex products as mobile phones, tailings are more similar to the ores they originate from due to the relatively low degree of processing [35,52]. In the assessment, established methods from the primary mining industry can be considered for mining, the valorisation process—including mineral processing, smelting and refining—deposit modelling, and economic evaluation [32,53]. However, these aspects are not to be considered in the screening phase as they require detailed knowledge on a mineral deposit [31]. In the case of natural mineral RMs assessment, there are two types of studies which are conducted independently: first, the geological exploration of mineral RMs, which can be divided into reconnaissance, prospection, general and detailed exploration [46]. Generally, the intensity in the applied techniques and efforts increases in each phase [47]. Second, the techno-economic scoping, pre-feasibility and feasibility studies. They accompany geological exploration once reasonable prospects for an eventual economic development can be assumed [31].

Differences: in contrast to natural mineral RM deposits where the site and size are unknown in an early exploration stage, TSFs are usually close to their geogenic origin since mining operators avoid transporting material they deem valueless over long distances [47]. Hence, possible locations of TSFs can be screened for using information on active and abandoned mines. Simple methods as a visual identification of TSFs on satellite images can help to obtain basic structural information. This procedure is not applicable for finding ore deposits. Tailings consist of similar minerals as the ores they originate from so that information on local geology or on ore deposits can be used to obtain a first indication of their composition. These aspects generally enable a remote localisation and screening.

For tailings characterisation, newly generated but also historical data can be used. The targeted minerals in TSFs can vary depending on market conditions and available recovery or processing technologies. Based on the generally available data, the assessment of TSFs can be viewed as brownfield exploration [47]. Environmental and/or social aspects can also influence tailings valorisation [54]. The state of target minerals can alter in relatively short time spans due to their exposition to biological, chemical and physical processes of the Earth's surface [15]. For instance, the local climate can influence the formation of secondary minerals inside TSFs [15] (p. 172). Due to the alteration process, an inventoried TSF might need to be re-assessed in the future regarding geological conditions, its economic relevance and interested stakeholders.

In comparison to ore deposits, TSFs have an inherent negative socio-environmental impact. The severity of the individual footprint varies, depends on the condition a TSF is in and must be assessed on site [54,55]. The involved actors in a tailings mining project are partly the same as in primary mining projects: they comprise investors, mining companies, geologists, mining engineers and metallurgists for instance. However, there can be additional actors, such as modern recycling companies [56]. Furthermore, TSF owners can be the landowners if TSFs are not monitored under the Mining Law anymore [57]. Public acceptance plays a major role and depends on factors as a local population's cultural experiences [56]. Regarding the legislation, the situation can be less clear than for natural mineral RMs. For instance, tailings are considered mineral waste, thus, they fall under the Circular Economy Act (KrWG) in Germany [57]. Hence, they can only be treated in certified waste disposal plants for RMs recovery unless their legal status can be changed to a mineral RM [57]. Obtaining permission for the disposal of new residues might be a challenge too [57]. Alternatively, if environmental considerations are a project's driver, a TSF not monitored under the Mining Law would be treated under the Soil Protection Act in Germany [57]. On this basis, the legal situation must be assessed individually in more advanced studies since the uncertainties make the permitting process more costly [21]. The above aspects illustrate the complexity of the operating environments TSFs are situated in. Consequently, the screening needs to consider potential TSF-related socio-environmental impacts and a broad stakeholder group to identify project benefits and risks.

Résumé: the comparison shows that TSFs can be assessed according to the principles of natural and anthropogenic mineral RMs (cf., summary in Table 1). The natural mineral RMs approach provides the necessary geological and techno-economic information. A major difference is that in the case of anthropogenic RMs, the environmental, social and legal aspects are taken into account at an early stage, which is uncommon in natural mineral RMs exploration. Furthermore, these aspects are equally important so that all aspects must be considered concurrently to provide a comprehensive picture of the potentials of and barriers to a RM's development. These requirements can be fulfilled by assessing tailings as anthropogenic mineral RMs under consideration of the UNFC principles.

Table 1. General features of tailings in the context of natural mineral raw materials, consequences for the assessment of tailings, and addressed UNFC axes.

Group & Factor	Feature	Considerations for Tailings Assessment	UNFC Axis [1]
similarities with the assessment of natural mineral raw materials			
mine planning			
mining methods	same as for ores	existing portfolio of proven methods to resort to	F
valorisation	same as for ores	existing portfolio of proven methods to resort to	F
deposit modelling	same as for ores	existing portfolio of proven methods to resort to	F
economic evaluation	same as for ores	existing portfolio of proven methods to resort to	E (econ.)
differences from the assessment of natural mineral raw materials			
project identification			
location	remnants of mining operations	mapped (non-)active mine sites can be investigated to locate TSFs	G
composition	similar to ore composition	first indication of tailings composition derivable from ore composition	G
TSF content			
characterisation	with historical & newly generated data	brownfield exploration: remote localisation & assessment of TSFs possible	G
target minerals	formerly & newly relevant raw materials	re-assessment of project viability might be necessary for inventoried TSFs	G
state of target minerals	can alter with time	geological re-assessment might be necessary for inventoried TSFs	G
project boundaries			
socio-environmental impact	inherent footprint of TSFs	not only geological data but also status quo impacts must be considered	E (env., soc.)
involved actors	broader scope of actors involved	broad stakeholder assessment necessary from screening phase on	E (soc.)
legislation	legal situation less clear	individual assessment necessary to clarify which laws are applicable	E (leg.)

[1] econ.: economic aspects, env.: environmental aspects, soc.: social aspects, leg.: legal aspects.

3. Development of a UNFC-Compliant Approach for Systematic TSF Screening

3.1. Concept for a Systematic TSF Screening

While investors seek economic benefits from a project, a lack of public acceptance can jeopardise project development. Therefore, a successful project implementation must include the interest of both investors and the public. In this context, a systematic approach was developed to quickly identify project potentials and barriers in 5 steps (cf., Figure 2, elaborated in Sections 3.2–3.6). The approach implements the discussion results of Section 2.3 which stipulate to consider the principles of natural and anthropogenic mineral RMs assessment in TSF assessment. After the initial collection of basic information, the order of the steps reflects an increasing effort to obtain information. Therefore, the collection of information can be interrupted before too much time and money are invested. A reiteration of each step can be performed if it is decided that more information is necessary or when new information on preceding steps becomes available. Table 2 shows an overview of the knowledge generated in each step, as well as general criteria for the case of a positive rating of each step.

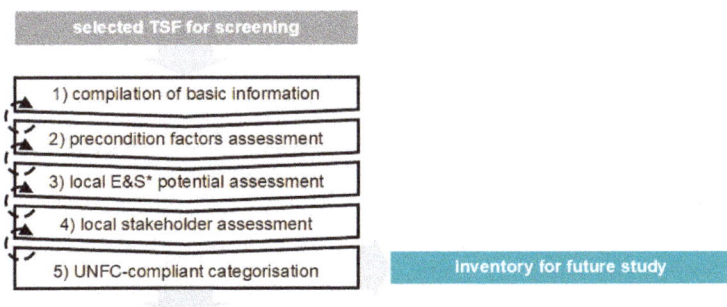

Figure 2. Quick and efficient UNFC-compliant approach for a systematic tailings storage facility (TSF) screening in 5 steps. The dotted lines indicate possible reiteration steps.

Table 2. Generated knowledge in each step and general criteria for the positive rating of each step.

Screening Step	Generated Knowledge	General Positive Rating Criteria
(1) basic TSF information compilation	overview is obtained, base for project definition created	all readily available basic information captured for later evaluation
(2) precondition factors assessment	potential project drivers identified, favourable technological & legal conditions identified	criteria of the G & F categories fulfilled, minimum one criterion of the E subcategories met
(3) local E&S potential assessment	possible environmental and social risks identified, potentials to reduce environmental risks and/or to create social benefits identified	minimum one conceivable positive environmental and/or social impact identified
(4) local stakeholder assessment	potentially affected stakeholders by TSF failure or raw materials recovery identified, potential social issues identified	all potentially affected stakeholder captured
(5) UNFC-compliant categorisation	generally favourable project conditions warrant on-site exploration	economic, environmental and/or social potentials/barriers identified

The 5 steps are: (1) Basic TSF information is compiled for a general project definition. Important aspects, such as project location, environment, contained RMs, TSF condition, and potential negative impacts, are investigated. (2) The general project conditions are assessed to determine whether economic, environmental and/or social aspects could be a project's driver, and if favourable technological and investment conditions for project execution can be assumed. (3) The potential to reduce environmental and/or social risks by removing the TSF is assessed. (4) Stakeholders directly affected by the TSF or its removal are assessed and captured for a consideration in later project planning phases. (5) The generated knowledge is reviewed and a decision regarding the further proceeding is justified with project potentials and barriers.

3.2. Basic TSF Information for Project Definition

Information on 3 categories is required to characterise a TSF for a first impression:

1. content
2. structure
3. location

Information on the *content* describes the geological and criticality characteristics of the tailings. It is the base to determine the tailings' economic relevance and enables an estimation of the tailings' valorisation for potentially interested stakeholders. Information on the *structure* describes a TSF's construction, technical features and current state. It is the base for a first estimation of a project's technical feasibility, expected operational risks and necessary rehabilitation measures. Information on the *location* is the basis to differentiate a TSF from its surroundings, which is necessary for describing environmental, social and infrastructural conditions.

Žibret et al. [21] propose a list of 19 factors as key basic parameters for the valorisation of mine waste, which they derived from literature and best practices reviews, as well as discussions in expert workshops. In this article, 21 basic factors are identified for the knowledge base of TSFs (cf., Table 6).

The following adaptations are made: the 21 factors are allocated to the above-described categories. A more complete impression on the TSF is obtained by adding the 12 factors raw materials, resource criticality, grade, mass, current use, local geology, topography, land use, climate, settlements, surface waters and infrastructure. At this stage, various aspects are excluded from the assessment as they require a detailed investigation (cf., Section 2.3). These are legal and permitting aspects, detailed knowledge on material- and mineral-centric valorisation parameters, and actual impacts. Potential environmental and social impacts are addressed in the environmental and social (E&S) potential assessment. The methodology of data collection and availability are presented in the case study in Section 4.

3.3. Precondition Factors Assessment to Identify Potential Project Drivers

Little information is available in the TSF screening phase. Therefore, the preconditions for project development are determined with 6 basic TSF information factors and 1 legal factor (investment conditions) (cf., Table 3). All factors can be allocated to the UNFC's G, F, and E categories.

Table 3. Precondition factors, assessed aspects, and addressed UNFC axes.

Precondition Factor	Assessed Aspect	UNFC Axis [1]
(1) TSF volume	justification for mid- to long-term investment	G
(2) local infrastructure	cost savings due to accessible infrastructure or incurred costs due to necessary disposal of existing infrastructure	F
(3) TSF condition	necessity of special safety measures during mining or extensive environmental rehabilitation due to contamination	F
(4) resource criticality	economic importance of targeted minerals	E (econ.)
(5) climatic conditions	enhanced environmental risks due to TSF's location	E (env.)
(6) proximity to human settlements	necessity of special protective measures during mining	E (soc.)
(7) investment conditions	general regulatory conditions in a country	E (leg.)

[1] econ.: economic aspects, env.: environmental aspects, soc.: social aspects, leg.: legal aspects.

G category: to attract investors, a project must be large enough to justify the investment. However, there is no empirical data available on required criteria for a tailings mining project to be viable. Hence, the factor TSF volume is chosen to address this aspect and the minimum volume is defined as 0.2 million m^3. It is derived from the assumption of a minimum Life of Mine (LOM) for mining metalliferous ores of 5 years (American plc

(2013) cited in Reference [58]), the tonnage according to the Taylor's rule [47] (p. 320) and an average tailings density of 2 t/m^3 [30]).

F category: the factor local infrastructure addresses locally available technology, buildings and transportation infrastructure, and the proximity to accessible utilities infrastructure. The aim is to identify potential cost savings due to accessible infrastructure or costs due to necessary asset disposal.

The factor condition addresses project risks associated with the TSF. The aim is to anticipate costs which might be incurred due to enhanced safety measures during RMs recovery or extensive environmental rehabilitation.

E category: the E subcategories economic, environmental, social and legal aspects are addressed separately; *legal aspects* being defined as a separate E subcategory in this article.

Resource criticality of target minerals is an important aspect to assess the tailings' economic relevance. Hence, information on Critical Raw Materials (CRMs) or other RMs with very high economic importance as defined by the European Commission (EC) [59] is sought. Such RMs are often used in high-technology industries, e.g., in decarbonised electric energy generation [3]. This factor is chosen since a reliable economic estimate cannot be made without detailed geological information.

The climatic conditions give an indication on environmental risks associated with a TSF. It is an important factor as it can aggravate already existing risks: for instance, dust emissions from a TSF are more likely in arid regions, and extreme weather occurrences, such as heavy rainfalls, can erode a TSF or increase the likelihood of TSF collapse especially in combination with seismic activities. This factor is also to be considered to reduce risks in case new residues need to be disposed of locally.

The factor proximity to human settlements gives an indication if special attention must be paid during mining to protect local population, e.g., from emissions. This factor needs to be considered in the context of the climatic and TSF conditions since both can increase potential risks.

The factor investment conditions is important to indicate if simple regulations and strong protection of property rights can be expected in a country. It is assessed with a country's rank on the Ease of Doing Business ranking by the World Bank [60]. The ranking covers 12 areas of business regulation, for instance getting electricity, getting credit, and enforcing contracts [60].

3.4. Local Environmental and Social Potential Assessment to Identify Benefits and Risks

In general, base metal grades in tailings are low and the processing is challenging so that potential projects can be economically unviable [30]. However, environmental and social benefits can be a key driver for developing a project in anthropogenic RMs recovery [42]. The removal of a high-risk TSF represents a social and environmental advantage since it usually incurs high ecological and social costs in the long run [61].

To reveal high-risk TSFs and to assess the benefits of their removal, a local E&S risk assessment is performed (cf., Table 4). It is based on the methodology of Owen et al. [38] which was developed to assess the vulnerability of the area surrounding a TSF to its potential failure. In this article, the methodology is applied to TSF removal. The reduction of identified E&S risks is regarded as a socially responsible action; hence, it produces benefits for society.

Table 4. Assessed environmental and social (E&S) categories, benefits derived from TSF removal, and addressed UNFC axes.

Category	Derived Benefits from TSF Removal	UNFC Axis [1]
(1) waste	reduced exposure to potential tailings flood by TSF collapse	E (env.)
(2) water	reduced risks to scarce water, aquatic ecosystems & drinking water	E (env.)
(3) landscape	reduced risk to ecosystems, aesthetically valuable lands & recreational lands	E (env.)
(4) biodiversity	reduced risk to nearby ecosystems	E (env.)
(5) land use	reduced social tensions due to land use conflicts	E (soc.)
(6) social vulnerability	reduced risk of harm to human health & social unrest	E (soc.)

[1] env.: environmental aspects, soc.: social aspects.

The categories waste, water, biodiversity, land use and social vulnerability are adopted; which are described in Owen et al. [38]. The category *landscape* is added due to the importance of protected landscapes for flora and fauna, their cultural-historical significance or their values for recreation [62].

The criteria seismic hazard, aqueduct water risk, Fragile States Index and human footprint are adopted (cf., Table 8). The criterion indigenous peoples is replaced by proximity to human settlements to consider the impacts on any local population. It provides an indication of the necessity to act to protect human health since local population may potentially or may already be affected by a TSF. The criterion nearby surface waters is added to consider their exposure to a potential TSF failure. The criteria nearby nature conservation areas, water protection areas and protected landscape areas are added to consider national environmental protection regulations.

3.5. Local Stakeholder Assessment to Identify Potential Social Issues

The increasing importance of stakeholders in mining projects and mine site remediation is generally acknowledged [63–65]. Even more, it is increasingly recognised that social conflicts can significantly increase costs and even impede project development [66,67]. The goal of the stakeholder assessment is to identify stakeholders who must be considered in further project planning. This aspect is particularly important for investors who must be aware of social conflict potentials. 5 stakeholder categories, adapted from Azapagic [63] and Valenta et al. [65], are considered (cf., Table 5).

Table 5. Stakeholder categories, their selection criteria, and addressed UNFC axes.

Category	Selection Criterion	UNFC Axis [1]
(1) nearby communities	potentially economically or physically affected by TSF failure or mining	E (soc.)
(2) TSF owner	approval required	E (soc.)
(3) local authorities	approval required, representing certain political interests which are relevant for tailings valorisation	E (soc.)
(4) NGOs [2]	representing environmental and/or social interest associated with TSF failure or tailings mining	E (soc.)
(5) other interested parties	any of the above	E (soc.)

[1] soc.: social aspects. [2] NGO: non-governmental organisation.

3.6. UNFC-Compliant Categorisation and Final Decision

For the categorisation of the project, the knowledge on the TSF, which is generated in the previous steps, is reviewed. The results are discussed on an individual basis depending on the user's point of view. For instance, a public entity might screen a particular region for

TSFs with high environmental impacts to appraise the required environmental remediation measures. The compilation of the identified potentials and barriers together with the criteria for the removal of the barriers serves as a decision-making aid for proceeding with a very preliminary assessment. There are 2 options for a first UNFC-compliant categorisation and classification:

Proceed with very preliminary study: if the criteria outlined in Table 2 are met, the project's further assessment is recommended and the project is classified as a 'Prospective Project' in the UNFC category E3F3G4 [42] (p. 5) (cf., Figure 1). Hence, the generation of further knowledge by on-site exploration is recommended.

Inventory for future study: however, if no further assessment is recommended, the project is inventoried with the classification as 'Remaining products not developed from prospective studies' in the UNFC categorisation E3F4G4 [42] (p. 5) (cf., Figure 1).

4. Case Study Results

The developed approach is tested with the case study TSF Bollrich near Goslar (Germany) (cf., Figure 3). The screening is undertaken for an area downstream of the TSF within a radius of 10 km around the TSF. It is assumed that this area would be immediately threatened in case of TSF failure [38]. Moreover, it is assumed that the TSF has not yet been explored. For this reason, the various scientific studies, media reports on the TSF, and on-site exploration results [56,68–70] are excluded.

Figure 3. Location of the TSF Bollrich and the associated disused processing plant (light shaded areas, bottom left pictures), and public infrastructure. The white lines represent public railway tracks, the red line represents the disused railway to the processing plant Bollrich, the yellow lines represent country roads, the orange line represents the 4-lane section of the federal highway B6, and the blue line represents the motorway A395 (adapted after Google Earth [71]).

4.1. Basic TSF Information

The results of step 1 are summarised in Table 6, showing that information for 20 out of 21 factors could be retrieved. The data quality is rated according to the following criteria: obvious or well-documented data is rated high quality, and remotely obtained data requiring exact data, speculative or indirect data is rated low quality. The primary sources of information are a combination of observations on Google Earth [71] and a Google search which evolved from the observations.

G category: regarding the geological evidence, most data is based on indirect evidence so the data quality is accordingly low: the presence of $BaSO_4$, Ag, Au, Cu, In, Pb, and Zn is only assumed based on the composition of the mined ores in References [72,73] and the description of ore processing by Eichhorn [74]. Several changes in the ore processing during the TSF's operation are described [74] so that variations in mineral quantity, quality and distribution can be expected. The TSF volume is a rough estimate based on a sketch with AutoCAD, the contents of the 3 ponds cannot be differentiated, and mineral quantities and qualities are missing. There is no information regarding the neutralised mine waters in the middle pond, and the quantity and composition of the discharged residues.

F category: public infrastructure, such as roads, motorways, highways, and railway tracks, are in near vicinity of the TSF. It is observable on Google Earth that the TSF is accessible via dirt roads and a disused railway track connects the processing plant Bollrich to the public railway network in Oker (cf., Figure 3). It can be observed that the railway track is partly overgrown by vegetation (cf., coordinates: 51°54′15.68″ N, 10°27′17.66″ E). This is confirmed by photos retrieved from an internet forum (http://www.goslarer-geschichten.de/showthread.php?2000-Regelspurige-Erzbahn-Bollrich-nach-Oker), also showing that the wooden railway sleepers are partly rotten. It is also observed that the buildings of the processing plant still exist.

Overall, most factors could be investigated with high quality data and only the factor *grade* lacks information. Based on the basic TSF information, the assessment is continued.

4.2. Precondition Factors Assessment

The results of step 2 are summarised in Table 7, showing that 6 out of 7 criteria are rated positive. The sources of information are scientific publications, public databases and observations on Google Earth [71].

It can be assumed that despite the simple estimation, the minimum TSF volume is exceeded 20-fold. Buildings, transportation and utilities infrastructure is present and it is assumed that all are accessible and might be reused. Erosion of the TSF or other problematic conditions, such as AMD, are not observable so that risks to a mining operation are assumed to be low and no major environmental rehabilitation measures can be anticipated. The presence of the CRMs $BaSO_4$ and In, and the economically highly important elements Cu, Pb, and Zn make the TSF economically interesting. A low climatic risk can be assumed so that related risks to a mining operation or the locally disposed of new residues are unlikely. The TSF's proximity of approximately 400 m to the nearest human settlement is rated critical. As for the investment conditions, Germany has a very high rating on the Ease of Doing Business ranking, so that favourable regulatory conditions for project execution are assumed. In summary, the project preconditions are rated favourable so that an investor's interest in the TSF can be justified.

4.3. Local Environmental and Social Potential Assessment

The results of step 3 are summarised in Table 8. The indicator thresholds are chosen conservatively to capture high risks only. The sources of information are public scientific and non-scientific databases, as well as published reports. An overview of the TSF's near environment in the context of environmentally sensitive areas is given in Figure 4.

A visual assessment of the tailings flow direction in case of a dam breach was performed with a topographic map (cf., Figure A1). It shows that the flanks of the valley in which the TSF has been built form a funnel which would direct the tailings towards the public railway tracks and the nearby industrial area in Oker. When conservatively assuming a flow rate of 5 km/h [38], the tailings would reach the nearest observable buildings on Google Earth and the public railway tracks in approximately 1.2 min (100 m distance) and 5.3 min (440 m distance), respectively. It is doubtable that the area could be evacuated in such a short time span so that harm to human health would be likely. A tailings spill could also affect the protected landscape area downstream of the TSF (cf., Figure 4).

The proximity to the river Gelmke, which flows immediately downstream of the TSF, is critical. Due to the river's small size, a tailings spill would completely fill up the river, destroy the aquatic ecosystem and deprive the river of its drain.

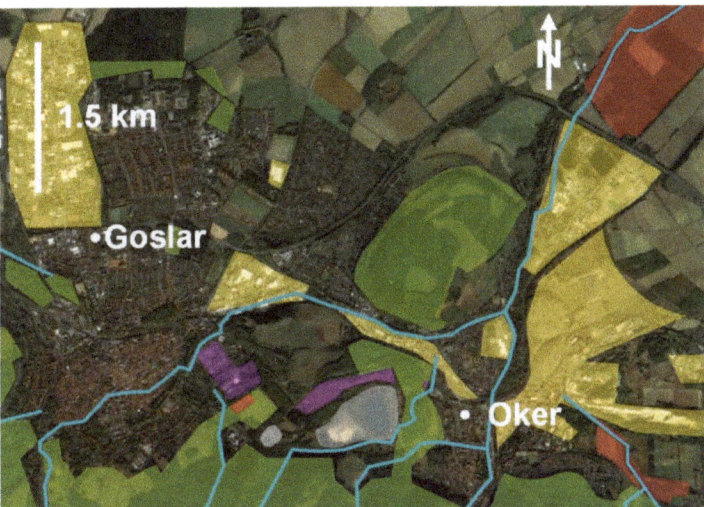

Figure 4. Simplified schematic illustration of the environment around the TSF Bollrich: the light grey shaded areas mark the TSF Bollrich (right area) and the associated disused processing plant (left area), the green shaded areas mark protected landscape areas, the red shaded areas mark nature conservation areas, the yellow shaded areas mark industrial and commercial areas, and the purple shaded areas mark sports areas close to the TSF. The blue lines represent rivers (adapted after District of Goslar | Environmental Service [75] and Google Earth [71]).

The area around the TSF is strongly affected by human activity with a Human Footprint Index of 60–80% so that there could be competing land use interests. The city administration of Goslar has the goal to develop the area around the TSF as an extensive natural and cultural landscape for calm recreation [76]. Hence, the removal of the TSF could contribute to fulfilling this goal, for instance by restoring a more natural environment.

In contrast, the seismic risk is relatively low and no signs of dam erosion are observable on Google Earth so that the risk of TSF failure is rated low. The water risk is low so that local water supply is assumed not to be endangered in case of TSF failure and a competition between different water users is unlikely. The spotted water protection and nature conservation areas downstream of the TSF is assumed not to be immediately threatened by a tailings spill due to the region's topography and the distance (cf., Figure 4). The social indicators give rise to the assumption that local communities would be able to cope with TSF failure.

Overall, the assessment of the environmental and social risk categories reveals that the TSF's environment is vulnerable to a possible TSF failure. Its removal would, therefore, generate benefits despite the low risk of failure. The proximity to human settlements and the complex surrounding environment are regarded as a necessity to act.

4.4. Local Stakeholder Assessment

The results of step 4 are depicted in Table 9, listing 17 stakeholders or stakeholder groups that could be identified. The primary sources of information are a combination of observations on Google Earth [71], a Google search with related and various other search terms, and a published report of an integrated development concept for the city Goslar by Ackers and Pechmann [76].

The largest stakeholder group consists of the citizens of Goslar and Oker with approximately 50,000 inhabitants in total [76]. Regarding local authorities, the State Office for Mining in Clausthal-Zellerfeld and various departments of the city administration of Goslar have to be considered for regulatory aspects. Three local environmental non-governmental organisations (NGOs) could be captured, counting more than 1800 members according to their websites. Their early consultation is crucial to obtain public acceptance for project implementation in an environment highly impacted by industrial activities. The German Railway and the company Oker-Chemie are directly threatened by a possible TSF failure and could advocate its removal. Farmers, foresters, and the air sports community surrounding the TSF need to be protected from negative impacts during mining. The development association of the Rammelsberg mine preserves the cultural heritage and it needs to be assured that a project would not contradict their interests. The Clausthal University of Technology and the Recycling Cluster Economically Strategic Metals (REWIMET e. V.) should be considered for their experience with recycling technologies and mine waste valorisation.

No information could be retrieved on the TSF's owner. 3 stakeholders could be captured but not specified: the responsible entity for the discharge of mine water from the Rammelsberg mine into the TSF, the owner of a tennis court and a company located approximately 230 m downstream of the TSF.

In summary, the assessment shows that the TSF is situated in a complex environment due to its proximity to agricultural, forest, industrial and commercial, nature and water protection, recreation, and residential areas (cf., Figure 4). In this context, a comprehensive stakeholder management is recommended if the project is to be continued.

4.5. UNFC-Compliant Categorisation

Overall, a further assessment within the scope of a very preliminary study is recommended due to the following aspects which are favourable for the development of a tailings mining project: assumed presence of CRMs and economically highly important metals, the identified potentials of environmental risk reduction and benefits of environmental rehabilitation, the potential to reduce land use-related social tension, favourable regulatory and infrastructure conditions, and a sufficiently large TSF volume. According to the UNFC [42], the project is classified as a 'Prospective Project' in the E3F3G4 categorisation (cf., Figure 1).

4.6. Path Forward for the Case Study Bollrich

In a very preliminary study, the following aspects should be addressed to remove the barriers for a higher classification as a 'Potentially Viable Project' (E2F2G3): the largest barrier is the lack of geological knowledge on the deposit so that, for instance, the quantities of products cannot be estimated. Hence, the next milestone is on-site exploration to determine material characteristics, such as the chemical and mineralogical composition of the tailings, their quantities and qualities, and their physico-chemical properties, as well as their distribution inside the TSF. Additionally, the TSF's geomechanical stability needs to be studied. Furthermore, the identification of the TSF's owner and a first assessment of the legal conditions for a project are important aspects to be clarified.

Table 6. Basic information on the TSF Bollrich. The green shaded and red shaded shaded areas indicate data of high and low quality, respectively.

Category & Factor	Data	Source & Data Quality
(1) content		
(i) raw materials	sulphates: $BaSO_4$; sulphides: Cu, Pb, Fe, Zn; others: Ag, Au, In	inferred from References [72,73]
(ii) resource criticality	$BaSO_4$, & In are Critical Raw Materials in the EU; Cu, Pb, & Zn of very high economic importance in the EU	[59]
(iii) grade	-	-
(2) structure		
(iv) history	start/end of operation in 1938/1988, froth flotation plant Bollrich closed in 1987, course of Gelmke was modified several times	[74]
(v) reasons for closure	closure of mine Rammelsberg in 1988 for economic reasons	[74]

Table 6. Cont.

Category & Factor	Data	Source & Data Quality
(vi) design	valley impoundment, 1 small pond & 2 large ponds, 1 main dam & 2 intermediate dams, estimated dam height 35 m	observed on Google Earth [71], cf., Figure 3
(xii) surface area	estimated 315,000 m^2	Ruler tool [74]
(xiii) volume	estimated 4.7 & 4 million m^3 (including & excluding main dam, respectively)	Ruler tool [74], AutoCAD (Autodesk Inc.)
(iv) mass	estimated 9.4 & 8 million t (including & excluding main dam, respectively)	assumed tailings density 2 t/m^2 [30]
(x) homogeneity	several changes of ore processing reported, heterogeneity of minerals inside TSF can be assumed	[74]
(xi) condition	partially dry but mostly covered with water, no observable signs of AMD, erosion or controlled reclamation	observed on Google Earth [71], cf.
(xii) current use	since 1988 neutralised mine waters from the closed mine Rammelsberg are discharged into the lower pond	observed on Google Earth [71,74]
(3) location		
(xiii) position	Goslar district (51°54'8.97" N, 10°27'47.31" E, Lower Saxony, Germany), 270 m above mean sea level	observed on Google Earth [71]
(xiv) local geology	folded & faulted Palaeozoic rocks of the Harz Mountains are uplifted & thrust over younger Mesozoic rocks of the Harz foreland along the Northern Harz Boundary fault leading to steeply tilting & partly inverted Mesozoic strata, Mesozoic rocks are largely composed of Triassic to Cretaceous sedimentary rocks of varying composition (i.e., mostly impure limestones, clastic sandstones (greywackes) & shales), younger Quaternary sediments are rare & locally limited	[77]
(xv) topography	at the foot of Harz Mountain range, max. 1141 m altitude with deep valleys	[78]
(xvi) land use	in near vicinity: agricultural, forest, industrial & commercial, & recreation & residential areas	observed on Google Earth [71]
(xvii) climate	moderately warm, temperature −0.7 to 16.3 °C (average 7.9 °C), average rain precipitation 768 mm/a	[79]
(xviii) settlements	nearest ~400 m E air-line distance downstream of main dam	observed on Google Earth [71], cf., Figure 3
(xix) surface waters	4 small rivers observed downstream of TSF within 1.5 km radius (Abzucht, Ammentalbach, Gelmke, Oker)	observed on Google Earth [71], cf., Figure 4
(xx) site accessibility	dirt roads, federal highway B6 ~1.6 km N air-line distance from TSF, public railway ~500 m E air-line distance from TSF, disused railway tracks from processing plant Bollrich to public railway network (estimated abandonment in 1988)	observed on Google Earth [71,74] cf., Figure 3
(xxi) infrastructure	disused processing plant Bollrich ~500 m W air-line distance from TSF, access to public electricity & water grid assumed	observed on Google Earth [71], cf., Figure 3

Table 7. Precondition factors, and the corresponding criteria and indicators for a TSF screening. ✓ indicates a fulfilled and ✗ a non-fulfilled criterion, respectively.

Factor	Criterion	Indicator	Result	Source	Rating	UNFC Axis [1]
(1) TSF volume	TSF volume (V) high enough for a LOM [2] of ≥ 5 years	$V \geq 0.2$ million m^3	4 million m^3 (excluding main dam)	estimated with Ruler tool in Google Earth [71] & AutoCAD (Autodesk Inc.)	✓	G
(2) infrastructure	buildings, transportation & utilities infrastructure present	observable	buildings, railway tracks, roads, highways, motorways & utilities infrastructure observable	assumption based on observation with Google Earth [71]	✓	F

Table 7. Cont.

Factor	Criterion	Indicator	Result	Source	Rating	UNFC Axis [1]
(3) TSF condition	erosion of TSF and/or emissions (e.g., AMD [3])	not observable	no signs of erosion and/or emissions observable	observation with Google Earth [71]	✓	F
(4) resource criticality	number (n) of elements or minerals that are CRMs [4] in EU or that are of very high economic importance	$n \geq 1$	$n = 4$ (BaSO$_4$, Cu, Pb & Zn expected to be present)	inferred from [73]	✓	E (econ.)
(5) climatic conditions	favourable climatic conditions with low probability of extreme climate or weather occurrences	moderate climate	moderately warm, average 7.9 °C, average rain precipitation 768 mm/a	[79]	✓	E (env.)
(6) human settlements	distance (d) to settlements	$d \leq 10$ km	$d \approx 400$ m E air-line	[71]	✗	E (soc.)
(7) investment conditions	good conditions as per Ease of Doing Business ranking	country rank ≤ 75	rank 22 (Germany)	[60]	✓	E (leg.)

[1] econ.: economic aspects, env.: environmental aspects, soc.: social aspects, leg.: legal aspects. [2] LOM: Life of Mine. [3] AMD: Acid Mine Drainage. [4] CRM: Critical Raw Mater.

Table 8. Results of the local E&S potentials assessment for the TSF Bollrich (modified after Owen et al. [38]). ✓ indicates a fulfilled and ✗ a non-fulfilled criterion, respectively.

Domain [1]	Category	Criterion	Indicator	Result	Source	Rating
env.	waste	seismic hazard	peak ground acceleration > 3.2 m/s^2	0.4 m/s^2	[80]	✗
	water	aqueduct water risk	overall water risk > 3 (high)	1–2 (low-medium)	[81]	✗
		nearby surface waters	downstream distance to TSF < 10 km	in near vicinity, cf., Figure 4	[71]	✓
		nearby water protection areas	downstream distance to TSF < 10 km	~7.3 km N-E of the TSF near Vienenburg	[75]	✓
	landscape	protected landscape areas	downstream distance to TSF < 10 km	nearest immediately at the foot of the dam, cf., Figure 4	[75]	✓
	biodiversity	nature conservation areas	downstream distance to TSF < 10 km	~3.5 km N-E of TSF, cf., Figure 4	[75]	✓
soc.	social vulnerability	proximity to human settlements	downstream distance to TSF < 10 km	nearest settlement Oker ~400 m E of main dam, potential flow path in direction of settlement, cf., Figure A1	[71,82]	✓
		Fragile States Index	country score ≥ 4 for social indicators	average score 2 (Germany)	[83]	✗
	land use	human footprint	Human Footprint Index > 40%	60–80% (area around the TSF)	[84]	✓

[1] env.: environmental potentials, soc.: social potentials.

Table 9. Potential stakeholders of a tailings mining project at the TSF Bollrich (stakeholder categories derived from Azapagic [63] and Valenta et al. [65]).

Stakeholder Category	Result	Source	Remark
nearby communities	(1) citizens of Goslar & its borough Oker	observation on Google Earth [71,76]	total population of ~50,000 inhabitants
TSF owner	(2) -	-	could not be clarified with internet search
local authorities	(3) Goslar administrative bodies	www.landkreis-goslar.de www.landkreis-goslar.de/eh-	Various departments, such as for Regional Economic Development or the Environment, the Circular Economy Department, are responsible for the disused landfill Paradiesgrund in near vicinity of the TSF
	(4) State Office for Mining, Energy & Geology Office Clausthal-Zellerfeld	www.lbeg.niedersachsen.de	~15 km S-W from TSF, included due to relevance for approval
NGOs	(5) German Federation for the Environment & Nature Conservation in the western Harz region (BUND)	www.bund-westharz.de	~600 members
	(6) Nature & Biodiversity Conservation Union (NABU)	www.nabu-goslar.de	~1000 members
	(7) Nature & Environmental Aid Goslar (NU)	www.nu-goslar.de	~200 members
other interested parties	(8) German Railway (DB)	observation on Google Earth [71]	connection to railway network would potentially have to be reactivated, a potential TSF failure might affect the railway
	(9) farmers	observation on Google Earth [71]	proximity to farmlands around the TSF
	(10) foresters	observation on Google Earth [71]	proximity to forests around the TSF
	(11) Development Association World Cultural Heritage Ore Mine Rammelsberg Goslar/Harz	https://foerderverein-rammelsberg.de	the association is responsible for the preservation of the World Heritage
	(12) Oker-Chemie GmbH	observation on Google Earth [71]	a potential TSF failure might affect the industrial site
	(13) Air Sports Community Goslar	www.segelfliegen-goslar.de	glider airfield in near vicinity of TSF
	(14) REWIMET e. V.—Recycling Cluster	www.rewimet.de	network of companies, scientific institutions & local authorities, promotes recycling from research up to the industrial scale
	(15) Clausthal University of Technology (TUC)	www.ifa.tu-clausthal.de	~14 km S-W from TSF, included due to regional knowledge & research experience on mineral wastes of >25 years

Table 9. Cont.

Stakeholder Category	Result	Source	Remark
	non-specifiable:		
	(16) responsible entity for mine water discharge into the TSF	observation on Google Earth [71]	could not be specified with internet search
	(17) owner of tennis courts downstream of the TSF	observation on Google Earth [71]	could not be specified with internet search
	(18) company downstream of the TSF	observation on Google Earth [71]	could not be specified with internet search

As for the technical feasibility, different valorisation scenarios should be investigated. Hence, the tailings' processability needs to be assessed together with a conceptual mine plan under consideration of various valorisation options. This includes an investigation of the decommissioned Bollrich processing plant, whether there is reusable machinery, and the condition of the road and railway access.

This article shows that a large, diverse and socially active stakeholder group is involved. Therefore, early proactive stakeholder engagement is recommended. Measures should be taken to avoid negative environmental impacts on local population during active mining to avoid social conflicts. A public discussion of the benefits and risks of the status quo of the TSF can help to promote public acceptance. A strong argument for removing the TSF is the risk of greater harm in the event of TSF collapse. In this case, an expansion of the 10 km screening radius could help to better estimate potential harm and determine whether additional stakeholders would be affected. A detailed survey of actual emissions from the TSF could provide additional arguments for its removal.

Economic and social aspects of the city administration's development goals can also contribute to the evaluation: strengthening the regional industrial and commercial role, creating high-value jobs, fostering cultural heritage and traditions, harnessing the cultural potential of the industrial history, and developing tourism [76]. The likelihood of obtaining political acceptance increases if possible RMs recovery scenarios do not contradict these goals. Additionally, the likelihood of obtaining political and public acceptance can be increased if part of the revenues from a project would be used for partial environmental rehabilitation of contaminated land in Oker [76].

5. Discussion

5.1. Limitations of the Developed Screening Approach

Data sources and quality: the developed approach is intended to enable a quick and cost-efficient TSF screening. This goal can be achieved as shown by the case study application. A low degree of data quality can be tolerated in the screening phase similar as in reconnaissance exploration [47]. However, one needs to be aware of the potential sources of error: the information's quality from publicly accessible sources can vary from speculative (e.g., private websites or internet forums) to scientifically proven (e.g., peer-reviewed articles), and cross-checking is not always possible. In internet forums and websites, participants generally do not reflect representative interest groups, and the opinions shared may be biased.

The methods used can generally be expected to provide low quality data on a TSF's content and structure. Tailings production records or exploration data are unlikely to be publicly accessible especially for older TSFs. The visual assessment of satellite images can only provide a rough estimate of a TSF's volume and water body. Statements about the dam material or other materials inside a TSF cannot be made.

Certain aspects can only be hardly or not evaluated at all with Google Earth, such as the condition of present infrastructure. A visual assessment and internet search are unsuitable for carrying out a comprehensive stakeholder analysis. In the case study, for instance, there are unspecifiable stakeholders, such as the TSF's owner. In the early stage of project development, no tendencies towards public acceptance can be anticipated.

The E&S potential assessment is generally expected to generate information of high quality since it relies on established scientific and public databases. However, information on certain factors might not always be available in the required quality, especially in remote areas. In addition, actual negative environmental impacts, such as emissions to air, need to be assessed on site.

Categorisation of screening results: the case study shows that the developed approach can be used to compile sufficient information for a TSF screening. The evaluation of the generated knowledge allowed to identify project potentials and barriers that need to be considered for further project development. According to the UNFC [42] (p. 5), only 2 categorisations are possible in the screening phase: a 'Prospective Projects' (E3F3G4) or 'Remaining products not developed from prospective projects' (E3F4G4).

A disadvantage of this limitation is that important aspects of a project's status cannot be communicated directly. It is, therefore, recommended to consider the following aspects for sustainable resource management of anthropogenic RMs: the sources of information, e.g., historical, indirect, or speculative, cannot be differentiated in the G subcategories. A differentiation could provide a quick overview of the information's quality. The F subcategories are not applicable since they focus on the degree of development of recovery technologies and neglect factors, such as already existing infrastructure. This is overcome in this article by assuming that present or absent observable infrastructure can be distinguished with the F3 and F4 categories. An according description should be added to the guideline for anthropogenic RMs. The E categorisation does not allow for a differentiated communication of a project's potentials and barriers in an appropriate level of detail since several dimensions are aggregated in the E category. There is the need to differentiate the E category by introducing the 4 separate subcategories economic, environmental, social, and legal aspects.

5.2. The Developed Screening Approach in a Global Raw Materials Classification Context

The conventional classification of tailings under the CRIRSCO has several shortcomings: first, early exploration focusses on geological aspects, such as mineral quality and quantity, in order to identify potentially economic mineral RMs [31]. Exploration Targets, Exploration Results, and non-economic mineral RM deposits are excluded from the classification [31]. However, metal grades in tailings are generally low [30] so that, from the CRIRSCO's perspective, the exploration of TSFs can be expected to be a priori unattractive due to the high costs. In contrast, the developed approach shows that it is possible to perform a quick TSF screening and a UNFC-compliant categorisation with little effort in order to determine a TSF's potentials. This quick and efficient approach can help make TSF exploration more attractive.

Second, the CRIRSCO generally focusses on providing information for investors [31]. Hence, the definition of a RM's potential under the CRIRSCO is limited to material and monetary aspects. However, when assessing mineral RMs, aspects other but purely economic ones are increasingly becoming important [65], and environmental, social, and legal aspects must be taken into account explicitly in the case of anthropogenic RMs. As shown by the case study, the latter aspects can even be decisive for the screening result. It also shows that these aspects can be assessed parallel to geological and economic ones unlike it is standard practice in natural mineral RMs assessment.

Third, under the CRIRSCO, sustainability aspects are discussed in Public Reports but they are not relevant at exploration stage and they are not part of the classification [31]. However, the physical risks of TSFs are often borne by local populations and the environment while mining companies mostly face financial risks only [38]. Communities living in near vicinity to TSFs are often unable to properly judge the risks associated with TSFs since these are rarely disclosed [38]. This is particularly important since the communities can usually not move away to avoid these risks [38]. Currently, resource management recognises that E&S risks can form a barrier to mining projects [65]. Therefore, the assessment of relevant stakeholders, including public institutions and local communities, is important even in an early exploration phase. Furthermore, identified TSF-related environmental and social impacts at local level must be incorporated into a resource strategy at an early stage. Additionally, the awareness of cultural factors must be included especially in the development of anthropogenic RMs since they can enable or prevent their valorisation [56]. Overall, these aspects would ensure a high level of transparency for all stakeholders.

Fourth, the environmental context a RM deposit is situated in is not considered when reporting Exploration Targets or Exploration Results [31]. This includes aspects such as already present infrastructure and risks. However, it could be shown in this article that information on these aspects is remotely retrievable with low effort and that it can provide important information on a project's potentials and barriers.

Fifth, the assessment and reporting criteria defined in the CRIRSCO apply to Exploration Results and more advanced studies only [31]. Several authors state that there is a lack of knowledge on TSFs and their properties [29,44,85] so that the overall risks and economic potentials remain generally unknown. In the EU, for instance, current mine waste inventories are not comprehensive: the 'Minerals4EU' Knowledge Data Platform (http://www.minerals4eu.eu) provides too little information, and national mine waste registries of the EU's countries are incomplete and focus on environmental aspects in most cases [21]. To remedy this shortcoming, a comprehensive analysis of the status is needed, including economic, environmental, social, legal, technological, and geological aspects altogether. A status analysis would enable one to filter for specific aspects, such as RMs recovery potentials. However, the CRIRSCO is unsuitable for a screening that fulfils these requirements. In contrast, the developed 5-step approach can enable a status analysis in a remote, quick, and, therefore, cost-efficient manner.

6. Conclusions and Recommendations

To recapitulate, anthropogenic RMs are becoming an increasingly important source of RMs and they are available in vast amounts: for instance, the world's largest waste stream, mineralised waste, is produced in an estimated range of 20–25 Gt/a [15]. The current knowledge gaps concerning their RM potential and the lack of comparability to other RMs impede valorisation. In the context of an expected increase of global demand for metallic and non-metallic RMs of 96% and 168%, respectively, between 2015 and 2050 [86], actions must be taken to include anthropogenic RMs in strategic resource management. One solution is a comprehensive investigation and sharing of the information with decision-makers and the public. Therefore, the potentials of and barriers to their development need to be mapped, comparable to current practice in natural mineral RMs assessment. This must be done in a quick and cost-efficient manner. This study uses tailings as an example to demonstrate how a UNFC-compliant approach can be used for a systematic TSF screening, similar to the concept of reconnaissance exploration for natural mineral RMs. The case study TSF Bollrich (Germany) is chosen to show how a TSF can be systematically screened in practice with the developed approach. Hence, the innovative approach contributes to a re-interpretation of the material value of tailings in terms of resource efficiency within a circular economy.

The research questions are answered: (1) Tailings are a suitable example to demonstrate the difference between natural and anthropogenic RMs. The CRIRSCO classification standards from the primary mining industry are designed for natural mineral RMs and they focus on material and monetary aspects. However, the case study shows that sustainability aspects together with legal aspects and the interest of stakeholders are of vital importance for tailings assessment. This corresponds to the general requirements for the classification of anthropogenic RMs. (2) For a comprehensive assessment of tailings, all relevant aspects must be considered including information on a TSF's contents, physical structure and surroundings. The concerns of local population on potential negative environmental and social impacts can be a major barrier to project development. Therefore, the stakeholders who are potentially affected by a TSF and its removal should be investigated even in a screening phase. (3) Applying a UNFC-compliant approach, which considers environmental and/or social aspects in addition to economic viability, can increase the chances for RMs recovery from TSFs. A categorisation of the retrieved information is possible and a classification of 'Prospective Projects' (categorisation E3F3G4) and 'Remaining products not developed from prospective studies' (categorisation E3F4G4) can be performed with the current UNFC concept.

The case study results are summarised: the case study application shows that a quick and remote identification of a potentially viable tailings mining project is possible. It is based on geological, techno-economic, environmental, social and legal aspects. Hence, a decision for further assessment can be made before costly on-site exploration is carried out. A particularity of the case study TSF is its embedding in a complex environment in near vicinity to local population. It highlights the importance of considering TSF-related environmental and social impacts on a local scale. Potentials for project development are that a potential source of economically highly important RMs is identified, the city administration's development goals can be supported, environmental rehabilitation can be promoted, and social risks can be reduced. Barriers are the lack of a conceptual mine plan including techno-economic feasibility, high uncertainties regarding data on mineral quantity and quality, and the lack of information on actual environmental and social impacts.

The following recommendations are made: for the case study TSF Bollrich, increase the degree of confidence in knowledge on geology and technical feasibility by on-site exploration. Identify the TSF's owner and determine legal conditions for mining. Investigate stakeholder opinions to anticipate conditions for public acceptance. Develop valorisation scenarios which consider the development goals for Goslar, environmental rehabilitation, and a wide variety of stakeholder interests. Systematic screening approach: identify the requirements for a project status as 'Potentially Viable'. Investigate if mineral- and structure-related information on TSFs can be obtained with spaceborne hyperspectral and radar measurements, respectively. Develop a standard at European level, including reporting guidelines, in order to comprehensively map the RM potential of tailings. Test if governance-related risks can be included in the screening. Develop more case studies to identify essential and universally applicable valorisation factors and assessment criteria including sustainability. Test the developed approach with other mine wastes. Anthropogenic resource management: implement a screening for potentially viable recovery projects. Break down the UNFC's E category into economic, environmental, social and legal aspects to visualise specific project potentials and barriers. Develop guidelines for rating data quality and uncertainty ranges in project development stages. Include stakeholder assessments in case studies to capture potential sources of social conflict. To enable an EU-wide comparison of the RM potential of tailings and to reveal barriers for project development, structure information in EU national mine waste registries in a UNFC-compliant manner.

Author Contributions: Conceptualisation, R.S.; methodology, R.S.; validation, R.S., S.H.-A.; resources, R.S.; data curation, R.S.; writing—original draft preparation, R.S.; writing—review and editing, R.S., S.H.-A.; visualisation, R.S.; project administration, R.S.; funding acquisition, R.S., S.H.-A. Both authors have read and agreed to the published version of the manuscript.

Funding: This research was funded by the German Ministry of Research and Education (BMBF) as part of the research project ADRIANA under the Client II programme, grant agreement number 033R213A-D.

Institutional Review Board Statement: Not applicable.

Informed Consent Statement: Not applicable.

Data Availability Statement: This research used publicly available data available in the referenced sources.

Acknowledgments: The authors are thankful to Bernd G. Lottermoser for his comments and to Johannes J. Emontsbotz for assisting with the CAD modelling. In addition, the authors would like to express their deep gratitude to 4 anonymous reviewers who helped to improve the manuscript.

Conflicts of Interest: The authors declare no conflict of interest. The funders had no role in the design of the study; in the collection, analyses, or interpretation of data; in the writing of the manuscript, or in the decision to publish the results.

Abbreviations

Abbreviation/Unit	Description
Ag	lat. *argentum* (silver)
Au	lat. *aurum* (gold)
$BaSO_4$	barium sulphate (barite)
Cu	lat. *cuprum* (copper)
Fe	lat. *ferrum* (iron)
In	indium
Pb	lat. *plumbum* (lead)
Zn	zinc
AMD	acid mine drainage
CRIRSCO	Committee for Mineral Reserves International Reporting Standards
CRM	Critical Raw Material
E	East
E&S	environmental and social
EC	European Commission
EU	European Union
LOM	Life of Mine
N	North
N-E	Northeast
NGO	non-governmental organisation
REWIMET e. V.	Recycling Cluster Economically Strategic Metals
RM	raw material
S-W	Southwest
TSF	tailings storage facility
UNFC	United Nations Framework Classification for Resources
UNFC E category	represents environmental-socio-economic viability
UNFC F category	represents technical feasibility
UNFC G category	represents degree of confidence in the geological estimate
W	West
°C	degree Celsius (unit of temperature on the Celsius scale)
Gt/a	gigatons per year (unit of mass flow, equivalent to 10^{12} kg per year)
km	kilometre (unit of length, equivalent to 1,000 metres)
m	metre (SI unit of length)
m/s^2	metre per square second (unit of acceleration)
m^3	cubic metre (SI-derived unit of volume)
mm/a	millimetres per year (annual rain precipitation)
t	metric tonne (unit of weight, equivalent to 1,000 kg)

Appendix A

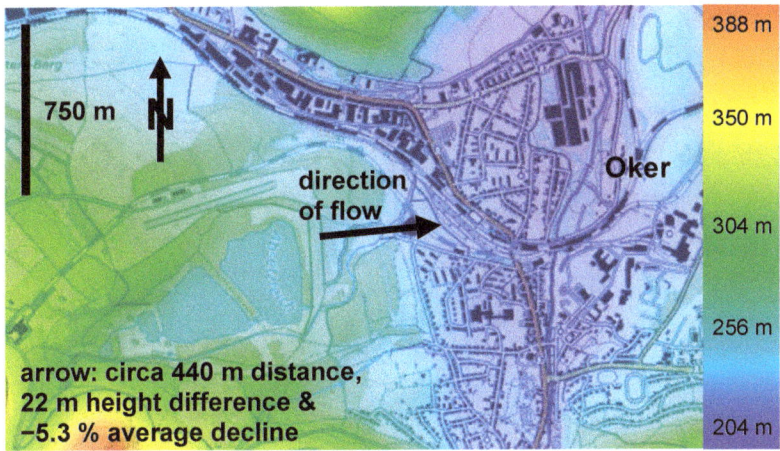

Figure A1. Contour map of the area around the TSF Bollrich and assessed direction of tailings flow in case of TSF failure (adapted after topographic-map.com [82]).

References

1. Singh, N.; Duan, H.; Yin, F.; Song, Q.; Li, J. Characterizing the Materials Composition and Recovery Potential from Waste Mobile Phones: A Comparative Evaluation of Cellular and Smart Phones. *ACS Sustain. Chem. Eng.* **2018**, *6*, 13016–13024. [CrossRef]
2. Valero, A.; Valero, A.; Calvo, G.; Ortego, A.; Ascaso, S.; Palacios, J.-L. Global material requirements for the energy transition. An exergy flow analysis of decarbonisation pathways. *Energy* **2018**, *159*, 1175–1184. [CrossRef]
3. Vidal, O.; Goffé, B.; Arndt, N. Metals for a low-carbon society. *Nat. Geosci.* **2013**, *6*, 894–896. [CrossRef]
4. Watari, T.; McLellan, B.C.; Giurco, D.; Dominish, E.; Yamasue, E.; Nansai, K. Total material requirement for the global energy transition to 2050: A focus on transport and electricity. *Resour. Conserv. Recycl.* **2019**, *148*, 91–103. [CrossRef]
5. Martins, F.F.; Castro, H. Raw material depletion and scenario assessment in European Union—A circular economy approach. *Energy Rep.* **2020**, *6*, 417–422. [CrossRef]
6. Bastian, D.; Brandenburg, T.; Buchholz, P.; Huy, D.; Liedtke, M.; Schmidt, M.; Sievers, H. *DERA List of Raw Materials*; Deutsche Rohstoffagentur (DERA) in der Bundesanstalt für Geowissenschaften und Rohstoffe (BGR): Berlin, Germany, 2019; p. 116. ISBN 978-3-943566-61-1. Available online: https://www.deutsche-rohstoffagentur.de/DE/Gemeinsames/Produkte/Downloads/DERA_Rohstoffinformationen/rohstoffinformationen-40.pdf?__blob=publicationFile (accessed on 28 August 2020). (In German)
7. Grilli, M.L.; Bellezze, T.; Gamsjäger, E.; Rinaldi, A.; Novak, P.; Balos, S.; Piticescu, R.R.; Ruello, M.L. Solutions for Critical Raw Materials under Extreme Conditions: A Review. *Materials* **2017**, *10*, 285. [CrossRef]
8. Rabe, W.; Kostka, G.; Smith Stegen, K. China's supply of critical raw materials: Risks for Europe's solar and wind industries? *Energy Policy* **2017**, *101*, 692–699. [CrossRef]
9. Sievers, H.; Tercero, L. European Dependence on and Concentration Tendencies of the Material Production. 2012. Available online: https://www.google.com/url?sa=t&rct=j&q=&esrc=s&source=web&cd=&ved=2ahUKEwjHtcbe453sAhWD3KQKHXxMAhIQFjAAegQIAxAC&url=https%3A%2F%2Fwww.isi.fraunhofer.de%2Fcontent%2Fdam%2Fisi%2Fdokumente%2Fccn%2Fpolinares%2FPolinares_WP_14_March_2012&usg=AOvVaw01huy0wOnuuqtkRHm0Q0K9 (accessed on 5 October 2020).
10. Burlakovs, J.; Kriipsalu, M.; Klavins, M.; Bhatnagar, A.; Vincevica-Gaile, Z.; Stenis, J.; Jani, Y.; Mykhaylenko, V.; Denafas, G.; Turkadze, T.; et al. Paradigms on landfill mining: From dump site scavenging to ecosystem services revitalization. *Resour. Conserv. Recycl.* **2017**, *123*, 73–84. [CrossRef]
11. European Commission (EC). Towards a Circular Economy: A Zero Waste Programme for Europe. 2014. Available online: https://ec.europa.eu/environment/circular-economy/pdf/circular-economy-communication.pdf (accessed on 4 August 2019).
12. Krook, J.; Baas, L. Getting serious about mining the technosphere: A review of recent landfill mining and urban mining research. *J. Clean. Prod.* **2013**, *55*, 1–9. [CrossRef]
13. United Nations Environment Programme (UNEP). *Critical Metals for Future Sustainable Technologies and their Recycling Potential: Sustainable Innovation and Technology Transfer Industrial Sector Studies*; United Nations Environment Programme & United Nations University: Tokyo, Japan, 2009; Available online: https://www.oeko.de/oekodoc/1070/2009-129-en.pdf (accessed on 6 October 2020).

14. Eurostat. Waste Statistics. 2020. Available online: https://ec.europa.eu/eurostat/statistics-explained/index.php/Waste_statistics (accessed on 25 October 2020).
15. Lottermoser, B. *Mine Wastes*, 3rd ed.; Springer: Berlin/Heidelberg, Germany, 2010; p. 400. [CrossRef]
16. Drif, B.; Taha, Y.; Hakkou, R.; Benzaazoua, M. Recovery of Residual Silver-Bearing Minerals from Low-Grade Tailings by Froth Flotation: The Case of Zgounder Mine, Morocco. *Minerals* **2018**, *8*, 273. [CrossRef]
17. López, F.; García-Díaz, I.; Rodríguez Largo, O.; Polonio, F.; Llorens, T. Recovery and Purification of Tin from Tailings from the Penouta Sn–Ta–Nb Deposit. *Minerals* **2018**, *8*, 20. [CrossRef]
18. Okereafor, U.; Makhatha, M.; Mekuto, L.; Mavumengwana, V. Gold Mine Tailings: A Potential Source of Silica Sand for Glass Making. *Minerals* **2020**, *10*, 448. [CrossRef]
19. Su, Z.; Chen, Q.; Zhang, Q.; Zhang, D. Recycling Lead–Zinc Tailings for Cemented Paste Backfill and Stabilisation of Excessive Metal. *Minerals* **2019**, *9*, 710. [CrossRef]
20. Tang, C.; Li, K.; Ni, W.; Fan, D. Recovering Iron from Iron Ore Tailings and Preparing Concrete Composite Admixtures. *Minerals* **2019**, *9*, 232. [CrossRef]
21. Žibret, G.; Lemiere, B.; Mendez, A.-M.; Cormio, C.; Sinnett, D.; Cleall, P.; Szabó, K.; Carvalho, M.T. National Mineral Waste Databases as an Information Source for Assessing Material Recovery Potential from Mine Waste, Tailings and Metallurgical Waste. *Minerals* **2020**, *10*, 446. [CrossRef]
22. Dong, L.; Deng, S.; Wang, F. Some developments and new insights for environmental sustainability and disaster control of tailings dam. *J. Clean. Prod.* **2020**, *269*, 122270. [CrossRef]
23. Dong, L.; Tong, X.; Li, X.; Zhou, J.; Wang, S.; Liu, B. Some developments and new insights of environmental problems and deep mining strategy for cleaner production in mines. *J. Clean. Prod.* **2020**, *210*, 1562–1578. [CrossRef]
24. Lee, E.-S.; Cho, S.-J.; Back, S.-K.; Seo, Y.-C.; Kim, S.-H.; Ko, J.-I. Effect of substitution reaction with tin chloride in thermal treatment of mercury contaminated tailings. *Environ. Pollut.* **2020**, *264*, 114761. [CrossRef]
25. Lyu, Z.; Chai, J.; Xu, Z.; Qin, Y.; Cao, J. A Comprehensive Review on Reasons for Tailings Dam Failures Based on Case History. *Adv. Civ. Eng.* **2019**, *2019*, 4159306. [CrossRef]
26. Wang, G.; Tian, S.; Hu, B.; Xu, Z.; Chen, J.; Kong, X. Evolution Pattern of Tailings Flow from Dam Failure and the Buffering Effect of Debris Blocking Dams. *Water* **2019**, *11*, 2388. [CrossRef]
27. Blasenbauer, D.; Bogush, A.; Carvalho, T.; Cleall, P.; Cormio, C.; Guglietta, D.; Fellner, J.; Fernández-Alonso, M.; Heuss-Aßbichler, S.; Huber, F.; et al. Knowledge Base to Facilitate Anthropogenic Resource Assessment: Deliverable of COST Action Mining the European Anthroposphere. 2020. Available online: https://zenodo.org/record/3739164#.X6WHdFBo3b0 (accessed on 6 November 2020).
28. Dino, G.A.; Mehta, N.; Rossetti, P.; Ajmone-Marsan, F.; Luca, D.A.d. Sustainable approach towards extractive waste management: Two case studies from Italy: Two case studies from Italy. *Resour. Policy* **2018**, *59*, 33–43. [CrossRef]
29. European Commission (EC). *Assessment of Member States' Performance Regarding the Implementation of the Extractive Waste Directive; Appraisal of Implementation Gaps and Their Root Causes; Identification of Proposals to Improve the Implementation of the Directive*; European Commission: Brussels, Belgium, 2017; Available online: https://ec.europa.eu/environment/waste/studies/pdf/KH-01-17-904-EN-N.pdf (accessed on 6 November 2020).
30. Kuhn, K.; Meima, J.A. Characterization and Economic Potential of Historic Tailings from Gravity Separation: Implications from a Mine Waste Dump (Pb-Ag) in the Harz Mountains Mining District, Germany. *Minerals* **2019**, *9*, 303. [CrossRef]
31. Committee for Mineral Reserves International Reporting Standards (CRIRSCO). International Mineral Reporting Template for the Public Reporting of Exploration Targets, Exploration Results, Mineral Resources and Mineral Reserves. 2019. Available online: http://www.crirsco.com/templates/CRIRSCO_International_Reporting_Template_November_2019.pdf (accessed on 9 June 2020).
32. Attila Resources. Attila to Acquire the Century Zinc Mine. 2017. Available online: https://www.newcenturyresources.com/wp-content/uploads/2018/01/170301-AYA-Acquisition-of-Century-ASX-Ann.pdf (accessed on 14 August 2019).
33. Campbell, M.D.; Absolon, V.; King, J.; David, C.M. Precious Metal Resources of the Hellyer Mine Tailings. 2015. Available online: http://www.i2massociates.com/downloads/I2MHellyerTailingsResourcesMar9-2015Rev.pdf (accessed on 14 August 2019).
34. Cronwright, M.; Gasela, I.; Derbyshire, J. Kamativi Lithium Tailings Project. 2018. Available online: http://sectornewswire.com/NI43-101TechnicalReport-Kamativi-Li-Nov-2018.pdf (accessed on 14 August 2019).
35. Johansson, N.; Krook, J.; Eklund, M.; Berglund, B. An integrated review of concepts and initiatives for mining the technosphere: Towards a new taxonomy. *J. Clean. Prod.* **2013**, *55*, 35–44. [CrossRef]
36. Hartman, H.L.; Mutmansky, J.M. *Introductory Mining Engineering*, 2nd ed.; Wiley: Hoboken, NJ, USA, 2002; p. 570.
37. Laurence, D. Establishing a sustainable mining operation: An overview. *J. Clean. Prod.* **2011**, *19*, 278–284. [CrossRef]
38. Owen, J.R.; Kemp, D.; Lèbre, É.; Svobodova, K.; Pérez Murillo, G. Catastrophic tailings dam failures and disaster risk disclosure. *Int. J. Disaster Risk Reduct.* **2020**, *42*, 101361. [CrossRef]
39. Reid, C.; Bécaert, V.; Aubertin, M.; Rosenbaum, R.K.; Deschênes, L. Life cycle assessment of mine tailings management in Canada. *J. Clean. Prod.* **2009**, *17*, 471–479. [CrossRef]
40. Roche, C.; Thygesen, K.; Baker, E.E. *Mine Tailings Storage: Safety Is No Accident: A UNEP Rapid Response Assessment*; United Nations Environment Programme and GRID-Arendal: Arendal, Norway, 2017; ISBN 978-827-701-170-7. Available online: https://www.grida.no/publications/383 (accessed on 10 January 2020).

41. Worrall, R.; Neil, D.; Brereton, D.; Mulligan, D. Towards a sustainability criteria and indicators framework for legacy mine land. *J. Clean. Prod.* **2009**, *17*, 1426–1434. [CrossRef]
42. United Nations Economic Commission for Europe (UNECE). *United Nations Framework Classification for Resources—Update 2019*; UNECE: Geneva, Switzerland, 2020; p. 20. Available online: https://www.unece.org/fileadmin/DAM/energy/se/pdfs/UNFC/publ/UNFC_ES61_Update_2019.pdf (accessed on 13 November 2020).
43. United Nations Economic Commission for Europe (UNECE). Specifications for the Application of the United Nations Framework Classification for Resources to Anthropogenic Resources. 2018. Available online: https://www.unece.org/fileadmin/DAM/energy/se/pdfs/UNFC/Anthropogenic_Resources/UNFC_Anthropogenic_Resource_Specifications.pdf (accessed on 4 October 2019).
44. Heuss-Aßbichler, S.; Kral, U.; Løvik, A.; Mueller, S.; Simoni, M.; Stegemann, J.; Wäger, P.; Horváth, Z.; Winterstetter, A. Strategic Roadmap on Sustainable Management of Anthropogenic Resources. 2020. Available online: https://zenodo.org/record/3739269#.X6WBG1Bo3b1 (accessed on 6 November 2020).
45. Suppes, R.; Heuss-Aßbichler, S. Resource potential of mine wastes: A conventional and sustainable perspective on a case study tailings mining project. *J. Clean. Prod.* **2021**. [CrossRef]
46. Ilich, M. On the use of geological exploration data in mineral projects. *J. Environ. Geol.* **2018**, *2*, 61–63.
47. Revuelta, M.B. *Mineral Resources*; Springer International Publishing: Cham, Switzerland, 2018; p. 653. [CrossRef]
48. Winterstetter, A.; Laner, D.; Rechberger, H.; Fellner, J. Evaluation and classification of different types of anthropogenic resources: The cases of old landfills, obsolete computers and in-use wind turbines. *J. Clean. Prod.* **2016**, *133*, 599–615. [CrossRef]
49. Huber, F.; Fellner, J. Integration of life cycle assessment with monetary valuation for resource classification: The case of municipal solid waste incineration fly ash. *Resour. Conserv. Recycl.* **2018**, *139*, 17–26. [CrossRef]
50. Mueller, S.R.; Kral, U.; Wäger, P.A. Developing material recovery projects: Lessons learned from processing municipal solid waste incineration residues. *J. Clean. Prod.* **2020**, *259*, 120490. [CrossRef]
51. Mueller, S.R.; Wäger, P.A.; Widmer, R.; Williams, I.D. A geological reconnaissance of electrical and electronic waste as a source for rare earth metals. *Waste Manag.* **2015**, *45*, 226–234. [CrossRef] [PubMed]
52. Dino, G.A.; Rossetti, P.; Perotti, L.; Alberto, W.; Sarkka, H.; Coulon, F.; Wagland, S.; Griffiths, Z.; Rodeghiero, F. Landfill mining from extractive waste facilities: The importance of a correct site characterisation and evaluation of the potentialities. A case study from Italy. *Resour. Policy* **2018**, *59*, 50–61. [CrossRef]
53. Wates, J.; Goetz, A. Practical Considerations in the Hydro Re-Mining of Gold Tailings. In *Gold Ore Processing*; Adams, M.D., Ed.; Elsevier Science: San Diego, CA, USA, 2016; pp. 729–738.
54. Sözen, S.; Orhon, D.; Dinçer, H.; Ateşok, G.; Baştürkçü, H.; Yalçın, T.; Öznesil, H.; Karaca, C.; Allı, B.; Dulkadiroğlu, H.; et al. Resource recovery as a sustainable perspective for the remediation of mining wastes: Rehabilitation of the CMC mining waste site in Northern Cyprus. *Bull. Eng. Geol. Environ.* **2017**, *76*, 1535–1547. [CrossRef]
55. Candeias, C.; Melo, R.; Ávila, P.F.; da Silva, E.F.; Salgueiro, A.R.; Teixeira, J.P. Heavy metal pollution in mine–soil–plant system in S. Francisco de Assis—Panasqueira mine (Portugal). *Appl. Geochem.* **2014**, *44*, 12–26. [CrossRef]
56. Bleicher, A.; David, M.; Rutjes, H. When environmental legacy becomes a resource: On the making of secondary resources. *Geoforum* **2019**, *101*, 18–27. [CrossRef]
57. Poggendorf, C.; Rüpke, A.; Gock, E.; Saheli, H.; Kuhn, K.; Martin, T. Nutzung des Rohstoffpotentials von Bergbau- und Hüttenhalden am Beispiel des Westharzes. 2015, p. 22. Available online: https://www.researchgate.net/profile/Tina_Martin5/publication/303941732_Nutzung_des_Rohstoffpotentials_von_Bergbau-_und_Huttenhalden_am_Beispiel_des_Westharzes/links/575fbf8d08aed884621bbfa3/Nutzung-des-Rohstoffpotentials-von-Bergbau-und-Huettenhalden-am-Beispiel-des-Westharzes.pdf (accessed on 13 November 2020).
58. Statista. Duration of the Extraction Period of a Mine by Selected Commodities. 2013. Available online: https://www.statista.com/statistics/255479/mine-life-per-commodity/ (accessed on 12 September 2020).
59. European Commission (EC). Critical Raw Materials Resilience: Charting a Path towards Greater Security and Sustainability. 2020. Available online: https://eur-lex.europa.eu/legal-content/EN/TXT/PDF/?uri=CELEX:52020DC0474&from=EN (accessed on 21 December 2020).
60. World Bank. *Doing Business 2020: Comparing Business Regulation in 190 Economies*; World Bank: Washington, DC, USA, 2020; p. 135. Available online: http://documents1.worldbank.org/curated/en/688761571934946384/pdf/Doing-Business-2020-Comparing-Business-Regulation-in-190-Economies.pdf (accessed on 13 November 2020).
61. Roemer, F.; Binder, A.; Goldmann, D. Basic Considerations for the Reprocessing of Sulfidic Tailings Using the Example of the Bollrich Tailing Ponds. *World Metall.* **2018**, *71*, 1–9.
62. Federal Agency for Nature Conservation (BfN). Landscape Protection Areas. 2019. Available online: https://www.bfn.de/en/activities/protected-areas/landscape-protection-areas.html (accessed on 11 November 2020).
63. Azapagic, A. Developing a framework for sustainable development indicators for the mining and minerals industry. *J. Clean. Prod.* **2004**, *12*, 639–662. [CrossRef]
64. European Commission. Establishment of Guidelines for the Inspection of Mining Waste Facilities, Inventory and Rehabilitation of Abandoned Facilities and Review of the BREF Document: Annex 3—Supporting Document on Closure Methodologies for Closed and Abandoned Mining Waste Facilities. 2012. Available online: https://ec.europa.eu/environment/waste/mining/pdf/Annex3_closure_rehabilitation%20.pdf (accessed on 23 September 2020).

65. Valenta, R.K.; Kemp, D.; Owen, J.R.; Corder, G.D.; Lèbre, É. Re-thinking complex orebodies: Consequences for the future world supply of copper. *J. Clean. Prod.* **2019**, *220*, 816–826. [CrossRef]
66. Environmental Law Alliance Worldwide. Guidebook for Evaluating Mining Project EIAs. 2010. Available online: https://www.elaw.org/files/mining-eia-guidebook/Full-Guidebook.pdf (accessed on 16 August 2020).
67. Franks, D.M.; Davis, R.; Bebbington, A.J.; Ali, S.H.; Kemp, D.; Scurrah, M. Conflict translates environmental and social risk into business costs. *Proc. Natl. Acad. Sci. USA* **2014**, *111*, 7576–7581. [CrossRef]
68. Goldmann, D.; Zeller, T.; Niewisch, T.; Klesse, L.; Kammer, U.; Poggendorf, C.; Stöbich, J. *Recycling of Mine Processing Wastes for the Extraction of Economically Strategic Metals Using the Example of Tailings at the Bollrich in Goslar (REWITA): Final Report*; TU Clausthal: Clausthal-Zellerfeld, Germany, 2019; Available online: https://www.tib.eu/de/suchen/id/TIBKAT:1688127496/ (accessed on 22 July 2020). (In German)
69. Roemer, F. Investigations on the Processing of Deposited Flotation Residues at the Tailings Pond Bollrich with Special Consideration of the Extraction of Economically Strategic Raw Materials. Ph.D. Thesis, Clausthal University of Technology, Clausthal, Germany, 28 November 2019. (In German).
70. Woltemate, I. Assessment of the Geochemical and Sediment Petrographic Significance of Drilling Samples from Flotation Tailings in Two Tailing Ponds of the Rammelsberg Ore Mine (German). Ph.D. Thesis, University of Hanover, Hanover, Germany, 5 November 1987. (In German).
71. Google Earth Pro 7. Available online: https://www.google.com/earth/ (accessed on 20 December 2020).
72. Bertrand, G.; Cassard, D.; Arvanitidis, N.; Stanley, G. Map of Critical Raw Material Deposits in Europe. *Energy Procedia* **2016**, *97*, 44–50. [CrossRef]
73. Large, D.; Walcher, E. The Rammelsberg massive sulphide Cu-Zn-Pb-Ba-Deposit, Germany: An example of sediment-hosted, massive sulphide mineralisation. *Miner. Depos.* **1999**, *34*, 522–538. [CrossRef]
74. Eichhorn, P. *Ore Processing Rammelsberg—Origin, Operation, Comparison*; Förderverein Weltkulturerbe Rammelsberg Goslar/Harz e.V.: Goslar, Germany, 2012; Available online: https://docplayer.org/16359673-Erzaufbereitung-rammelsberg.html (accessed on 30 August 2020). (In German)
75. District of Goslar, Environmental Service. Map of Nature Conservation Areas. 2020. Available online: https://webgis.landkreis-goslar.de/MapSolution/apps/map/client/Umweltinformation?view=[Umweltinformation][true][3] (accessed on 15 August 2020). (In German).
76. Ackers, W.; Pechmann, S. *Integriertes Stadtentwicklungskonzept Goslar 2025*; GOSLAR Marketing Gmbh: Goslar, Germany, 2011; Available online: https://www.goslar.de/stadt-buerger/stadtentwicklung/isek-2025 (accessed on 28 July 2020). (In German)
77. Mohr, K. *Geology and Mineral Deposits of the Harz Mountains: With 37 Tables in Text and on 5 Folded Inserts and 2 Overview Tables on the Inside Pages of the Cover*, 2nd ed.; Schweizerbart: Stuttgart, Germany, 1993; p. 496. (In German)
78. Liessmann, W. *Historical Mining in the Harz Mountains*, 3rd ed.; Springer: Berlin/Heidelberg, Germany, 2010; p. 470, (In German). [CrossRef]
79. Climate-Data.org. Climate Goslar (Germany). Available online: https://de.climate-data.org/europa/deutschland/niedersachsen/goslar-22981/ (accessed on 23 August 2020).
80. Giardini, D.; Grünthal, G.; Shedlock, K.M.; Zhang, P. The GSHAP Global Seismic Hazard Map. In *International Handbook of Earthquake & Engineering Seismology*; Lee, W., Kanamori, H., Jennings, P., Kisslinger, C., Eds.; Academic Press: Amsterdam, The Netherlands, 2003; pp. 1233–1239.
81. Hofste, R.W.; Kuzma, S.; Walker, S.; Sutanudjaja, E.H.; Bierkens, M.F.P.; Kuijper, M.J.M.; Sanchez, M.F.; van Beek, R.; Wada, Y.; Rodriguez, S.G.; et al. *Aqueduct 3.0: Updated Decision-Relevant Global Water Risk Indicators*; World Resources Institute: Washington, DC, USA, 2019; Available online: https://www.wri.org/publication/aqueduct-30 (accessed on 11 June 2020).
82. Topographic-map.com. Topography of Germany. Available online: https://en-gb.topographic-map.com/maps/d93/Germany/ (accessed on 23 August 2020).
83. The Fund for Peace. Frage States Index. 2020. Available online: https://fragilestatesindex.org/ (accessed on 16 August 2020).
84. Wildlife Conservation Society (WCS) and Center for International Earth Science Information Network (CIESIN). *Last of the Wild Project, (LWP-2): Global Human Footprint Dataset (Geographic)*; NASA Socioeconomic Data and Applications Center (SEDAC): Palisades, NY, USA, 2005. [CrossRef]
85. Lèbre, É.; Corder, G. Integrating Industrial Ecology Thinking into the Management of Mining Waste. *Resources* **2015**, *4*, 765–786. [CrossRef]
86. European Union. *Raw Materials Scoreboard 2018: European Innovation Partnership on Raw Materials*; Publications Office of the European Union: Luxembourg, 2018. [CrossRef]

Article

How to Identify Potentials and Barriers of Raw Materials Recovery from Tailings? Part II: A Practical UNFC-Compliant Approach to Assess Project Sustainability with On-Site Exploration Data

Rudolf Suppes [1,2,*] and Soraya Heuss-Aßbichler [3]

1. Institute of Mineral Resources Engineering (MRE), RWTH Aachen University, Wüllnerstr. 2, 52064 Aachen, Germany
2. CBM GmbH—Gesellschaft für Consulting, Business und Management mbH, Horngasse 3, 52064 Aachen, Germany
3. Department of Earth and Environmental Sciences, Ludwig-Maximilians-Universität München, Theresienstr. 41, 80333 Munich, Germany; soraya@min.uni-muenchen.de
* Correspondence: rudolf.suppes@rwth-aachen.de or suppes@cbm-ac.de

Citation: Suppes, R.; Heuss-Aßbichler, S. How to Identify Potentials and Barriers of Raw Materials Recovery from Tailings? Part II: A Practical UNFC-Compliant Approach to Assess Project Sustainability with On-Site Exploration Data. *Resources* 2021, 10, 110. https://doi.org/10.3390/resources10110110

Academic Editors: Andrea Thorenz and Armin Reller

Received: 31 July 2021
Accepted: 15 October 2021
Published: 29 October 2021

Publisher's Note: MDPI stays neutral with regard to jurisdictional claims in published maps and institutional affiliations.

Copyright: © 2021 by the authors. Licensee MDPI, Basel, Switzerland. This article is an open access article distributed under the terms and conditions of the Creative Commons Attribution (CC BY) license (https://creativecommons.org/licenses/by/4.0/).

Abstract: A sustainable raw materials (RMs) recovery from waste requires a comprehensive generation and communication of knowledge on project potentials and barriers. However, a standardised procedure to capture sustainability aspects in early project development phases is currently missing. Thus, studies on different RM sources are not directly comparable. In this article, an approach is presented which guides its user through a practical interpretation of on-site exploration data on tailings compliant with the United Nations Framework Classification for Resources (UNFC). The development status of the overall project and the recovery of individual RMs are differentiated. To make the assessment results quickly comparable across different studies, they are summarised in a heat-map-like categorisation matrix. In Part I of this study, it is demonstrated with the case study tailings storage facility Bollrich (Germany) how a tailings mining project can be assessed by means of remote screening. In Part II, it is shown how to develop a project from first on-site exploration to a decision whether to intensify costly on-site exploration. It is concluded that with a UNFC-compliant assessment and classification approach, local sustainability aspects can be identified, and a commonly acceptable solution for different stakeholder perspectives can be derived.

Keywords: anthropogenic raw materials; sustainability assessment; tailings recycling

1. Introduction

A growing world population, the growth of emerging economies, and the global transition to a decarbonised energy supply lead to an increasing demand for mineral raw materials (RMs) [1–4]. For more than a century, the annual average increase in global mineral RM demand is reported to be 3% [1], and a 2- to 3-fold increased global demand for Al, Cu, Fe, Mn, Ni, Pb and Zn is expected between 2010 and 2050 [5,6]. Due to net stock additions and low recycling rates, the primary mining industry is expected to remain an important supplier of RMs in the foreseeable future [6,7].

In mining, valuable RMs are extracted from ores by separating wanted from unwanted minerals. A common method to do so is froth flotation, which requires the ores to be finely ground to a particle size of typically 10–200 μm [8]. The unwanted minerals are rejected as tailings, and they are usually stored in tailings storage facilities (TSFs). The global annual tailings production is estimated to lie in the range of 5–14 Gt [9], and it is estimated that in China alone some 12,000 TSFs exist [10]. Globally, ore grades are decreasing and ore complexities are increasing [11] so that the amount of produced tailings and energy spent per unit of produced commodity are increasing.

Despite continuous improvements in the construction and management of TSFs, they can be regarded as legacies with long-lasting environmental impacts, such as the occupation of large surface areas, and high external costs [12–16]. Risks associated with TSFs comprise the contamination of soil and water with acidic leachates or heavy metals, especially in the case of sulphidic tailings [13,17–19]. Other risks include dam stability issues which, on average, cause 2 to 3 annual TSF failures, leading to a contamination of large areas and threatening human lives [20,21]. The environmental impact of TSFs has increased public pressure on the primary mining industry to act more environmentally friendly [6,22,23].

At the same time, tailings contain usable RMs due to former processing inefficiencies or an emerging demand for RMs which were not exploitable in the past [24]. The active promotion of sustainability in RM sourcing in the past decade by institutions such as the European Commission (EC) has initiated a paradigm shift so that formerly regarded waste is now becoming interesting for valorisation [25–27]. Scientists have investigated the recovery of metalliferous or industrial minerals from tailings [28–30], or an alternative valorisation, e.g., in construction materials [31–33] or glass making [34–36].

A comprehensive exploration is required to identify if tailings can be valorised. However, conventional case studies under consideration of the Committee for Mineral Reserves International Reporting Standards (CRIRSCO) classification principles from the primary mining industry usually target single RMs and neglect other contained RMs (cf., References [37–39]). Hence, the knowledge on their RM potential is incomplete. Usually, economic aspects are mainly considered in the primary mining industry [8,40], while environmental and social aspects of RMs recovery are mostly neglected or ignored; only recently have sustainability aspects been given greater attention [41].

The United Nations Sustainable Development Goals aim at a worldwide sustainable extraction of natural RMs [42]. Therefore, the prospects of mineral RMs recovery requires environmental and social aspects to be regarded as equal to economic ones. As a result, these aspects must be assessed concurrently with geological, technological, and legal aspects to obtain comprehensive exploration results [43]. This is possible when applying the United Nations Framework Classification for Resources (UNFC) principles, which are based on the 3 categories: *degree of confidence in the estimates* (G category), *technical feasibility* (F category), and *environmental-socio-economic viability* (E category) [44]. In this way, decision-makers in RM management can get an overview of the potentials and barriers of mineral RMs recovery from tailings and its competitiveness across different RM sources.

In mineral RM exploration in the primary mining industry, a mineral deposit is first identified with remote techniques [8,45]. It is then investigated on site with intensified techniques to obtain data for a first techno-economic assessment, termed a scoping study [8,45]. Despite the many recent case studies on anthropogenic RMs developed in analogy to natural RMs [46], a standardised procedure is missing. Existing case studies provide a snapshot of a specific stage of project development in the RMs recovery chain [47], e.g., the remote exploration [48]. Hence, there is a research gap in the development of case studies which outline the progression of RMs recovery project development [47].

This study addresses the lack of a standardised procedure to explore tailings as anthropogenic RMs. It is the first to demonstrate how a UNFC-compliant tailings mining project assessment and classification can evolve from a first remote TSF screening (Part I [43]) to a consecutive interpretation of on-site exploration data (Part II). In this article, a systematic and practical UNFC-compliant approach is developed for a very preliminary assessment and classification of tailings mining projects based on on-site exploration data. It is tested to what extent an overview of project potentials and barriers can be obtained. The research questions are: (1) is it possible to reconcile different stakeholder interests with a UNFC-compliant approach or must different perspectives be considered on their own merits? (2) which aspects should be considered in very preliminary UNFC-compliant assessments? (3) can a UNFC-compliant approach be used to identify site-specific project potentials and barriers?

The approach focuses on metalliferous tailings from industrial processes. A project's development status is differentiated in terms of geological, technological, economic, environmental, social, and legal aspects. Beside the rating of the overall project, each contained RM is rated individually as a separate subproject. The rating is performed in a categorisation matrix in a heat map-like style. In this way, driving factors as well as barriers can be identified quickly. The approach is tested with the case study TSF Bollrich (Germany) from a public decision-maker's perspective, considering the interests of local environmental non-governmental organisations (NGOs), private investors, and the city administration of Goslar. The TSF was chosen since it is a potential source of economically highly relevant RMs, it is situated in a complex environment with several stakeholders, and there is a potential to relieve the burden on the environment and society [43].

The article is structured as follows: (i) outline of the frame conditions for the further development of the case study Bollrich, (ii) proposal of a UNFC-compliant anthropogenic RMs assessment and classification approach, (iii) development of a categorisation matrix for a UNFC-compliant rating of the overall project and subprojects for individual RMs, (iv) case study application, and (v) discussion of the developed approach.

2. Terms and Methods

2.1. Key Words and Definitions

TSF: physical structure for tailings storage. *Deposit*: potential RM source. *Target minerals:* minerals wanted for valorisation. *Other minerals*: unwanted minerals. *Recovery:* physical extraction process. *Material recovery:* extraction of minerals to be used in construction materials. *Tailings mining:* process from exploration, recovery, and processing to rehabilitation. A *very preliminary study* is regarded as an analogue to a scoping study from the primary mining industry [45] (p. 31), and it is defined as follows: *it is the first quantification of a tailings mining project's potentials and barriers with respect to geological, technological, economic, environmental, social, and legal aspects. The degree of uncertainty in the estimates is high. The study is based on directly generated project data, for instance from on-site exploration or information from other sources such as from the literature and model assumptions based on similar projects. Technological considerations are based on conceptual foundations.*

2.2. Considerations for the Development of the Case Study TSF Bollrich

This case study is based on the screening results from Reference [43], where the following potentials are identified: an economic interest in the TSF is justified due to its size and the presumably contained critical raw materials (CRMs) $BaSO_4$ and In, as well as the highly economically relevant RMs Ag, Au, Cu, Pb, and Zn. The development costs are expected to be low since buildings, transportation, and utilities infrastructure are present in the near vicinity. As Germany has a high rating on the ease of doing business ranking, favourable regulatory conditions for an investment can be assumed. The TSF's environment is vulnerable to a potential TSF failure: the nearest human settlement is located ~400 m downstream of the TSF, and the high score on the Human Footprint Index indicates that land-use-related social tension with competing interests can be expected in the area. Therefore, a removal of the TSF would reduce the potentially severe risks of a TSF failure.

The following barriers are identified [43]: the TSF is located in a challenging environment with a potential for social conflicts due to agricultural, forest, industrial and commercial, nature and water protection, recreation, and residential areas in the near vicinity. A diverse and socially active stakeholder group of a minimum of 18 parties could be identified, which may potentially form a strong base for a project rejection. Amongst others, these include environmental NGOs, the Development Association Cultural Heritage Ore Mine Rammelsberg, and the Air Sports Community Goslar. The geological knowledge on the deposit is limited due to unknown RM quantities and qualities. Furthermore, potentially contained RMs are presumed based on literature on mined ores and their processing. Knowledge on the TSF's geomechanical stability is missing. Valuable ecosystems with protected species have formed as a result of ecological succession. To overcome these barriers, on-site exploration and evaluating techno-economic feasibility is required; local stakeholders' environmental, social, and economic interests must be considered; and advantages and disadvantages of RMs recovery need to be weighed against each other.

2.3. UNFC-Compliant Anthropogenic Raw Materials Assessment and Classification Approach

The assessment and classification approach from Heuss-Aßbichler et al. [47] (p. 17) was adopted and modified by adding sub-steps and assigning assessment methods. The modified approach consists of 3 phases (cf., Figure 1), which can be reiterated when additional information is required or when new information on preceding steps is generated:

1. Definition of project and generation of information.
2. Assessment of project's development status.
3. UNFC-compliant categorisation of criteria and project classification.

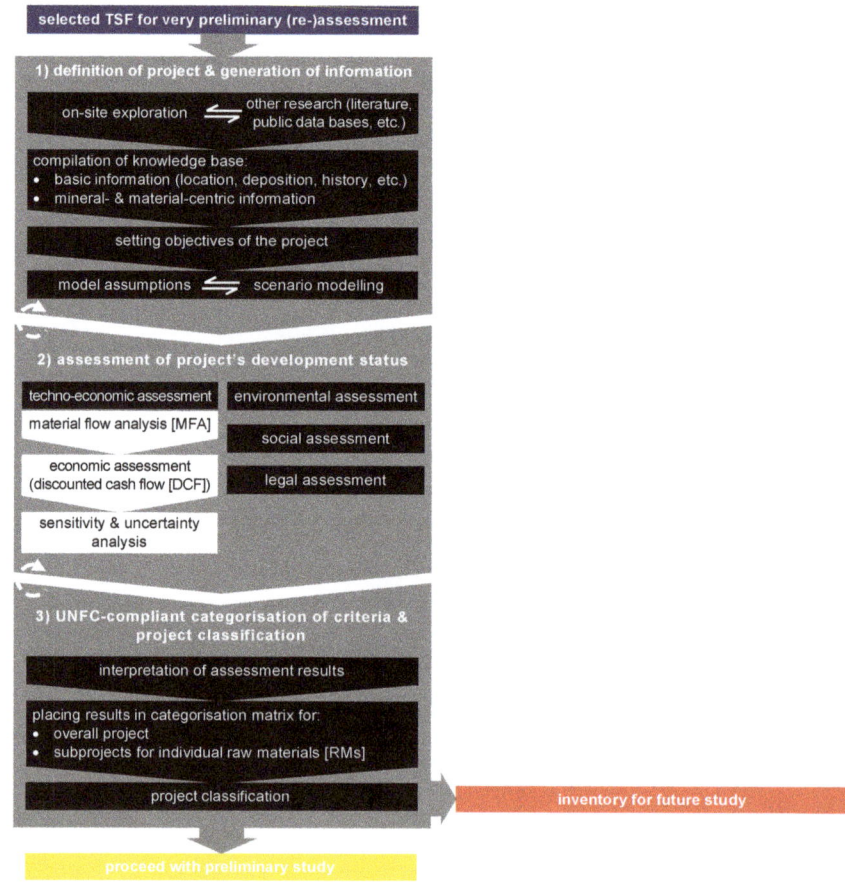

Figure 1. Practical UNFC-compliant approach for a systematic assessment and classification of mineral RMs recovery from tailings at very preliminary level. The leftwards arrow over rightwards arrow indicates mutual influence, and the dotted circles indicate possible reiteration steps.

2.4. Case Study Assessment Methods

2.4.1. Environmental Assessment

TSF-related risks can have a great influence on the classification result of a tailings mining project [49]. Based on data from scientific literature, publicly accessible sources, and observations on Google Earth [50], a status quo risk assessment is performed. The TSF's stability and its impacts on the surrounding environment is assessed, including the following subjects of protection (adopted from Reference [51]): air, flora and fauna, ground, groundwater, human health, landscape, and surface water.

2.4.2. Social Assessment

Investors are recognising that ignoring social aspects in project development can create barriers to RMs recovery [6]. Amongst others, it is therefore important to consider the attitudes of local stakeholders such as communities towards a possible RMs recovery. From the stakeholders identified in Reference [43], this study focused on administrative bodies, industry, and local environmental NGOs as proxies for concerned citizens. Due to a lack of data, only basic tendencies on stakeholder attitudes are assessed. The assessment is based on an internet search and the study of Bleicher et al. [52] who interviewed stakeholders on a potential RMs recovery from mine waste in the Harz region including the TSF Bollrich. They focused on stakeholders from non-specified local and regional environmental NGOs, industry, administrative bodies, and scientific institutions, and they considered secondary sources such as public media.

2.4.3. Material Characterisation and Material Flow Analysis

The drill core sampling campaigns on the TSF Bollrich for tailings characterisation are described in References [53,54]. 3 scenarios are developed: no RMs recovery (NRR0), conventional RMs recovery (CRR1), and enhanced RMs recovery (ERR2). The amount and composition of generated commodities and residues are evaluated with a material flow analysis (MFA) according to Reference [55] under consideration of available recovery technologies:

1. Scenario definition and selection of relevant processes and mass flows.
2. Mass flow quantification with published and estimated data, and model assumptions for unavailable data.
3. Mass flow visualisation with Sankey diagrams.

2.4.4. Economic Assessment

The economic viability is assessed with a discounted cash flow (DCF) analysis to determine the net present value (NPV) before taxes, considering internal costs and revenues. The NPV is estimated with the open-source software R (www.r-project.org, accessed on 16 January 2021) after

$$NPV = -I_0 + \sum_{i=1}^{t} (I_i/(1+r)^r), \quad (1)$$

where I_0 is the initial investment [€] in year 0, I_i is the net cash flow [€] in the i-th year, r is the discount rate [-], and t is the project's duration [a]. Given estimated figures for target mineral masses, prices and recovery rates are rounded down; they are rounded up for costs to estimate conservatively as per CRIRSCO [45].

2.4.5. Sensitivity and Uncertainty Analysis

To increase the reliability of the assessment, sensitivity and uncertainty analyses is performed [56]. The sensitivity analysis is performed by varying input factors to determine how the outputs depend on them. The uncertainties are assessed with dynamic price forecasts by applying autoregressive functions to historical price data of metals, minerals, diesel, and electric energy (cf., Supplementary Materials, Figures S1–S9).

2.4.6. Legal Assessment

The legal aspects right of mining, environmental protection, and water protection are considered. Due to a lack of data, the state of development of legal aspects are assessed by making basic considerations based on data from Reference [53].

2.5. Development of a Categorisation Matrix for a UNFC-Compliant Project Rating

In the categorisation matrix, the overall project and subprojects for individual RMs are differentiated. The UNFC's G, F, and E categories are addressed. The E category is subdivided into economic (a), environmental (b), social (c), and legal (d) aspects, the latter being defined as a distinct subcategory in this article. For the project categorisation and classification, an exemplary 35 factors for the rating of the overall project and 9 factors for the rating of the subprojects for individual RMs are assessed. They are adapted and modified after a literature search on established assessment factors from the primary mining industry, literature on sustainability in mining, case studies, and our own reasoning. Table 1 provides an overview of the chosen factors, their allocation to groups, and the rationale for choosing them based on their influence on a project. A proposal is made for a UNFC-compliant rating with *descriptive indicators* to describe a state and *performance indicators* to quantitatively compare the status quo with target values. For better legibility, the categorisation matrix is divided into separate tables (cf., Appendix A, Tables A1–A10). With the above nomenclature, an exemplary rating in the social subcategory might look like E3.1c or E1c. Factors with high uncertainty remain in the 3rd UNFC subcategorisation (3.1, 3.2, 3.3), while more developed factors can be rated as high as in the 1st UNFC category (1, 2, 3). For a quick overview of project potentials and barriers, an individual colour is assigned to each rating. In the discussion in Section 4.1, the rating results are presented in a heat-map-like style for a quick overview.

Table 1. Categorisation matrix: assessed factors and rationale behind their application based on their influence on a project.

Category & Factor	Influence on	UNFC Axis [1]
overall project rating		
geological conditions (relevant for project development)		
(1) quantity, (2) quality, (3) homogeneity	potential profitability, mine planning, overall uncertainty	G
TSF condition & risks (relevant for project development)		
(4) ordnance	exploration costs, overall project safety	F
mine planning considerations (relevant for project execution)		
(5) mine/operational design, (6) metallurgical testwork, (7) water consumption	reliability of the financial analysis, efficiency of the operation, environmental footprint	F
infrastructure (relevant for project development)		
(8) real estate, (9) mining & processing, (10) utilities, (11) transportation & access	project viability, ramp-up time	F
post-mining state (relevant for future impacts)		
(12) residue storage safety, (13) rehabilitation	necessary aftercare measures, public acceptance	F
microeconomic aspects (relevant for project development)		
(14) economic viability, (15) economic uncertainty	potential returns, investor interest	E a
financial aspects (relevant for project development)		
(16) investment conditions, (17) financial support	potential returns, investor interest, security of investment	E a
environmental impacts during project execution		
(18) air emission, (19) liquid effluent emission, (20) noise emission	mine planning, local population, local ecosystems	E b
environmental impacts after project execution		
(21) biodiversity	quality of ecosystem after the project	E b
(22) land use	land which can be repurposed	
(23) material reactivity	aftercare measures, local ecosystems	
social impacts during project execution		
(24) local community, (25) health & safety, (26) human rights & business ethics	social acceptance, peace & wellbeing, (unforeseeable) costs for compensation	E c

Table 1. Cont.

Category & Factor	Influence on	UNFC Axis [1]
social impacts due to project execution (27) wealth distribution, (28) investment in local human capital (29) degree of RM recovery, (30) RM valorisation	social peace & wellbeing, employment of local population, valuable legacy for workers & society after mine closure amount of new residues, ecological risks, effort for & efficiency of future RMs recovery	E c
social impacts after project execution (31) aftercare, (32) landscape	social risks, social wellbeing, external costs	E c
legal situation (relevant for project development) (33) right of mining, (34) environmental protection, (35) water protection	project feasibility, social acceptance, effort for formal project planning	E d
subproject for individual RMs rating		
geological conditions (relevant for project development) (36) quantity, (37) quality, (38) homogeneity	potential profitability, mine planning, RM uncertainty	G
mine planning considerations (relevant for project execution) (39) recoverability	efficiency of the operation, amount of new residues	F
microeconomic aspects (relevant for project development) (40) demand, (41) RM criticality, (42) price development	project viability, investor interest, overall project risk	E a
impacts after project execution (43) solid matter, (44) eluate	environmental risks of new deposition, aftercare measures	E b

[1] a: economic aspects, b: environmental aspects, c: social aspects, d: legal aspects.

3. Results

3.1. Definition of the Project and Generation of Information

3.1.1. Knowledge Base on the Case Study Deposit

The tailings deposit Bollrich (cf., Figure 2) near Goslar was part of the Rammelsberg mining operation [57]. It contains $BaSO_4$, Co, Ga, and In, which are CRMs in the European Union (EU), and the elements Cu, Pb, and Zn, which are economically highly important in the EU [58]. The deposit is nationally relevant as it is one of the few possible CRM sources [59]. The first exploration with a focus on geological aspects took place in 1983 before its abandonment in 1988 after ca. 50 years of operation [54]. In the 2010s, the exploration's main focus was on mineral processing. Geological, technological, environmental, legal, [53] and social aspects [52] were also investigated. A comprehensive assessment of a potential tailings mining project has not been carried out.

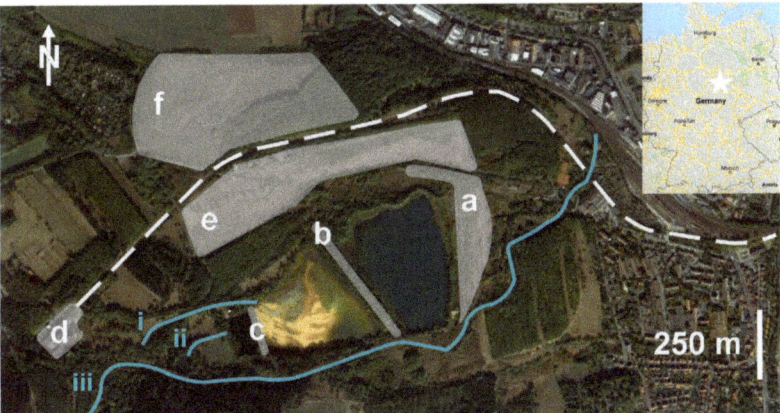

Figure 2. Schematic illustration of the TSF Bollrich's near environment: (a) marks the main dam, (b) the middle dam, (c) the water retention dam, (d) the disused processing plant, (e) a glider airfield, and (f) the disused landfill Paradiesgrund. The neutralisation sludge between the dams (b, c) is yellowish. The white dotted line marks the disused railway connection from Oker to the processing plant, (i) the stream of neutralised mine water, (ii) the connection between the pond Gelmketeich and the water retention pond, and (iii) the river Gelmke. Adapted after Google Earth [50].

In this study, the deposit in its current condition is assessed and classified from a sustainability viewpoint, considering the area around the TSF within a radius of 10 km. Information was derived from the existing scientific studies on the deposit in References [52–54,60] and from publicly available data sources. The knowledge base on the deposit is summarised in Table A11. The material flows and economics are evaluated quantitatively based on published data and model assumptions for unavailable data (cf., Table 2).

Table 2. Summary of model assumptions for the case study TSF Bollrich.

Model Assumption
(1) for in-situ rehabilitation, TSF abandonment is performed as for DK II class landfills [1] under the German Landfill Regulation (DepV) [61].
(2) mass of dam material is neglected in mineral RMs recovery scenarios alongside its further treatment.
(3) freight costs for commodities & residues to downstream processes are neglected.
(4) all equipment can be used over the whole life of mine (LOM) without renewal except for the pipelines & pumps, which are exchanged in year 6 of the mining operation due to abrasive wear.
(5) processing plant Bollrich: assets can be used (for operation, administration, etc.), processing machinery can be reactivated, & the $BaSO_4$ concentrate can be conditioned on site; basic infrastructure is in place.
(6) experimental tailings recovery rates from lower pond applicable to tailings from upper pond, neglecting the influence of neutralisation sludge on processing.
(7) no losses & dilution of tailings occur during mining & transport.
(8) the processing plant produces 3 types of products: (i) a pure industrial mineral concentrate ($BaSO_4$), (ii) a mixed sulphide concentrate ($CuFeS_2$, PbS, ZnS) including all high-technology metals (Co, Ga, In), & (iii) mixed residues due to inefficiencies in mineral processing.
(9) smelters pay for the recoverable Co, Ga, & In content in the mixed sulphide concentrate based on a recovery with ammonia leaching as specified in Reference [60].
(10) a discount rate of 15% is chosen to reflect a high risk investment [8].

[1] Above-ground landfill for contaminated but non-hazardous waste such as pre-treated domestic waste or commercial mineral waste. Geological base and surface sealing is required.

3.1.2. Setting Objectives of the Project

Based on current research, the TSF Bollrich offers the potential for action by a public decision-maker at national level seeking a sustainable solution at reasonable costs. Based on the stakeholder considerations (cf., Section 3.2.2), 3 relevant stakeholder perspectives are considered: NGOs with environmental concerns due to TSF-related risks, private investors seeking economic opportunities, and the city administration of Goslar seeking an opportunity to create high-value jobs and to establish a regional recycling industry.

The selected scenarios' objectives are: no RMs recovery (NRR0)—a physically and chemically stable, maintenance-free structure is created. Environmental and social risks are minimised by preventing the release of contaminants due to recovery and by avoiding the transport of hazardous material in a vulnerable region. The environment is rehabilitated, and the current landform is retained. RMs recovery (CRR1)—application of conventional technologies with off-site residue disposal. The original landform is restored, and the area is rehabilitated. RMs recovery (ERR2)—the same processes as in CRR1 but the produced residues are sold to a local recycling company.

3.1.3. Scenario Modelling

In the rehabilitation scenario (NRR0), a leachate collection system is installed, the TSF is stabilised by in-situ concrete injection, its surface is sealed, and leachates are captured and treated on site in a 5-year closure phase. In a 30-year aftercare phase, emissions and the

TSF's stability are monitored. Reference data is used for the techno-economic assessment (cf., Tables A12 and A13). No historical data is available for a price forecast.

Figure 3 outlines the general project for CRR1 and ERR2 from a material flow perspective. Geotechnical and mine planning considerations are conceptual. The low mineral content estimated in Reference [53] is adopted to estimate conservatively (cf., Table A11). A homogeneous deposit is assumed. The tailings are mined in a dredging operation (cf., Figure S10) and processed on site in the existing processing plant at a constant rate over a 10-year period, followed by a 1-year rehabilitation period. The products leave the system boundaries at the mineral processing plant's outlet where the reference point is set. The target minerals are extracted with a multi-stage froth flotation as specified by Roemer [60] (cf., Table A16) based on a sampling campaign on the lower pond [53]. A pure industrial mineral concentrate (BaSO$_4$), a mixed sulphide concentrate containing base metals (Cu, Pb, Zn) and high-technology metals (Co, Ga, In), and mixed residues are produced. Tailings, commodity, and residue masses are estimated as dry matter.

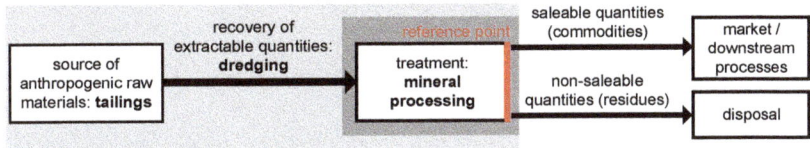

Figure 3. Tailings mining project Bollrich for the mineral RMs recovery scenarios (CRR1, ERR2) from a material flow perspective. The light grey and dark grey shaded fields illustrate the spatial and mineral processing system boundaries, respectively.

The database with fixed and variable parameters for the techno-economic assessment is given in Tables A14–A16. Energy flows are considered for tailings recovery and processing. Initial and intermediate investment costs for mining and processing equipment, and infrastructure, are included in the capital expenditure (CAPEX). Variable costs for mining, processing, electric and mechanical maintenance, administration, and general services are included in the operating expenditure (OPEX). Revenues are realised immediately. In ERR2, the mixed residues are sold to a recycling company for an application in construction materials. Mine site preparation costs are estimated to be low due to the simple mine plan, good mine site accessibility by road, and the availability of buildings for the processing plant and the operation's administration. Mine site rehabilitation costs such as for revegetation and environmental monitoring are considered. Assets and machinery are liquidated at the operation's end at a residual value of 10%.

Certain relevant aspects are out of the scope of this study: costs for preventing emissions during development, mining, transport and processing, for renewing the railway access, for removing roads and railway at mine closure, for treating and disposing of water from mining and processing, and downstream processing.

The uncertainty analysis comprises 3 price forecasts: pessimistic (p), mean (m), and optimistic (o), after which the respective scenarios are named (CRR1p, CRR1m, etc.). The pessimistic and optimistic forecasts refer to the lower and upper limits of the 95% confidence interval, respectively. CuFeS$_2$, PbS, and ZnS concentrate prices are estimated [62]. Prices for selling and costs for disposing of residues are fixed due to a lack of data. The mean price forecast (m), representing the most realistic case, is focussed. Material flow uncertainties are neglected as the dependence on price and cost variations is focussed.

3.2. Case Study Assessment

3.2.1. Environmental Assessment: Status Quo Risks

The area around the TSF is contaminated with heavy metals such as As, Cd, and Pb, which partially exceed the concentration threshold values for soil in parks and recreational areas in Germany [63,64]. However, the source of pollution could also be the former transport of ores via the Bollrich area to smelters in Oker [65]. Hence, the TSF's contribution to the pollution is unknown.

No data is available on the TSF's impact on human health, local flora and fauna, and surface and groundwater as there currently is no monitoring in place [53]. Dust emissions from the TSF can be excluded due to the wet tailings storage. The neutralisation sludge is unlikely to emit dust as it hardens when being exposed to air [54]. Heavy-metal-laden seepage is collected at the foot of the dam and returned into the TSF [53]. However, the unsealed TSF base constitutes a risk for the release of contaminants [53]. A general safety concern is that the TSF is freely accessible (observed on Google Earth [50]), and there are several trails around the TSF (https://regio.outdooractive.com/oar-goslar/de/touren/#filter=r-fullyTranslatedLangus-,sb-sortedBy-0&zc=15,10.46323,51.90085, accessed on 16 January 2021). Hence, people who are not familiar with the area may come in direct contact with the TSF.

The main dam's stability in its current state and in the case of extreme rainfalls could be confirmed by conservative calculations [66]. However, 2 sinkholes in karstified zones in near vicinity to the TSF were reported [53]. The knowledge on the karstified zones is limited [53] so that the long-term risk for the TSF's stability is currently unknown.

3.2.2. Social Assessment: Stakeholder Considerations

The Harz region has an ore mining history ranging from the Middle Ages to the 1980s [52]. Today, the region is facing the challenges of demographic change, young people's emigration, a weak economy, and environmental burdens from former mining [52,65]. A particularity is the Goslar community's and city administration's strong awareness of the region's mining history, which is regarded as a cultural heritage and an important factor for tourism [52,65]. This can be observed in public social media such as the Goslar Tales forum: the category *Mines and Smelters* has 70 topics from 2011 to 2019 with 925 contributions (http://www.goslarer-geschichten.de/forum.php, accessed on 26 September 2020). The TSF's history, basic knowledge, opinions, and safety concerns on water quality are discussed, and photos and videos are shared.

The results of Bleicher et al. [52] are summarised: generally, RMs recovery from mine waste is regarded as a development opportunity for the Harz region, and the trust in scientists and the industry is shared by public media. Scientific institutions and the industry are identified as the current regional drivers of CRMs recovery from mine waste. All interviewed stakeholders were in favour of developing knowledge and technologies for mine waste valorisation, with the exception of minor criticism from an environmental activist about the presumption of scientists that good ideas are approved by everyone. However, environmental NGOs see RMs recovery from mine waste as an opportunity to at least partially rehabilitate the environment. The city's administration is interested in RMs recovery from mine waste since the establishment of a recycling industry might attract highly skilled workers, and the possible knowledge transfer with scientific institutions and the opportunity to test novel technologies is seen as one of the region's strengths.

3.2.3. Techno-Economic Assessment: Material Flow Analysis

No material flow takes place in NRR0 due to in-situ stabilisation. Figure 4 depicts the specific material flows for the RMs recovery scenarios (CRR1, ERR2) (cf., Figure A1 for a detailed production breakdown). Over a 10-year period, 7.1 million t of tailings are mined and processed. In CRR1, 2.7 million t of commodities (i.e., 38 wt% of total tailings), and 4.4 million t of mixed mineral residues are produced. The commodities consist of an industrial mineral and a mixed sulphide concentrate. In ERR2, all tailings are valorised. The commodities (CRR1, ERR2) leave the system boundaries for off-site conditioning.

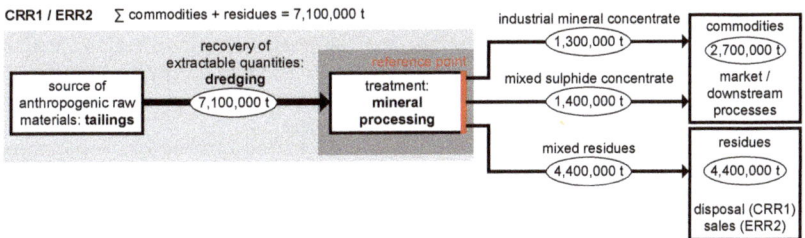

Figure 4. Material flow systems and 5-year material flows for the mineral RMs recovery scenarios (CRR1, ERR2). The light grey and dark grey shaded fields illustrate the spatial and mineral processing system boundaries, respectively. All figures were rounded to the sixth digit.

3.2.4. Techno-Economic Assessment: Discounted Cash Flow Analysis

Table 3 summarises the results of the DCF analysis (cf., Figures S15–S17). Generally, mineral RMs recovery is economically viable (CRR1m, ERR2m) under the project's current state of assessment. The DCF analysis yields positive NPVs in ERR2 regardless of the price forecast. The NPV in CRR1 becomes negative in the pessimistic forecast (CRR1p). The NPVs of NRR0, CRR1m, and ERR2m are EUR −124.5 million, EUR 73.9 million, and EUR 172.5 million, respectively. 98% of all costs in the rehabilitation scenario (NRR0) are attributed to the 5-year closure and leachate phase. In the mineral RMs recovery scenarios (CRR1m, ERR2m), the largest share of revenues is attributed to $BaSO_4$ with a 49% and 47% contribution, respectively, and a share of the total commodity masses of 64.4 wt% and 24.5 wt%, respectively. The second highest revenues are attributed to Zn with a contribution of 27% and 25%, respectively, and a ZnS share of the total commodity masses of 5.5 wt% and 2.1 wt%, respectively. The high-technology metals Co, Ga, and In contribute least to the revenues from RMs sales with a combined share of ca. 2% of total revenues and a combined share of total commodity mass of 0.6% and 0.02%, respectively.

Table 3. Results of the DCF analysis. The rehabilitation scenario (NRR0) has a project duration of 35 years. The RMs recovery scenarios (CRR1, ERR2) has a project duration of 11 years. The left column shows cost and revenue factors of the NPVs. Figures are given in millions of EUR.

	Scenarios [1]						
	NRR0	CRR1p	ERR2p	CRR1m	ERR2m	CRR1o	ERR2o
NPV Factor							
total NPV	−124.6	−16.6	82.0	73.9	172.5	164.4	263.1
costs							
CAPEX	-	−14.6	−14.6	−14.6	−14.6	−14.6	−14.6
OPEX	-	−29.1	−29.1	−29.1	−29.1	−29.1	−29.1
diesel	-	−3.4	−3.4	−5.1	−5.1	−6.9	−6.9
electric energy	-	−1.2	−1.2	−1.2	−1.2	−1.2	−1.2
residue disposal	-	−87.7	-	−87.7	-	−87.7	-
rehabilitation	-	−4.0	−4.0	−4.0	−4.0	−4.0	−4.0
closure & leachate phase	−122.0	-	-	-	-	-	-
aftercare phase	−2.6	-	-	-	-	-	-
revenues							
$BaSO_4$	-	92.1	92.1	106.2	106.2	120.4	120.4
Cu	-	9.4	9.4	14.9	14.9	20.3	20.3
Pb	-	14.1	14.1	30.5	30.5	47.0	47.0
Zn	-	6.1	6.1	58.2	58.2	110.2	110.2
Co	-	0.7	0.7	2.6	2.6	4.6	4.6
Ga	-	0.3	0.3	0.7	0.7	1.0	1.0
In	-	0.2	0.2	2.1	2.1	4.1	4.1
asset liquidation	-	0.1	0.1	0.1	0.1	0.1	0.1
residue sales	-	-	11.0	-	10.9	-	10.9

[1] p: pessimistic price forecast (lower limit of 95% confidence interval), m: mean price forecast, o: optimistic price forecast (upper limit of 95% confidence interval).

Residue disposal is the highest cost factor in CRR1m with a share of 62% of total costs. The OPEX is the second highest cost factor in CRR1m and the highest in ERR2m with a share of total costs of 21% and 58%, respectively. In both scenarios, the smallest cost factor is electric energy consumption with a share of 0.8% and 2.4%, respectively.

3.2.5. Techno-Economic Assessment: Sensitivity and Uncertainty Analysis

The NPV is most sensitive to $BaSO_4$ price variations (cf., Figures A2 and A3). In CRR1m and ERR2m, a decreased $BaSO_4$ price by 69% and 100% yields an NPV decrease of 100% and 62%, respectively. In CRR1m, decreased Pb and Zn prices by 100% yields an NPV decrease of 42% and 79%, respectively. In ERR2m, a decreased Zn price by 100% yields an NPV decrease of 34%. The NPV is relatively insensitive to other price variations.

Residue disposal was the most influential cost factor in CRR1m, with a price increase of 84% yielding an NPV of zero. CAPEX and OPEX increases of 504% and 253% (CRR1m), respectively, and 1178% and 592% (ERR2m), respectively, yields NPVs of zero.

3.2.6. Legal Assessment: Basic Considerations

The legal aspects for a possible project execution have not been considered so far. The TSF is still monitored under Mining Law (State Office for Mining Energy and Geology (LBEG), personal communication, 16 September 2020). As for the right of mining, it needs to be assessed if the mining or waste legislation applies [67]. Goldmann et al. [53] rate the legal aspects for environmental protection as follows: strict legal restrictions and high efforts to achieve legal consent are expected since heterogeneous and high-quality flora and fauna ecosystems were identified during preliminary on-site inspections. It is likely that an environmental impact study and a concept to protect the ecosystems and/or to remediate impacts upfront are necessary. Potential impacts on the surrounding protected natural areas and landscapes need to be assessed. As for water protection, potential impacts on the river Gelmke in near vicinity (cf., Figure 2) and the nearby Ammentalbach need to be assessed. Potential impacts on groundwater are unclarified.

4. Discussion

4.1. Interpretation of the Case Study Results

The rating results are summarised in the categorisation matrix in Tables 4 and 5. The justification for the rating is given in Tables A17–A26. As no RMs are recovered in the rehabilitation scenario (NRR0), only the overall project is rated. The lowest rating in a category is chosen for the rating of the overall category (cf., Reference [68] (p. 37)).

For NRR0, the categorisation matrix shows that the knowledge on the TSF's geology has medium confidence (G2). The rehabilitation scenario's state of technological development has a low overall rating (F3) due to the uncertainty regarding possible ordnance, the conceptual operational design, the unclarified usability of TSF water, and the unclarified long-term storage safety. The infrastructural conditions (F1–F2) and rehabilitation planning (F2) are rated high. As only costs are incurred and as there currently is no knowledge on a potential financial support, the economics are rated low (E3.3a). As for the environmental aspects, the unclarified potential dust emission and in-situ cementation of reactive material lead to a low rating (E3.3b). As for the social aspects, only the retained landscape is rated positively (E2c). The legal aspects are generally underdeveloped (E3.3d).

In CRR1m and ERR2m, the project can be expected to be economically viable (E3.1a). However, the NPV in the pessimistic forecast for CRR1 is negative. ERR2 is more resilient in this respect due to the sales of the new residues. The favourable economics of ERR2 are highlighted in the overall category rating (E.3.1a) as opposed to CRR1 (E3.3a) due to the higher uncertainty in the pessimistic price forecast. The driving revenue factor is the $BaSO_4$ sales due to its relatively high grade (24.5 wt%), its high price compared to the other commodities, its high recovery rate (74%), and the forecasted price increase. The $BaSO_4$ price is relatively stable, with the largest price drop being ca. 17% in the past 20 years (cf., Figure S3). CRR1m is relatively insensitive to $BaSO_4$ price variations with the NPV becoming negative at a decreased $BaSO_4$ price by 69%. ERR2 is more resilient with a $BaSO_4$ price drop to EUR 0, leading to a decreased NPV of 38%. In general, the presence of real estate, transportation, and utilities infrastructure reduces the mine development costs.

Residue disposal is the greatest cost factor in CRR1 with 64% of all costs, and it is the greatest economic risk with a price increase of 93% leading to a negative NPV. A price increase is possible if a further conditioning is necessary to meet the criteria of disposal sites. Regarding CAPEX and OPEX, CRR1m and ERR2m are relatively insensitive to cost variations, and they are regarded as economically viable given that the estimates are in the accuracy and contingency range for scoping studies of 50% and 30%, respectively [45].

For the upper pond, there is high uncertainty regarding geological knowledge on the neutralisation sludge, as well as the Co, Ca, and In contents (G3). The TSF's volume, and the $BaSO_4$ and base metal contents are well known (G2). Metallurgical testwork on the tailings from the upper pond is missing (F3), and it is unknown if the neutralisation sludge could be valorised in ERR2. These tailings might be difficult to process due to the high sulphate ion content [54]. If they need to be disposed of too, the disposal costs would increase in both scenarios (CRR1, ERR2). RMs recovery has a higher rating regarding environmental aspects as compared to rehabilitation only (NRR0). However, planning considerations such as the resettlement of rare flora and fauna still requires fundamental work (E3.3d), and the RMs efficiency (E3.3c) and preservation of RMs for future generations (E3.2c) in CRR1 could be improved. In contrast, the complete tailings valorisation (E1c) and high RM efficiency (E3.1c) are positively highlighted in the categorisation matrix. The development status of social aspects is generally low, just as for legal aspects (E3.3d).

For the individual RMs, a clear distinction in the geological and technological categories between the development status for $BaSO_4$ (G2F2), base metals (G2F2), FeS_2 (G2F1), and inert material (G2F1) can be seen as compared to the high-technology metals (G3F3). The development status for economic and environmental aspects is heterogeneous. Most RMs have a high economic importance or are CRMs in the EU, and all except for FeS_2 and inert material have a clear demand. The mean RM price forecast yields increasing $BaSO_4$, Co, and In prices (E3.1a); stagnant Pb and Zn prices (E3.2a); and decreasing Cu and Ga prices (E3.3a). For the new residues, the Pb solid matter content and dissolved Pb in leachate impede a disposal as inert waste (DK 0 class) (E3.2b) [61]. On the extreme ends, Ga and FeS_2 has the lowest (G3F3E3.3a) and highest (G2F1E3.2a) rating, respectively.

In sum, all 3 scenarios are rated equally in the overall rating in terms of the degree of confidence in the geological estimates and technical feasibility (G2F3). The scenarios differ in the economic performance with rehabilitation incurring costs only, and CRR1 having a higher uncertainty as compared to ERR2. Considering the proposed differentiation of the E category, the scenarios are categorised as G2/F3/E3.3a/E3.3b/E3.3c/E3.3d (NRR0), G2/F3/E3.3a/E3.2b/E3.3c/E3.3d (CRR1), and G2/F3/E3.1a/E3.2b/E3.3c/E3.3d (ERR2). The conversion into the current official UNFC categorisation yields G2F3E3 for all 3 scenarios. There is currently no class for this categorisation [44]. In comparison to the categorisation of G4F3E3 in the preceding screening study [43], only the G category could be improved.

Table 4. Categorisation matrix for the overall project rating of the rehabilitation scenario (NRR0) and the mineral RMs recovery scenarios (CRR1, ERR2).

Factor	Scenario		
	NRR0	CRR1	ERR2
	UNFC G Category		
geological conditions (relevant for project development)			
(1) quantity	G2	G2	G2
(2) quality	G2	G2	G2
(3) homogeneity	G2	G2	G2
	UNFC F Category		
TSF condition & risks (relevant for project development)			
(4) ordnance	F3	F3	F3
mine planning considerations (relevant for project execution)			
(5) mine/operational design	F3	F3	F3
(6) metallurgical testwork	-	F3	F3
(7) water consumption	F3	F1	F1
infrastructure (relevant for project development)			
(8) real estate	F1	F1	F1
(9) mining & processing	-	F3	F3
(10) utilities	F2	F2	F2
(11) transportation & access	F2	F2	F2
post-mining state (relevant for future impacts)			
(12) residue storage safety	F3	F3	F3
(13) rehabilitation	F2	F2	F2
	UNFC E Category [1]		
microeconomic aspects (relevant for project development)			
(14) economic viability	E3.3a	E3.1a	E3.1a
(15) economic uncertainty	-	E3.3a	E3.1a
financial aspects (relevant for project development)			
(16) investment conditions	-	E3.1a	E3.1a
(17) financial support	E3.3a	E3.1a	E3.1a
environmental impacts during project execution			
(18) air emission	E3.3b	E3.1b	E3.1b
(19) liquid effluent emission	E3.1b	E3.1b	E3.1b
(20) noise emission	E3.2b	E3.2b	E3.2b
environmental impacts after project execution			
(21) biodiversity	E3b	E3b	E3b
(22) land use	E3.2b	E3.2b	E3.2b
(23) material reactivity	E3.3b	E3.1b	E3.1b
social impacts during project execution			
(24) local community	E3.3c	E3.2c	E3.2c
(25) health & safety	E3.3c	E3.3c	E3.3c
(26) human rights & business ethics	E3.3c	E3.3c	E3.3c
social impacts due to project execution			
(27) wealth distribution	E3.3c	E3.3c	E3.3c
(28) investment in local human capital	E3.3c	E3.3c	E3.3c
(29) degree of RM recovery	E3.3c	E3.2c	E1c
(30) RM valorisation	E3.3c	E3.3c	E3.1c
social impacts after project execution			
(31) aftercare	E3c	E1c	E1c
(32) landscape	E2c	E1c	E1c
legal situation (relevant for project development)			
(33) right of mining	E3.3d	E3.3d	E3.3d
(34) environmental protection	E3.3d	E3.3d	E3.3d
(35) water protection	E3.3d	E3.3d	E3.3d
total rating	G2	G2	G2
	F3	F3	F3
	E3.3a	E3.3a	E3.1a
	E3.3b	E3.2b	E3.2b
	E3.3c	E3.3c	E3.3c
	E3.3d	E3.3d	E3.3d

[1] a: economic aspects, b: environmental aspects, c: social aspects, d: legal aspects.

Table 5. Categorisation matrix for the subproject rating for individual RMs (CRR1, ERR2).

Factor	Subprojects for RMs								
	BaSO$_4$	Cu	Pb	Zn	Co	Ga	In	FeS$_2$	Inert Material [1]
	UNFC G Category								
geological conditions (relevant for project development)									
(36) quantity	G2	G2	G2	G2	G3	G3	G3	G2	G2
(37) quality	G2	G2	G2	G2	G3	G3	G3	G2	G2
(38) homogeneity	G2	G2	G2	G2	G3	G3	G3	G2	G2
	UNFC F Category								
mine planning considerations (relevant for project execution)									
(39) recoverability	F2	F2	F2	F2	F3	F3	F3	F1	F1
	UNFC E Category [2]								
microeconomic aspects (relevant for project development)									
(40) demand	E3.1a	E3.1a	E3.1a	E3.1a	E3.1a	E3.1a	E3.1a	E3.2a	E3.3a
(41) RM criticality	E1a	E2a	E2a	E2a	E1a	E1a	E1a	E2a	E3a
(42) price development	E3.1a	E3.3a	E3.2a	E3.2a	E3.1a	E3.3a	E3.1a	-	-
impacts after project execution									
(43) solid matter	-	E3.1b	E3.2b	E3.1b	-	-	-	-	E1b
(44) eluate	E3.1b	E3.1b	E3.2b	E3.1b	-	-	-	-	E1b
total rating	G2	G2	G2	G2	G3	G3	G3	G2	G2
	F2	F2	F2	F2	F3	F3	F3	F1	F1
	E3.1a	E3.3a	E3.2a	E3.2a	E3.1a	E3.3a	E3.1a	E3.2a	E3.3a
	E3.1b	E3.1b	E3.2b	E3.1b	-	-	-	-	E1b

[1] Wissenbach shales & ankerit. [2] a: economic aspects, b: environmental aspects, c: social aspects, d: legal aspects.

4.2. Reconciliation of Stakeholder Perspectives with an Application of the UNFC Principles

Environmental NGOs' perspective: the TSF Bollrich constitutes an ecological burden in a sensitive environment with high potential long-term environmental and social risks [43]. Indeed, the TSF's current geomechanical state is stable, but it requires constant maintenance such as the removal of large trees and assuring seepage in the main dam [66]. The TSF is an upstream dam type, which is the most vulnerable type [16,20]. The lacking knowledge on the karstified zones in the area and the former occurrence of sinkholes near the TSF are currently rated as non-problematic [53]. However, for a conservative approach, the risk must be rated high due to the uncertainty. A sudden release of the contained masses and toxic elements would cause widespread environmental destruction and social issues, and would threaten human lives [43]. Therefore, the long-term physical and chemical risks and associated legacy costs are regarded as a necessity to act. Hence, early actions are preferable, and the rehabilitation costs (NRR0) can be seen as external costs borne by society to prevent harm. As the TSF is integrated well into the landscape, being visible only from nearby hills or from close up, the benefit of NRR0 is that the current landscape is mostly retained. On top, NRR0 has a relatively short duration of perceptible works on the TSF of 5 years. Hence, negative environmental and social impacts due to project execution are kept at a minimum as compared to RMs recovery (CRR1, ERR2). However, stabilising the tailings impedes a future RMs recovery. On top, rehabilitation incurs costs only so that a combination with RMs recovery (CRR1, ERR2) is preferable. Since the new residues in CRR1 consume land due in a disposal site and since future emissions cannot be excluded as the storage conditions are currently unclear, ERR2 is preferable.

Private investors' perspective: TSF rehabilitation (NRR0) generates relatively high revenues. However, the TSF Bollrich is an economically viable source of important RMs. Since a domestic RMs recovery can contribute to reducing RM supply risks by diversifying the sourcing of CRMs on a national level, a private company could benefit from a positive public perception when engaging in RMs recovery. As CRR1 and ERR2 include environmental rehabilitation, they reduce the anthropogenic footprint. As the highest revenues of all scenarios are generated in ERR2, and as there is a certain economic risk in CRR1 shown with the pessimistic price forecast, ERR2 is preferable economically.

Goslar city administration's perspective: NRR0 is in line with the city development goals [65] by restoring the recreational qualities of the TSF area in a relatively short period.

However, the anthropogenic footprint is not reduced and the tailings' long-term stability is unclear [69] so that future measures might be necessary. With RMs recovery (CRR1, ERR2), the city administration saves rehabilitation expenses. An intensified interaction of industry and scientific institutions could strengthen the region in the long run. However, the short duration of active works (CRR1) thwart the goal to establish long-term high-quality jobs and to attract investors who seek long-term opportunities [65]. Such opportunities are created in ERR2 so that the Harz region's challenge of a weak economic structure and emigration of young people can be tackled [52], and an innovative recycling industry can be established [65]. Dealing with the region's environmental legacy from former mining is seen by the city administration of Goslar as a key challenge for a sustainable development [65] so that negative impacts of new residues must be avoided (ERR2).

Résumé: with the application of the UNFC-principles, the advantages and disadvantages of all 3 scenarios could be made visible for all 3 stakeholders. The overview of all factors shows that all 3 stakeholder interests are best fulfilled with the RMs recovery scenario ERR2 in which most benefits are generated, namely, environmental rehabilitation, economic revenues, and long-term regional development. In the assessed constellation, the city administration of Goslar would be a particularly eligible main project driver under compulsory consideration of the enablers environmental NGOs and private investors.

4.3. Path Forward for the Case Study Bollrich

For the RMs recovery scenarios (CRR1, ERR2), a higher rating of the project as *potentially viable* (G2F2E2) requires the following aspects to be addressed: the extent of karstified zones needs to be investigated to better assess the risk of a potential damage to the TSF. The amount of dam material, and the amount, composition, distribution and valorisability of neutralisation sludge need to be investigated. Furthermore, a solution is required for the discharge of the Rammelsberg mine water, preferably with a recovery of RMs such as Zn. The costs for residue disposal (CRR1) and conditioning for an application in construction materials (ERR2) needs to be investigated. To enhance RM efficiency, a potential concentrate buyer needs to be willing to valorise the FeS_2 and to recover the high-technology metals. It should be investigated if all residues in ERR2 can be valorised. The recoverability of As, Cd, Cr, Ni, and Tl needs to be investigated as they are important in high-technology applications, e.g., robotics or decarbonised energy production [70].

A milestone is the determination of site-specific processing costs for which reference values are used in this article. An economic estimation after taxes and other governmental charges are required to make it comparable across country borders [71]. An uncertainty analysis on tailings mass could account for errors in the geological estimates.

In terms of legal aspects, fundamental work must be carried out such as the estimation of costs and the duration of clarifying legal barriers, the engagement of authorities, and the drafting of applications. As for environmental aspects, the present flora and fauna needs to be inventoried in detail; measures for the compensation of environmental impacts need to be drafted; and rehabilitation, environmental monitoring, and post-closure land use plans need to be conceptualised. For the endorsement of a project plan, a disposal site for residues needs to be determined, and a transportation concept must be developed.

A comprehensive systematic stakeholder assessment is required. The process should be transparent and clearly structured to enable a fact-based discussion at all times. For all scenarios, the TSF's long-term risks need to be weighed against the temporary disturbance of local nature and communities, potential long-term regional benefits such as environmental rehabilitation, and the local recruitment of workforce.

4.4. Integrating Sustainability Aspects into Raw Materials Classification

RMs recovery from tailings can have certain benefits: processing the already ground tailings is less energy-intense than processing ores under similar conditions [72]. The potential savings are high since ore crushing and grinding are the most energy-intense processes with ca. 40% of a mine's energy consumption [73,74]. Moreover, it is increasingly acknowledged that aspects other than the RMs have to be considered in present-day RMs assessments [52]. RMs recovery from tailings offers the opportunity to rehabilitate the environment [12,75], which can reduce environmental and social risks. Hence, tailings can be regarded as a secondary RM source with a lower social conflict potential than ores [11].

The challenge is to identify and communicate these potential benefits, especially for environmental and social aspects [46]. Indeed, geological and techno-economic aspects can be assessed with established methods from the conventional CRIRSCO classification [45], but it is unsuitable for capturing sustainability aspects [43,49]. In contrast, the UNFC recognises environmental and social aspects as potential driving factors, integrating them into the classification [44]. Current shortcomings of the UNFC are its lacking practicability [8], user guidance [43,49], specification of knowledge which must be generated in very preliminary studies [49], and standardised assessment and classification template for anthropogenic RMs including key factors which must be considered [47,49]. This article demonstrates how one can be guided through a practical UNFC application. Established methods from the conventional mineral RMs classification are combined with methods to account for environmental and social benefits. With the following aspects, the developed approach supports the integration of sustainability aspects into RMs classification:

First, the report of on-site exploration data by Goldmann et al. [53] on the TSF Bollrich documents relevant aspects extensively but it lacks a frame for an overall rating. In their report, a techno-economic classification of the tailings in terms of conventional *resources* or *reserves* as well as the determination of cut-off grades was not possible due to the geological uncertainties [53]. Environmental and legal aspects are discussed separately, but they do not contribute to the classification. This is common in current classification practice, which focusses on economic aspects [16,40]. Therefore, current practice cannot fully reflect a project's potentials. In contrast, the presented UNFC-compliant assessment and classification approach provides a comprehensive framework to communicate the development status of the TSF Bollrich case study by considering all relevant geological, technological, and environmental-socio-economic aspects on site during exploration.

Second, mining companies worldwide are increasingly recognising that their economic interests need to be aligned with social values for long-term success [6,23,76]. However, the reinterpretation of waste as a RM source requires a change of mindset [52]. In this context, a challenge is to create a common understanding of sustainable acting as local stakeholders' perspectives on sustainable mining often diverge [77]. Hence, the sustainable prospects of a potential project need to be communicated transparently to local communities in the project development phase to create a common understanding. Thus, the developed assessment and classification approach offers the opportunity to integrate a stakeholder assessment in the decision-making process. The needs of local stakeholders are particularly addressed in terms of impacts related to land use, the environment, and health.

Third, the example of the Harz region highlights the importance of including social aspects such as involving local communities in the development of RMs recovery projects and transparently communicating potential long-term impacts on former contaminated sites: although the Mansfeld area is comparable to the Goslar area, the local population is sceptical about RMs recovery due to dishonest communication and selfish behaviour of potential project developers in the past [52]. Especially in densely populated areas, social conflicts can arise. The inclusion of local values, such as those expressed by the town council as the elected representative of local citizens, can help to improve the sustainability of a project and influence a project assessment in terms of enhancing the common good [77].

Fourth, the developed categorisation matrix addresses several issues: in the classification of tailings with conventional practice, the RM potential beside the target RM potential is usually not captured, e.g., References [37–39]. This means that part of the RM potential remains unassessed. The distinct classification of the individual RMs in the categorisation matrix highlights the potentials of and barriers to their recovery. The heat map-like visualisation of the categorisation enables a quick comparison of all aspects with each other, promoting a transparent communication of the assessment results. For instance, in each of the scenarios, the impairment of local ecosystems around the TSF Bollrich are captured in the categorisation matrix. Consequently, a project developer is required to comment on how further measures can be taken to overcome the scenario-specific barriers. As another example, even a longer duration of the RMs recovery scenarios (CRR1, ERR2) could be considered more favourable than the relatively short impairment caused by the rehabilitation scenario (NRR0) due to the long-term benefits resulting from the risk reduction associated with the removal of the tailings. In a stakeholder assessment, all relevant stakeholders can question the factors considered in order to reach a mutually agreed decision. In the course of the study, consensus building can be documented and evaluated.

Fifth, the case study shows how the application of the UNFC principles can reconcile 3 different stakeholder perspectives: the TSF-related long-term risks are identified as the main project drivers. Considering the remediation costs as external costs borne by society enables a comparison of the monetary impacts of the TSF in case of rehabilitation (NRR0) with those of the other scenarios (CRR1, ERR2). Scrutinising the considered stakeholder perspectives leads to the following common values: minimisation of physico-chemical risks associated with the TSF, minimisation of emissions to the environment during any operation, achievement of a long-term aftercare-free state after project execution, and the preservation of the area's recreational value and ecosystem quality. On this basis, the RMs recovery scenario ERR2 should be prioritised since it addresses all common values.

4.5. Development Potential of the Assessment and Classification Approach

A comparison of the classification result from the screening of the TSF Bollrich (G4F3E3) in Reference [43] to the result from this article (G2F3E3) shows that the improvements in the E and F categories are not reflected in the overall rating. This can be explained with the selected factors and indicators to measure the development status, especially for the social and legal aspects. A comparison of the factors and indicators applied in this study with other case studies could show if they all suit the scope of a very preliminary study or if some of them should be applied in more developed studies. Additionally, the low rating in the E and F categories can be explained with the procedure to choose the lowest rating in a category as the overall rating. An example is the rating of economic aspects for the RM Cu: despite the favourable rating of the demand (E3.1a) and RM criticality (E2a), the low rating of the forecasted decreasing price development (E3.3a) is determinant. This issue could be resolved by weighting factors for instance. It is worth noting that there is currently no class defined for a rating as G2F3E3. A proposal is made for a possible description: *based on very preliminary results, a prospective project has been identified as a potential source of RMs for which further studies are required to justify further development.*

Factors related to the impact on global warming are not considered in this study. This could be remediated by performing a life cycle assessment (LCA). It enables the consideration of external costs, and it was also used in conjunction with the UNFC [78]. Another advantage is that it allows for a comparison to projects from primary mining [78]. Regarding tailings, the LCA has been used to assess aspects such as environmental impacts in early phases of mine planning [79], and TSF site management and closure scenarios [80]. For RMs recovery from tailings, an LCA should provide decision-makers with information on environmental impacts which could be compared with primary mining. In general, the LCA requires site-specific data for a detailed analysis of processes and their impacts [81]. The LCA performed by Goldmann et al. [53] for the conceptualised dredging system shows that an LCA in very preliminary studies can be applied to assess different mining options. The use of LCAs in early project development phases on aspects such as mineral processing and a possible contribution to the classification must yet be examined.

5. Conclusions and Recommendations

To recapitulate, the deposition of tailings in TSFs impacts the environment and local communities and can even threaten human health [16]. These impacts could be aggravated in the future due to a climate-change-induced increased likelihood of extreme weather occurrences [20]. At the same time, the global tailings production is increasing due to an increasing demand for highly important RMs, which are forecasted to at least double between 2010–2050 [4,5]. The increasing RM demand could partially be met by using the RM potential of tailings: 10–20% of all technospheric metal RMs are estimated to be deposited in landfills and TSFs; metal grades in tailings can be as high as in ores [40]. Technological advancements enable the exploitation of the residual metals content [29,82] or the valorisation in construction materials [83,84]. RMs recovery from tailings can also be an opportunity to reduce the environmental and social impacts of TSFs [75]. For the re-interpretation of tailings as a source of RMs, the potential benefits of and barriers to their exploitation need to be captured and assessed holistically. The assessment shows that the TSF Bollrich is an economically interesting source of $BaSO_4$; the base metals Cu, Pb, and Zn; and the high-technology metals Co, Ga, and In. Removing the TSF has positive long-term environmental impacts. However, there is high uncertainty regarding geological knowledge and technological extractability of the CRMs. An issue is that the applied social and legal factors are generally underdeveloped.

The research questions are answered: (1) the tailings deposit Bollrich is an example of a RMs recovery project which takes place in a complex environment where the influence of various site-specific stakeholders needs to be considered. With a UNFC-compliant approach, different stakeholder perspectives can be addressed in order to derive a commonly acceptable solution. In the case study, the enhanced mineral RMs recovery scenario ERR2 aligns the interests of environmental NGOs, private investors, and the city administration of Goslar: environmental rehabilitation to protect the TSF's vulnerable environment, the generation of profits, and a long-term regional development. It can therefore be concluded that a UNFC-compliant assessment is suitable for identifying areas of conflict between economic, environmental and social interests, and for achieving a generally acceptable solution. (2) It is suggested that for very preliminary studies, aspects relevant for project development and execution, impacts due to project execution, and impacts after project execution should be considered. Furthermore, the availability of primary on-site exploration data and secondary research data could be regarded as a prerequisite for a very preliminary study on tailings. As tailings usually contain multiple RMs, a comprehensive overview of the RM potential with differentiation of individual RMs is required. The data must allow for an initial assessment of the following aspects: (i) characterisation and quantification of the total and individual RM content, (ii) laboratory investigation of processability, (iii) technological conceptualisation of project execution and aftercare measures, (iv) DCF analysis, (v) inventory on present rare flora and fauna, (vi) status quo environmental risk assessment, and (vii) identification of relevant stakeholders. After a clarification of these aspects, a project can be advanced to a *preliminary study*. (3) The identification and communication of sustainability aspects in RMs classification poses a challenge. Despite a project's impact on its local environment and communities, related site-specific project potentials and barriers are usually not considered. The example of the Harz region demonstrates that, in addition to conventional economic interests, a site-specific approach is essential from the beginning of project development. The example of the tailings deposit Bollrich shows that an integration of local sustainability aspects into the assessment, represented by the development goals of the city administration of Goslar, can give a strong impulse for project development: strengthening the regional industrial role, creating high-value jobs, and developing tourism. The developed UNFC-compliant categorisation matrix captures the development status of specified factors and communicates the results in a quickly understandable manner in a heat-map-like style. Hence, it enables a point-by-point comparison of different scenarios so that the individual potentials and benefits become clear. In this way, the most auspicious option can be quickly identified, and its development can be justified.

Recommendations made: as for the case study TSF Bollrich, enhance the geological knowledge on the metalliferous CRMs; investigate the processability of the neutralisation sludge; assess the recoverability of As, Cd, Cr, and Tl; and consider a direct valorisation of RMs in the Rammelsberg mine water. If the RMs recovery project is executed, the city administration's tax revenues could be used to rehabilitate other contaminated areas from former mining activities. In this way, the local community hosting the mining activity can benefit directly from it, which is uncommon in current practice [77]. Thus, RMs recovery from the TSF Bollrich could serve as a role model for a sustainable development of the Harz region. As for the developed approach, investigate if all selected factors and indicators, especially those for social and legal aspects, are suitable for very preliminary studies. Correspondingly, determine which factors are necessary and which are optional in very preliminary studies. Since the overall rating does not properly reflect the improvements made and deficits encountered in the course of several studies, introduce a reporting to support decision-making. As for the development of an anthropogenic RMs management, a database for the assessment of the global anthropogenic RM potential needs to be established. For this, waste producers could be obligated by law to report on all contained RMs in their wastes. Lastly, UNFC-compliant case studies on anthropogenic RMs are currently very labour-intensive due to a lack of experience. More UNFC-compliant case studies are needed to derive a reference base of project potentials and barriers. This would provide future studies with a benchmark for a quick recognition of a project's prospects of reaching the next level of maturity.

Supplementary Materials: Figure S1: Results of autoregressive electric energy price forecast based on yearly historical data from 2014 to 2020 from Statista [85]. The blue line on the right-hand side depicts the mean price forecast, and the blue and grey areas represent the 95% and 75% confidence intervals, respectively, Figure S2: Results of autoregressive diesel price forecast based on yearly historical data from 1950 to 2020 from Statista [86]. The blue line on the right-hand side depicts the mean price forecast, and the blue and grey areas represent the 95% and 75% confidence intervals, respectively, Figure S3: Results of autoregressive BaSO4 price forecast based on yearly historical data from 2011 to 2020 from the USGS [87–90]. The blue line on the right-hand side depicts the mean price forecast, and the blue and grey areas represent the 95% and 75% confidence intervals, respectively, Figure S4: Results of autoregressive Co price forecast based on yearly historical data from 1996 to 2020 from the USGS [87,89–93]. The blue line on the right-hand side depicts the mean price forecast, and the blue and grey areas represent the 95% and 75% confidence intervals, respectively, Figure S5: Results of autoregressive Cu price forecast based on monthly historical data from 1999 to 2021 from IndexMundi [94]. The blue line on the right-hand side depicts the mean price forecast, and the blue and grey areas represent the 95% and 75% confidence intervals, respectively, Figure S6: Results of autoregressive Ga price forecast based on yearly historical data from 1999 to 2020 from the USGS [87,89–93]. The blue line on the right-hand side depicts the mean price forecast, and the blue and grey areas represent the 95% and 75% confidence intervals, respectively, Figure S7: Results of autoregressive In price forecast based on yearly historical data from 1999 to 2020 from the USGS [87,89–93]. The blue line on the right-hand side depicts the mean price forecast, and the blue and grey areas represent the 95% and 75% confidence intervals, respectively, Figure S8: Results of autoregressive Pb price forecast based on monthly historical data from 1999 to 2021 from IndexMundi [95]. The blue line on the right-hand side depicts the mean price forecast, and the blue and grey areas represent the 95% and 75% confidence intervals, respectively, Figure S9: Results of autoregressive Zn price forecast based on monthly historical data from 1999 to 2021 from IndexMundi [96]. The blue line on the right-hand side depicts the mean price forecast, and the blue and grey areas represent the 95% and 75% confidence intervals, respectively, Figure S10: Conceptual mine plan and processing schematic. The light grey shaded field indicates the spatial system boundaries and the dark grey shaded fields indicate products (adapted after Goldmann et al. [53]), Figure S11: Results of the sensitivity analysis of the conventional mineral RMs recovery scenario (CRR1p) with pessimistic price forecast and a discount rate of 15%, Figure S12: Results of the sensitivity analysis of the conventional mineral RMs recovery scenario (CRR1o) with optimistic price forecast and a discount rate of 15%, Figure S13: Results of the sensitivity analysis of the enhanced mineral RMs recovery scenario (ERR2p) with pessimistic price forecast and a discount rate of 15%, Figure S14: Results of the sensitivity analysis of the enhanced mineral RMs recovery scenario (ERR2o) with optimistic price forecast and a discount rate of 15%, Figure S15: Comparison of costs, revenues and NPVs for the mean price forecast of the 3 scenarios with no mineral RMs recovery (NRR0), conventional mineral RMs recovery (CRR1m) and enhanced mineral RMs recovery (ERR2m). With a discount rate of 15%, NRR0 is discounted over a period of 35 years, and CRR1m and ERR2m over a period of 11 years, Figure S16: Comparison of costs, revenues and NPVs for the pessimistic price forecast of the 3 scenarios with no mineral RMs recovery (NRR0), conventional mineral RMs recovery (CRR1p) and enhanced mineral RMs recovery (ERR2p). With a discount rate of 15%, NRR0 is discounted over a period of 35 years, and CRR1p and ERR2p over a period of 11 years, Figure S17: Comparison of costs, revenues and NPVs for the optimistic price forecast of the 3 scenarios with no mineral RMs recovery (NRR0), conventional mineral RMs recovery (CRR1o) and enhanced mineral RMs recovery (ERR2o). With a discount rate of 15%, NRR0 is discounted over a period of 35 years, and CRR1o.

Author Contributions: Conceptualisation, R.S.; methodology, R.S.; validation, R.S., S.H.-A.; resources, R.S.; writing—original draft preparation, R.S.; writing—review and editing, R.S., S.H.-A.; visualisation, R.S.; project administration, R.S.; funding acquisition, R.S., S.H.-A. All authors have read and agreed to the published version of the manuscript.

Funding: This research was funded by the German Ministry of Research and Education (BMBF) as part of the research project ADRIANA (Client II programme), grant agreement number 033R213A-D.

Institutional Review Board Statement: Not applicable.

Informed Consent Statement: Not applicable.

Data Availability Statement: This research used publicly available data available in the referenced sources. The database can be found in the Appendix A and supplementary materials.

Acknowledgments: The authors are thankful to Bernd G. Lottermoser for his comments and to Jonas Krampe for providing the R code. In addition, the authors would like to express their deep gratitude to two anonymous reviewers who helped to improve the manuscript.

Conflicts of Interest: The authors declare no conflict of interest. The funders had no role in the design of the study; in the collection, analyses, or interpretation of data; in the writing of the manuscript; or in the decision to publish the results.

Abbreviations

Abbreviation/Unit	Description
Ag	lat. *argentum* (silver)
Al	aluminium
Au	lat. *aurum* (gold)
$BaSO_4$	barium sulphate (barite)
Cd	lat. *cadmia* (cadmium)
Co	cobalt
Cu	lat. *cuprum* (copper)
$CuFeS_2$	copper iron disulphide (chalcopyrite)
Fe	lat. *ferrum* (iron)
FeS_2	iron disulphide (pyrite)
Ga	lat. *gallia* (gallium)
In	indium
Mn	manganese
Mo	molybdenum
Ni	nickel
Pb	lat. *plumbum* (lead)
PbS	lead sulphide (galena)
Tl	lat. *tellus* (tellurium)
Zn	zinc
ZnS	zinc sulphide (sphalerite)
ADRIANA	Airborne spectral Detection of Reusable Industry mAterials in taiLiNgs fAcilities

Abbreviation	Definition
BMBF	German Ministry of Research and Education
CAPEX	capital expenditure
CL:AIRE	Contaminated Land: Applications in Real Environments
CRM	Critical Raw Material
DCF	discounted cash flow
E	East
EC	European Commission
EU	European Union
LOM	Life of Mine
N	North
NPV	net present value
OPEX	operating expenditure
Qty.	quantity
RM	raw material
TSF	tailings storage facility
UNECE	United Nations Economic Commission for Europe
UNFC	United Nations Framework Classification for Resources
UNFC E category	represents environmental-socio-economic viability
UNFC F category	represents technical feasibility
UNFC G category	represents degree of confidence in the geological estimate
USGS	U.S. Geological Survey
W	West
°C	degree Celsius (unit of temperature on the Celsius scale)
μm	micrometre (unit of length, equivalent to 10^{-6} metres)
a	year
km	kilometre (unit of length, equivalent to 10^{3} metres)
kW	kilowatt (SI-derived unit of power)
kWh	kilowatt-hour (SI-derived unit of energy)
l	litre (SI-derived unit of volume, equivalent to 10^{-3} m^3)
m	metre (SI unit of length)
m^2	square metre (SI-derived unit of surface)
m^3	cubic metre (SI-derived unit of volume)
mm	millimetre (unit of length, equivalent to 10^{-3} metres)
t	metric tonne (unit of weight, equivalent to 1000 kilograms)

Appendix A

Table A1. Degree of confidence in the geological estimates (G) for the overall project rating with the UNFC-compliant categorisation matrix.

Factor	Explanation	Dependence on	Modification after	Indicator & UNFC Rating
Geological conditions (relevant for project development)				
(1) quantity	amount of target RMs	ore quality, former processing efficiency, deposit volume	[45]	degree of geological certainty: high (G1) medium (G2) low (G3)
(2) quality	physico-chemical properties of target RMs	former processing, storage conditions	[45]	degree of geological certainty: high (G1) medium (G2) low (G3)
(3) homogeneity	distribution of target RMs inside the deposit	manner of former deposition	[24]	degree of geological certainty: high (G1) medium (G2) low (G3)

Table A2. Technical feasibility (F) for the overall project rating with the UNFC-compliant categorisation matrix.

Factor	Explanation	Dependence on	Modification after	Indicator & UNFC Rating
TSF condition & risks (relevant for project development)				
(4) ordnance	unexploded ordnance from armed conflicts	regional history, former searching activities	–	degree of knowledge: non-existence proven (F1) existence proven (F2) unclarified (F3)
Mine planning considerations (relevant for project execution)				
(5) mine/operational design	optimising RMs recovery under consideration of strategic goals & restrictions	geological knowledge on deposit, project planning phase, quality of model assumptions, legal restrictions	[45]	level of detail of planning: extended (incl. detailed operational factors) (F1) advanced (incl. pit configuration & processing scheme) (F2) basic (conceptual) (F3)
(6) metallurgical testwork	investigation of possible methods for mineral processing	sampling techniques, representativeness of test feed, testing techniques	[45]	degree of research on mineral processability: industrial scale (F1) pilot scale (F2) laboratory scale (F3)
(7) water consumption	demand of fresh water supply for mining & processing	available water resources, water efficiency of mining system	[13,97,98]	percentage of recycled water: high (>80%) (F1) medium (50–80%) (F2) low (<50%) (F3)
Infrastructure (relevant for project development)				
(8) real estate	availability of land & reusability of buildings	former mine closure, current land use, time lapsed after abandonment	[45]	condition of infrastructure: highly developed (fully reusable) (F1) acceptable (usable after upgrade) (F2) bleak (requires (re-)construction) (F3)
(9) mining & processing	reusability of equipment related to general services, mining & processing	former mine closure, current land use, time lapsed after abandonment	[45]	condition of equipment: highly developed (fully reusable) (F1) acceptable (usable after upgrade) (F2) bleak (requires new acquisition) (F3)
(10) utilities	access to utilities supply lines (e.g., electricity)	mine closure & time lapsed after abandonment, current land use, proximity to human settlements	[45]	condition of infrastructure: highly developed (full access) (F1) acceptable (access after upgrade) (F2) bleak (requires (re-)construction) (F3)
(11) transportation & access	access to mine & markets via air, road, railway, or waterway	topography, former mine closure, current land use, time lapsed after mine abandonment, proximity to human settlements	[45]	condition of infrastructure: highly developed (fully reusable) (F1) acceptable (usable after upgrade) (F2) bleak (requires (re-)construction) (F3)
Post-mining state (relevant for future impacts)				
(12) residue storage safety	ability of new storage facility to safely store new residues for an indefinite time period	amount of new residues, topography, type of construction, climate, regional seismic activity	[13,98–100]	suitability of new disposal site for safe storage: high degree of safety proven (F1) preliminary assertion of safety (F2) unsafe or unclarified (G3)
(13) rehabilitation	process of recontouring, revegetating, & restoring the water & land values	residue characteristics, local ecosystem, landscape, environmental laws, local climate	[101]	level of detail of planning: concrete (F1) conceptual (F2) none (F3)

Table A3. Economic viability (E a) for the overall project rating with the UNFC-compliant categorisation matrix.

Factor	Explanation	Dependence on	Modification after	Indicator & UNFC Rating
Microeconomic aspects (relevant for project development)				
(14) economic viability	economic returns from project	mine planning, RMs prices, costs of input factors (labour, energy, materials), payments to public sector (e.g., taxes)	[45,97]	discounted cash flow over projected LOM: positive (NPV >> 0€) (E3.1a) neutral (NPV~0€) (E3.2a) negative (NPV << 0€) (E3.3a)
(15) economic uncertainty	overall uncertainty of economic estimates	degree of detail in planning, data quality of economic estimate	[45]	uncertainty of cash flow in pessimistic scenario: low (NPV >> 0€) (E3.1a) medium (NPV~0€) (E3.2a) high (NPV << 0€) (E3.3a)
Financial aspects (relevant for project development)				
(16) investment conditions	conditions concerning taxes, royalties, & other financial regulations, which are a precondition for decision makers with respect to location & investment	country-specific regulations, condition of financial market, social considerations, environmental considerations	[45,68]	country rank on the ease-of-doing-business index: country rank < 75 (E3.1a) country rank 75–125 (E3.2a) country rank > 125 (E3.3a)
(17) financial support	financial support from political institutions for innovative projects such as loans, equity financing, or guarantees can incentivise RMs from mineral waste	active socio-political support	[102]	probability of approval: high (E3.1a) medium (E3.2a) low (E3.3a)

Table A4. Environmental viability (E b) for the overall project rating with the UNFC-compliant categorisation matrix.

Factor	Explanation	Dependence on	Modification after	Indicator & UNFC Rating
Environmental impacts during project execution				
(18) air emission	risk of tailings being eroded by wind	particle size, TSF cover, local climate, wind conditions, pit configuration	[13,98]	risk of dust emission: low (<80%) (E1b) medium (50–80%) (E2b) high (>50%) (E3b)
(19) liquid effluent emission	effluents from tailings can contaminate soil & surface water	soil liner, drainage system, wet tailings storage, local environment, tailings' chemical properties	[13,98]	risk of groundwater contamination: low (E1b) medium (E2b) high (E3b)
(20) noise emission	noise & vibrations during mining; transport & processing can cause disturbances of local communities determined by individual & collective perception	mine planning, protective measures, topography, proximity to human settlements	[97]	expected degree of impact: low (E1b) medium (E2b) high (E3b)
Environmental impacts after project execution				
(21) biodiversity	influence on habitats & species	local ecosystem, mining system, landscape, rehabilitation measures	[97]	total number of protected species that are affected by mining activities & that will be resettled on post-mining land: all (100%) (E1b) some (1–99%) (E2b) none (0%) (E3b)
(22) land use	land requirement after mine closure	amount of new residues, type of disposal, rehabilitation, land development opportunities	[97]	freely available post-mining land: most (>80%) (E1b) some (50–80%) (E2b) little (<50%) (E3b)
(23) material reactivity	capability of contained minerals to produce AMD	target minerals, concentration of sulphidic minerals	[13,103]	reduction of reactive material's mass: high (>80%) (E1b) medium (50–80%) (E2b) low (<50%) (E3b)

Table A5. Social viability (E c) for the overall project rating with the UNFC-compliant categorisation matrix.

Factor	Explanation	Dependence on	Modification after	Indicator & UNFC Rating
Social impacts during project execution				
(24) local community	commitment beyond formal regulatory requirements, the recognition of diverse values, & the right to be informed about issues & conditions that influence lives	communication with stakeholders, proximity to human urban, protected, or culturally relevant areas, participation of local communities in decision-making	[68,97,104]	probability of approval through active commitment: high (>80%) (E3.1c) medium (50–80%) (E3.2c) low (>50%) (E3.3c)
(25) health & safety	protection of workers & local communities from injuries & diseases, & environmental pollution	mining system, local health & safety standards, corporate values for the establishment of a safe work environment & lively safety culture	[97]	total number of complaints or prosecutions for non-compliance in planning phase: none (plans have been communicated publicly) (E3.1.c) more than 1 (plans have been communicated publicly) (E3.2c) none (plans have not been communicated publicly) (E3.3c)
(26) human rights & business ethics	degree to which a mining company values ethically correct behaviour	wages, right to organise trade unions, bribery & corruption, violation of human rights, forcefully gained control over land, a country's governance	[97]	total number of complaints or prosecutions for non-compliance in planning phase: none (plans have been communicated publicly) (E3.1.c) more than 1 (plans have been communicated publicly) (E3.2c) none (plans have not been communicated publicly) (E3.3c)
Social impacts due to project execution				
(27) wealth distribution	distribution of earning between mining company, local communities, & government	a country's governance, choice of suppliers, & contractors; percentage of locally hired workers; wages	[97]	total number of complaints or prosecutions for non-compliance in planning phase: none (plans have been communicated publicly) (E3.1.c) more than 1 (plans have been communicated publicly) (E3.2c) none (plans have not been communicated publicly) (E3.3c)
(28) investment in local human capital	fostering personal skill development & capacity-building of employees by education & skill development	percentage of locally hired workers, offering higher education & training & transferable skill development; degree to which work is contracted out	[97]	percentage of employees sourced from local communities: high (>80%) (E3.1c) medium (50–80%) (E3.2c) low (<50%) or unclarified (E3.3c)
(29) degree of RM recovery	RMs can become inaccessible for recovery for future generations	disposal of new residues, mineral processing, residue stabilisation, residue characteristics	-	residue disposal: complete residue valorisation (E1c) separate disposal (E3.1c) mixed disposal (E3.2c) sterilisation (E3.3c)
(30) RM valorisation	utilising a RM in a sustainable manner to limit the impact of its recovery on the environment	target minerals, maturity of valorisation technologies, potential markets, RMs prices	[97]	total mass reduction as percentage of original tailings mass: high (>80%) (E1c) medium (50–80% (E2c) low (<50%) (E3c)
Social impacts after project execution				
(31) aftercare	level of commitment & necessary measures on post-mining land	land management, national regulations, rehabilitation measures	-	duration of aftercare measures: short-term (<5 years) (E1c) mid-term (5–30 years) (E2c) long-term (>30 years) (E3c)
(32) landscape	mining activities can cause a visual impact by transforming landscapes	topography, local ecosystem, mine planning, local climate	[97]	impact on the environment: positive (E1c) neutral (E2c) negative (E3c)

Table A6. Legal viability (E d) for the overall project rating with the UNFC-compliant categorisation matrix.

Factor	Explanation	Dependence on	Modification after	Indicator & UNFC Rating
Legal situation (relevant for project development)				
(33) right of mining	regulations affecting project planning & realisation	supranational, national, & regional laws & rules	[45]	state of development: application in development (E3.1d) authorities engaged (E3.2d) application not begun or unclarified (E3.3d)
(34) environmental protection	regulations affecting project planning & realisation	supranational, national, & regional laws & rules	[45,53,97]	state of development: application in development (E3.1d) authorities engaged (E3.2d) application not begun or unclarified (E3.3d)
(35) water protection	regulations affecting project planning & realisation	supranational, national & regional laws & rules	[45]	state of development: application in development (E3.1d) authorities engaged (E3.2d) application not begun or unclarified (E3.3d)

Table A7. Degree of confidence in the geological estimates (G) for the rating of individual RMs with the UNFC-compliant categorisation matrix.

Factor	Explanation	Dependence on	Modification after	Indicator & UNFC Rating
Geological situation (relevant for project development)				
(36) quantity	amount of target RMs	ore quality, former processing efficiency, deposit volume	[45]	degree of geological certainty: high (G1) medium (G2) low (G3)
(37) quality	physico-chemical properties of target RMs	former processing, potential revenues	[45]	degree of geological certainty: high (G1) medium (G2) low (G3)
(38) homogeneity	distribution of target RMs inside the deposit	mine planning, mineral feed grade, timing of revenues	[45]	degree of geological certainty: high (G1) medium (G2) low (G3)

Table A8. Technical feasibility (F) for the rating of individual RMs with the UNFC-compliant categorisation matrix.

Factor	Explanation	Dependence on	Modification after	Indicator & UNFC Rating
Mine planning considerations (relevant for project execution)				
(39) recoverability	ability to extract a wanted RM from the tailings	technological development, state of metallurgical testing, equipment availability, state of target RM	-	percentage of RM which is extracted from the tailings: high (>80%) (F1) medium (50–80%) (F2) low (<50%) (F3)

Table A9. Economic viability (E a) for the rating of individual RMs with the UNFC-compliant categorisation matrix.

Factor	Explanation	Dependence on	Modification after	Indicator & UNFC Rating
Microeconomic aspects (relevant for project development)				
(40) demand	existence of a current practical use for the RM & absence of geological, technological, economic, environmental, social, &/or legal objections against its recovery	market, price, available technology, public acceptance, regulations	-	favourable conditions for RM extraction: yes (E3.1a) conditionally (E3.2a) no (E3.3a)
(41) RM criticality	importance of a RM in an industry or economy	economic importance, supply risk, substitutability	[59]	allocation to EC's criticality assessment: CRM (E1a) high economic importance or supply risk (E2a) no criticality (E3a)
(42) price development	forecasted RM price behaviour	demand, supply risk, quality, & quantity of historical data	-	forecasted mean price development over the project's duration: positive trend (E3.1a) stagnant trend (E3.2a) negative trend (E3.3a)

Table A10. Environmental viability (E b) for the rating of individual RMs with the UNFC-compliant categorisation matrix.

Factor	Explanation	Dependence on	Modification after	Indicator & UNFC Rating
Impacts after project execution				
(43) solid matter	a RM's potential to harm human health, flora, &/or fauna	concentration, toxicity, valorisation path	[13,105,106]	concentration of RM solid matter in new residues to qualify for class DK 0 (inert waste) according to German Landfill Regulation DepV [61]: non-hazardous material (E1a) threshold value not exceeded (E3.1a) threshold value exceeded (E3.2a) unclarified (E3.3a)
(44) eluate	a RM's potential to harm human health, flora, &/or fauna	concentration, toxicity, valorisation path, solubility	[13,105,106]	concentration of RM in eluate from new residues to qualify for class DK 0 (inert waste) according to German Landfill Regulation DepV [61]: non-hazardous material (E1a) threshold value not exceeded (E3.1a) threshold value exceeded (E3.2a) unclarified (E3.3a)

Table A11. Knowledge base on the Bollrich tailings deposit for project definition. The dark grey shaded fields indicate data associated with high uncertainties, while the light grey shaded fields indicate data associated with moderate uncertainties, and the dashes indicate factors for which no information is available.

Category & Factor	Data	Sources	UNFC Axis [1]
(A) type of study	very preliminary study	-	
(B) basic information			
(a) geography			
(i) location	Goslar district, Lower Saxony (Germany) (51°54′8.97″ N, 10°27′47.31″ E), 270 m above mean sea level nearest human settlement ~400 m E air-line distance downstream of main dam	[50]	
(ii) topography	at the foot of Harz mountain range, up to 1141 m altitude with deep valleys	[107]	
(iii) local geology	folded & faulted Paleozoic rocks of the Harz Mountains are uplifted & thrust over younger Mesozoic rocks of the Harz foreland along the Northern Harz Boundary fault leading to steeply tilting & partly inverted Mesozoic strata; Mesozoic rocks are largely composed of Triassic to Cretaceous sedimentary rocks of varying composition (i.e., mostly impure limestones, clastic sandstones (greywackes) & shales); younger Quaternary sediments are rare & locally limited	[108]	

Table A11. Cont.

Category & Factor	Data	Sources	UNFC Axis [1]
(iv) land use	in near vicinity: agricultural, forest, industrial & commercial, & recreation & residential areas	observed on Google Earth [50]	
(v) surface waters	Four small rivers observed downstream of TSF within a 1.5 km radius (Abzucht, Ammentalbach, Gelmke & Oker)	observed on Google Earth [50]	
(vi) climate	moderately warm, temperature −0.7 to 16.3 °C (average 7.2 °C), average rain precipitation 911 mm/a, average climatic water balance 366 mm/a	[109,110]	
(b) geogenic deposit			
(i) mineralisation	two strongly deformed lens-shaped main ore bodies (high & low grade), sedimentary exhalative deposit (SedEx), fine grained (10–30 μm) principle sulphide minerals sphalerite ((Zn,Fe)S) & pyrite (FeS_2), less amounts of galena (PbS) & chalcopyrite ($CuFeS_2$), Ag, Au, (average estimated grades 14 wt% Zn, 6 wt% Pb, 2 wt% Cu, 140 g/t Ag & 1 g/t Au), barite ($BaSO_4$) (average grade 20 wt%)—additionally ca. 30 trace elements such as Co, Ga, & In, hosted by Middle Devonian Wissenbach shales	[50,107,111]	
(ii) former mining	underground mine, closed for economic reasons in 1988 after >1000 years of operation, now UNESCO World Heritage site located ~3 km W air-line distance from second processing plant Bollrich & TSF	[50,107,111]	
(c) tailings deposit			
(i) data collection methods	scientific publications or publicly accessible data, assumptions based on scientific publications, &/or own reasoning	-	
(ii) history	was in operation for ~49 years, decommissioned in 1987; supplied by processing plants Rammelsberg (into upper pond, 1938–1987) & Bollrich (into lower pond, 1956–1987); course of river Gelmke was changed several times	[53,57,107]	
(iii) recoverability			
• target minerals	previously & non-previously mined minerals	-	G
• quantity & quality	$V_{tailings}$ = 2,030,000 m^3, m_{dry} = 7,100,000 t, ρ = 3.5 t/m^3 (weighted mean value), $\rho_{neutralisation\ sludge}$ = 2.3 t/m^3	[53,54]	
	exploration of deposit: (i) 10 drill cores (17–28 m) taken in upper pond along main dam & parallel to main dam in the middle of the pond, analysis of 16 elements; (ii) 90 water depth metering points	[53]	G
	26 drill cores taken in upper & lower ponds, analysis of 4 elements & 3 minerals	[54]	
	low degree of alteration associated with oxidation	[53]	
• TSF structure	valley impoundment, estimated surface area 315,000 m^3 consists of 3 ponds: (i) lower pond (west, 74 vol% of TSF, ρ = 3.0 t/m^3, max. water depth 4 m, average water depth 2 m), (ii) upper pond (middle, 26 vol% of TSF, ρ = 3.7 t/m^3, max. water depth 0.5 m, average water depth 0.4 m), (iii) water retention pond (East) consists of 3 dams: (i) main dam (max. 33 m height, max. 18° slope, raised 6 times, up-stream), (ii) middle dam (max. 19 m height), (iii) water retention dam (max. 8 m height)	[53,66], Ruler Tool [50], average water depth estimated with data from Reference [53]	F
• homogeneity	drill core data of upper pond shows relatively homogeneous deposit with slightly increasing Ba grades with depth; deposit modelled based on historical & current terrain models, water depth measurements, historical & current core data; validation by comparison to production records	[53]	G, F
• safety considerations	dam stability: occurrence of sinkhole at northern part of TSF documented in 1986 & several sinkholes near TSF reported in the past, which are associated with karstified geological structures nearby; expertise from 1986 concludes that TSF is not imminently threatened; confirmed by current calculations; unexploded ordnance: existence of WWII [2] ordnance cannot be excluded based on historical data so it needs to be investigated prior to mining	[53]	F
(iv) rehabilitation	not rehabilitated, left to ecological succession, no signs of AMD [3] or erosion observable	[53], observed on Google Earth [50]	

Table A11. Cont.

Category & Factor	Data	Sources	UNFC Axis [1]
(v) assessment status			
• maturity level	research work	-	
• characterisation	complete for lower pond	[53]	
	partial for upper pond; not all elements/minerals analysed; amount, composition, & shape of deposition of mine water neutralisation sludge in upper & lower pond roughly estimated		
• evaluation	partial	-	
• classification	prospective project (E3F3G4)	[43]	
(vi) economics			
• RM criticality	BaSO$_4$, Co, Ga, & In are CRMs in EU with very high economic importance; Cu, Pb, & Zn have high economic importance in EU	[112]	E a
• further valorisation	industrial & metalliferous minerals of interest, use of residues in construction materials conceivable	-	E a
(vii) social impacts			
• health protection	no apparent imminent hazards known; negative impacts through dermal contact, ingestion or inhalation not given; risk assessment not performed	[53]	E c
• scientific interest	first scientific exploration shortly in 1983 before TSF abandonment in 1988; one recent research project (REWITA) with focus on mineral RMs recovery (2015–2018); proposal for follow-up project (REMINTA) on material extraction submitted	[53,54], www.cutec.de/fileadmin/Cutec/documents/cutec-news/2020/new58_dezember2020.pdf (accessed on 24 February 2021)	E c
• SLO [4]	positive perception of project idea by administrative bodies, environmental NGOs, & scientists	[52]	E c
	local population's perception of project idea unknown	-	
(viii) environmental impacts			
• pollution	possible negative impacts unknown; disused landfill "Paradiesgrund" located 250 m N air-line distance from TSF; possible influence on landfill when mining the TSF needs to be investigated	[53]	E b
	TSF's base not sealed & in direct contact with tailings		
• landscape	integrated into landscape (visible only from up close or from hills); environment has been adapting through natural succession; active gilder airfield ~100 m N air-line distance from TSF; hiking trails next to TSF & biking Euroroute R1 near TSF	cf., Figure 2	E b
• current status	on-site inspection of the TSF showed that rare flora, & aerial & soil fauna colonise the site	[53]	E b
• protected areas	conservation areas & protected landscapes nearby, protected species of flora & fauna sighted in area around TSF	[53]	E b
• secondary use	since 1966, neutralised mine water from the Rammelsberg mine has been discharged into the TSF (mainly upper pond, currently ~450,000 to 900,000 m^3/a); overlay of tailings and neutralisation sludge	[54]	E b

Table A11. Cont.

Category & Factor	Data	Sources	UNFC Axis [1]
(d) technology			
(i) mine planning	mine planning considerations on conceptual basis (dredging)	-	F
(ii) processing	extraction of $BaSO_4$, Co, Cu, Ga, In, Pb, Zn, & inert residues evaluated in discontinuous laboratory experiments on tailings from lower pond, processing sequences: (i) sulphide separation together with contaminants (rougher+cleaner+leaching), (ii) $BaSO_4$ separation (rougher+cleaner+scavenger+conditioning); recovery rates (tested on material from lower pond; ammonia leaching route for sulphides): $BaSO_4$ (74%), Co (12%), Cu (74%), Ga (2%), In (26%), Pb (65%), Zn (72%) & inert material (93%) processing tests on tailings from upper pond not performed; precipitation of SO_4 ions in multiple stages necessary to recover metals	[60]	F
(e) infrastructure			
(i) real estate	buildings & land from former processing available	[53]	F
(ii) mining & processing	former processing plant available ~550 m E air-line distance from TSF	[53]	
(iii) utilities	access to public electricity, gas, & water grid assumed	based on observation on Google Earth [50]	F
(iv) transportation & access	dirt roads, federal highway B6 ~1.6 km N air-line distance from TSF & public railway ~500 m E air-line distance from TSF; disused railway tracks from processing plant Bollrich to public network (estimated abandonment in 1988)	[53], observed on Google Earth [50]	F
(f) politics			
(i) political willingness	-	-	E c
(g) legislation/licensing			
(i) ownership	Bergbau Goslar GmbH (address: Bergtal 18, 38640 Goslar, Germany)	[53]	E d
(ii) legal exploration framework	currently supervised under German Federal Mining Act (BBergG)	[53]	E d
(iii) legal mining framework	-	-	E d
(iv) operating license	-	-	E d
(v) contracts	-	-	E d
(C) mineral- & material-centric information			
(a) chemical & mineralogical composition			
(i) elements	Ba (14.4), Cu (0.15), Fe (12.5), Pb (1.2), Zn (1.3) [mean, wt%]; Ag (-), As (700), Cd (30), Co (185), Ga (23), In (5.9), Tl (70) [mean, µg/g]	[53]	G
(ii) minerals			G
• main mineral groups (& associated elements)	silica-based: Al, Si, K, Ni, Ga carbonate: Ca, Mn, Fe, (Mg), (Co) sulphidic: Fe, Co, Cu, Zn, Pb, As, Cd, In, Tl sulphate: Ba, Ca	[53,54]	
• quantities:	estimated cumulated minerals content (total dry mass/share of tailings' mass)	[53]	
• $BaSO_4$	1,739,000 t/24.5 wt% (monomineralic)		
• $CuFeS_2$	31,000 t/0.44 wt%		
• FeS_2	1,086,000 t/15.3 wt% (7.1 wt% Fe in tailings)		
• PbS	85,000 t/1.2 wt%		
• ZnS	149,000 t/2.1 wt%		

Table A11. Cont.

Category & Factor	Data	Sources	UNFC Axis [1]
• Wissenbach shales	2,350,000 t/33.1 wt%		
• ankerit	1,611,000 t/22.7 wt%		
• main minerals in neutralisation sludge:	masses unknown; high & low concentrations of Zn & BaSO$_4$, respectively	[54]	
• carbonate	CaCO$_3$		
• clay minerals	Al$_2$O$_3$		
• zinc hydroxide	Zn(OH)$_2$		
• quartz	SiO$_2$		
• gypsum	CaSO$_4$·2 H$_2$O		
(b) physico-chemical properties			
• particle size distribution	tailings: very fine, 90% of particles < 60 µm, predominantly 2–60 µm & partially >20% below 3 µm, analysed with 4 samples from 2 drill coresneutralisation sludge: very fine, ~80% of particles < 20 µm	[53,54]	G
• geomechanical properties	classified into geomechanical category GK III according to DIN 1054: highly difficult regarding the interaction of structure & subsoil	[113]	G
• abrasiveness	expected to be abrasive (30 wt% abrasive material in tailings)	[53]	G
• water content	29 wt%, estimated mean water content	[53]	G
• toxic elements	no valorisation as soil possible due to heavy metal concentration (As, Cd, Cr, Cu, Hg, Ni, Pb, Tl, & Zn) according to guideline "LAGA TR Boden" (note: tailings are not soil per definition); classified as DK IV hazardous waste according to Landfill Regulation DepV; As, Cd, & Tl mainly associated with sulphides (As mainly with FeS$_2$ & Cd mainly with ZnS)	[53,114]	G

[1] econ.: economic aspects, env.: environmental aspects, soc.: social aspects, leg.: legal aspects. [2] WWII: Word War II. [3] AMD: acid mine drainage. [4] SLO: social license to operate.

Table A12. Basic data for the in-situ rehabilitation scenario NRR0.

Parameter	Unit	Value	Source	Remarks
surface area	m^2	315,000	estimated with Google Earth [50]	-
duration of closure & leachate phase	a	5	following scenario B in Reference [51] (p. 104)	leachate emission constant; influx assumed only to occur in closure phase until in-situ stabilisation is completed & influx of rainwater or groundwater is phase neglected
duration of aftercare phase	a	30	Landfill Ordinance DepV [61]	minimum duration according to Landfill Ordinance DepV [61]
average emission of leachate	m^3/a	39,000	average water depth for lower & upper ponds calculated based on 82 out of 90 measurements taken from Reference [53]; visible water surface measured with Google Earth [50]	based on the assumption of a constant leachate flow & that only the standing water is drained
leachate treatment	-	-	assumption	active on-site treatment unit

Table A13. Economic parameters for closure and aftercare in the in-situ stabilisation and rehabilitation scenario NRR0. A conversion rate GBP-EUR of 0.9 is assumed as per 14 August 2020 [115] and rounded up. From the referenced sources, the maximum values are chosen for a conservative approach.

Parameter	Unit	Value	Source	Remarks
In-situ Stabilisation & Surface Sealing				
final surface cover including infrastructure	€/m^2	100	[51]	closure & leachate phase
concrete injection	€/m^3	68	[69] (p. 77)	closure & leachate phase
Leachate treatment				
active on-site treatment	€/m^3	50	[51]	closure & leachate phase
Other Costs				
maintenance & repair of leachate collection system	€/(a m^2)	0.6	[51]	closure & leachate phase
monitoring of leachates	€/(a m^2)	0.4	[51]	closure & leachate phase
monitoring of groundwater	€/(a m^2)	0.3	[51]	closure & leachate phase
insurances	€/(a m^2)	0.4	[51]	closure & leachate phase
maintenance of surface sealing	€/(a m^2)	1.0	[51]	aftercare phase
maintenance of infrastructure	€/(a m^2)	0.6	[51]	aftercare phase
monitoring of settlement	€/(a m^2)	0.1	[51]	aftercare phase
monitoring of environment including weather	€/(a m^2)	0.2	[51]	aftercare phase
aftercare management, reports, & documentation	€/(a m^2)	0.6	[51]	aftercare phase

Table A14. Fixed economic and technological parameters for the techno-economic assessment of the mineral RMs recovery scenarios CRR1 and ERR2. A conversion rate USD–EUR of 0.85 is assumed as per 4 August 2020 [116].

Parameter	Unit	Value	Source	Remarks	Qty.
CAPEX					
Mining					
dredger (including cutterhead)	€	1,579,000	[117] (p. SU 12), www.cat.com/en_US/products/new/power-systems/marine-power-systems/commercial-propulsion-engines/18493267.html (accessed on 14 March 2021)	230 kW ship engine (d) [1], 272 kW cutterhead (d–e) [2], Caterpillar C18 ACERT engine used as reference	1
excavator	€	160,000	www.cat.com/en_US/products/new/equipment/excavators/medium-excavators/1000032601.html (accessed on 14 March 2021)	CAT 320 GC, 1 m^3 bucket capacity, (d)	1
wheel loader	€	269,000	[117] (p. SU 22)	157 kW (d), 3.8 m^3 bucket capacity	1
bulldozer (with ripper)	€	145,000	[117] (p. SU 28)	-	1
dump truck	€	384,000	[117] (p. SU 34)	6x6 traction, 15 m^3 loading capacity, (d)	1
rubber boat (incl. engine)	€	4800	www.marine-sales.de (accessed on 14 March 2021)	transport of crew & light material to dredger, (d)	2
twin silo (2 × 810 m^3)	€	343,000	[117] (p. Misc 92)	ensuring continuous processing plant feed & contingency for feed stream disruptions; integrated stirring function assumed to keep tailings suspended	1
slurry pump	€	24,000	[117] (p. misc 56)	41 kW (e) [3], 40 m head @ 90 m^3/h, redundant system foreseen	6

Table A14. Cont.

Parameter	Unit	Value	Source	Remarks	Qty.
pipeline	€/m	1350	[118] (p. 42)	300 mm nominal diameter, assumed to be suitable for offshore & onshore application; 800 m one-way, redundant system foreseen; water recirculation included	267
floating bodies for pipeline	€	8750	[118] (p. 46)	longest distance to cover from landing site at northern part of middle dam to bottom right corner of lower dam (480 m)	40
Processing					
processing plant reactivation	€	6,000,000	[119] (p. 13)	low value is chosen since assets & machinery were assumed to be in place & reusable	-
Infrastructure					
mine site development (paving roads, reactivating railway, etc.)	€	1,300,000	[119] (p. 13)	low value chosen due to simple mine plan, good mine site accessibility & available buildings	-
reclamation				-	
removal of assets, surface rehabilitation, & environmental monitoring	€/$t_{tailings}$	2	[101] (p. 117)	mean value assumed due to relatively small reclamation area & off-site residue disposal	-
Other Fixed Economic Parameters					
discount rate	%	15	[8] (p. 297)	low value chosen to reflect very high risk	-
contingency factor	%	30	[45] (p. 58)	accounts for required non-specified assets	-
liquidating value	%	10	[120] (p. 16)	applied to assets & machinery under mining to estimate residual value	-
mine life	a	11	estimated with Taylor's Rule [62] (p. 80)	reclamation & asset liquidation only in year 11	-
run-of-mine (ROM)	t/h	170	assumption	-	-
working days administration	d/a	260	assumption	-	-
working days mining	d/a	260	assumption	-	-
working days processing	d/a	365	assumption	-	-
shift system mining	shifts/d	2	assumption	8 h per shift	-
shift system processing	shifts/d	3	assumption	8 h per shift	-
working hours administration	h/d	8	assumption	-	-
working hours mining	h/d	16	assumption	-	-
working hours processing	h/d	24	assumption	-	-
%-NSR_{Cu} (Europe)	%	65	[62] (p. 75)	percentage of net smelter return for Cu	-
%-NSR_{Pb}	%	65	[62] (p. 75)	percentage of net smelter return for Pb	-
%-NSR_{Zn}	%	50	[62] (p. 75)	percentage of net smelter return for Zn	-

Table A14. *Cont.*

Parameter	Unit	Value	Source	Remarks	Qty.
Technological Parameters					
tailings mass	t	7,100,000	[53] (p. AP1/75)	low value chosen for conservative approach	-
pump head	m	55	[53] (p. AP5/19)	-	-
r_{Ba} [4]	%	74	[60] (p. 254)	-	-
r_{Co}	%	12	[60] (p. 254)	for ammonia leaching path of sulphides	-
r_{Cu}	%	74	[60] (p. 176)	-	-
r_{FeS2}	%	87	[60] (p. 176)	-	-
r_{Ga}	%	2	[60] (p. 254)	for ammonia leaching path of sulphides	-
r_{In}	%	26	[60] (p. 254)	for ammonia leaching path of sulphides	-
$r_{inert\ material}$	%	93	[60] (p. 254)	-	-
r_{Pb}	%	68	[60] (p. 176)	-	-
r_{Zn}	%	70	[60] (p. 176)	-	-

[1] (d): diesel engine. [2] (d–e) diesel-electric engine. [3] (e): electric engine. [4] r: recovery rate.

Table A15. Variable economic parameters for the techno-economic assessment of the mineral RMs recovery scenarios CRR1 and ERR2. A conversion rate USD–EUR of 0.85 are assumed as per 4 August 2020 [116]. Data adopted from Reference [117] if not stated otherwise.

Machine/Item	Energy Consumption [l_{diesel}/h]	Energy Consumption [$kW_{electricity}$]	Maintenance & Overhaul [€/h]	Remarks
dredger	125	-	112	fuel consumption @ 502 kW approximated based on specification sheet & CAT engine assumed to constantly deliver 502 kW, http://s7d2.scene7.com/is/content/Caterpillar/LEHM0004-00 (accessed on 15 March 2021)
excavator	13	-	13	-
wheel loader	24	-	20	-
bulldozer (with ripper)	21	-	16	-
dump truck	15	-	13	-
rubber boat (including engine)	2	-	-	no data could be retrieved for maintenance & overhaul, negligible due to expected low value
twin silo (2 × 810 m^3)	-	-	5.8	-
slurry pump	-	41	3.2	-

Table A16. Variable economic parameters for the techno-economic assessment of the mineral RMs recovery scenarios CRR1 and ERR2. A conversion rate USD–EUR of 0.85 is assumed as per 4 August 2020 [116] if not stated otherwise.

Parameter	Unit	Value	Source	Remarks	Qty.
OPEX					
mining					
machine operating costs	€/h	200	derived from Reference [117]	overhaul, maintenance, lubricants, & wear	-
diesel consumption	l/h	202	derived from Reference [117]	-	-
electric energy consumption	kW	246	derived from Reference [117]	-	-
shift supervisor	€/(a person)	78.4	based on Reference [120]	including assumed employers' share of 40%	2
machine driver	€/(a person)	58.8	based on Reference [120]	including assumed employers' share of 40%	10
metal worker	€/(a person)	70.0	based on Reference [120]	including assumed employers' share of 40%	2
processing					
processing costs	€/$t_{metal\ recovered}$	7.2	[119]	-	-
machine operating costs	€/$t_{metal\ recovered}$	10.7	[119]	electric energy only	-
shift supervisor	€/(a person)	78.4	[120]	including assumed employers' share of 40%	3
control panel operator	€/(a person)	58.8	[120]	including assumed employers' share of 40%	3
machine operator	€/(a person)	58.8	[120]	including assumed employers' share of 40%	3
metal worker	€/(a person)	70.0	[120]	including assumed employers' share of 40%	3
services & administration					
general services	€/d	5210	[119]	-	-
administrative services	€/d	1310	[119]	-	-
RM prices					
electricity	€/kWh	cf., Figure S1	raw data from Reference [85]	forecast based on yearly average prices in Germany for commercial customers from 2014–2019	-
diesel	€/l	cf., Figure S2	raw data from Reference [86]	forecast based on yearly average prices in Germany from 1950–2020	-
BaSO$_4$	€/$t_{tailings}$	cf., Figure S3	raw data from References [87–90]	forecast based on yearly BaSO$_4$ prices from 2011–2020 [1]	-
Co	€/$t_{tailings}$	cf., Figure S4	raw data from References [87,89–93]	forecast based on yearly Co prices from 1996–2020 [1]	-
Cu	€/$t_{tailings}$	cf., Figure S5	raw data from Reference [94]	forecast based on monthly Cu prices from November 1999–March 2021 [1] & price per tonne tailings estimated after Wellmer et al. [62] (p. 47 ff.)	-
Ga	€/$t_{tailings}$	cf., Figure S6	raw data from References [87,89–93]	forecast based on yearly Ga prices from 1999–2020 [1]	-
In	€/$t_{tailings}$	cf., Figure S7	raw data from References [87,89–93]	forecast based on yearly In prices from 1999–2020 [1]	-
Pb	€/$t_{tailings}$	cf., Figure S8	raw data from Reference [95]	forecast based on monthly Pb prices from November 1999–March 2021 [1] & price per tonne tailings estimated after Wellmer et al. [62] (p. 74 ff.)	-
Zn	€/$t_{tailings}$	cf., Figure S9	raw data from Reference [96]	forecast based on monthly Zn prices from November 1999–March 2021 [1] & price per tonne tailings estimated after Wellmer et al. [62] (p. 74 ff.)	-
residue sales	€/t	5.0	assumption	intended valorisation as filler in construction materials; reference value for high-quality sand in Goslar is EUR 19.5 (www.recyclingpark.de/startseite.html, accessed on 2 June 2021); lower price assumed to estimate conservatively due to lack of information on effort to condition residues	-
residue disposal	€/t	40.0	[53] (p. AP7-9/58)	high value chosen to estimate conservatively	-

[1] under consideration of monthly/yearly USD–EUR conversion rates.

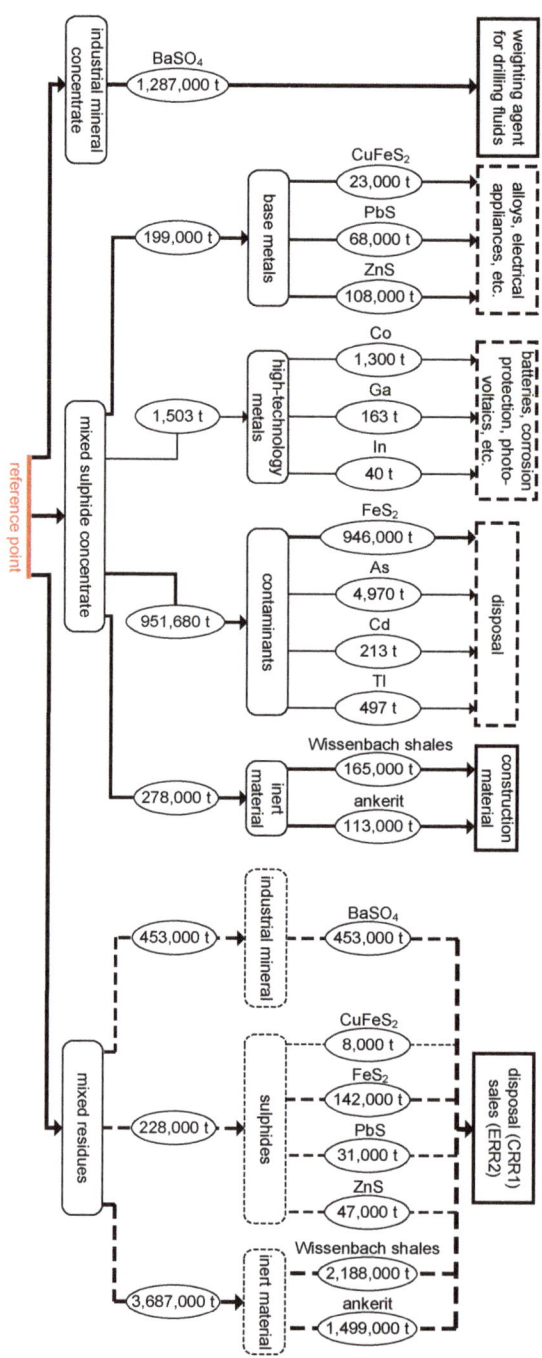

Figure A1. Detailed production breakdown of 10-year material flows for the RMs recovery scenarios (CRR1, ERR2).

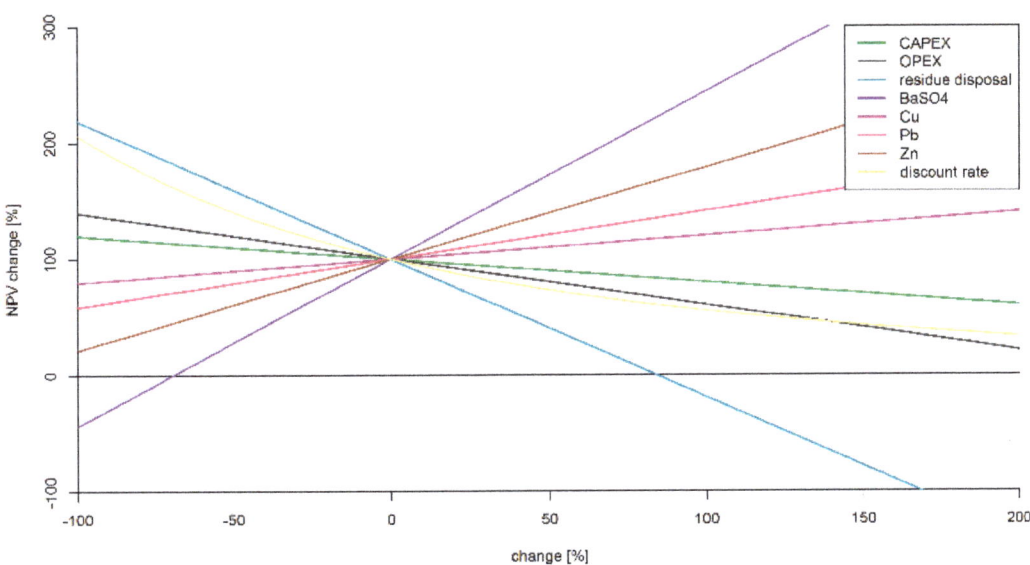

Figure A2. Results of the sensitivity analysis of the conventional mineral RMs recovery scenario (CRR1m) with mean price forecast and a discount rate of 15%.

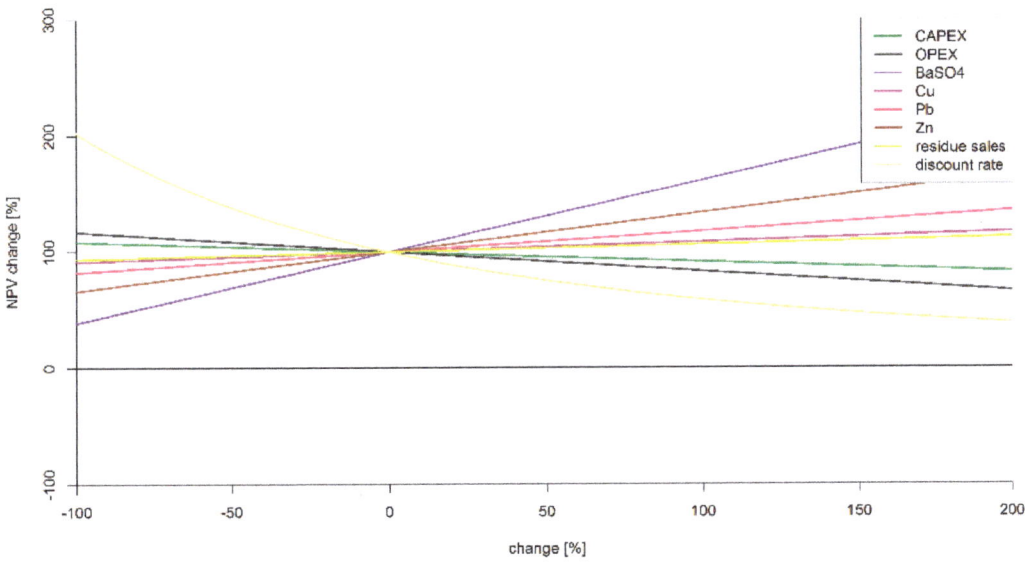

Figure A3. Results of the sensitivity analysis of the enhanced mineral RMs recovery scenario (ERR2m) with mean price forecast and a discount rate of 15%.

Table A17. Overall project rating with the UNFC-compliant categorisation matrix of the degree of confidence in the geological estimates (G).

Factor	Indicator	UNFC Rating	Justification	Source
Geological conditions (relevant for project development)				
(1) quantity	degree of geological certainty: medium	G2	NRR0, CRR1, & ERR2: deposit modelled based on direct data on 10 drill cores from lower pond, and pre-processed historical data on 14 & 12 drill cores from lower & upper pond, respectively. Model was validated with historical production data. Extension & volume of TSF known with medium confidence. Overall knowledge on mineral quantity with medium confidence in both ponds. Knowledge gap on quantity of neutralisation sludge & other dumped material.	[53]
(2) quality	degree of geological certainty: medium	G2	NRR0, CRR1, & ERR2: physico-chemical properties known with medium confidence.	[53]
(3) homogeneity	degree of geological certainty: medium	G2	NRR0, CRR1, & ERR2: mineral distribution in lower pond known with medium confidence. Knowledge gap on distribution of tailings & neutralisation sludge in both ponds.	[53,54]

Table A18. Overall project rating with the UNFC-compliant categorisation matrix for the technical feasibility (F).

Factor	Indicator	UNFC Rating	Justification	Source
TSF condition & risks (relevant for project development)				
(4) ordnance	degree of knowledge: unclarified	F3	NRR0, CRR1, & ERR2: existence cannot be excluded based on historical data. Requires clarification.	[53]
Mine planning considerations (relevant for project execution)				
(5) mine/ operational design	level of detail of planning: basic	F3	NRR0, CRR1, & ERR2: conceptual planning.	-
(6) metallurgical testwork	degree of research on mineral processing: -	-	NRR0: factor not applicable.	-
	laboratory scale	F3	CRR1 & ERR2: extraction of $BaSO_4$, Co, Cu, Ga, In, Pb, Zn, & inert material (Wissenbach shales, ankerit) evaluated in discontinuous laboratory experiments on tailings from lower pond.	[60]
(7) water consumption	percentage of recycled water: high (>80%)	F1	CRR1 & ERR2: water recirculated in dredging operation. Processing water can be recirculated, too.	[53]
	unclarified	F3	NRR0: unclear if TSF water can be used for making concrete.	-

Table A18. Cont.

Factor	Indicator	UNFC Rating	Justification	Source
Infrastructure (relevant for project development)				
(8) real estate	condition of infrastructure:			
	highly developed	F1	NRR0, CRR1, & ERR2: buildings & land from former processing available.	[53]
(9) mining & processing	condition of equipment:			
	-	-	NRR0: not applicable since specialised non-mining equipment is required.	-
	bleak	F3	CRR1 & ERR2: unclarified.	-
(10) utilities	condition of infrastructure:			
	acceptable	F2	NRR0, CRR1, & ERR2: access to public electricity, gas, & water grid assumed.	based on observation on Google Earth [50]
(11) transportation & access	condition of infrastructure:			
	acceptable	F2	NRR0, CRR1, & ERR2: dirt roads, federal highway B6 ~1.6 km N air-line distance from TSF & public railway ~500 m E air-line distance from TSF, disused railway tracks from processing plant Bollrich to public network (estimated abandonment in 1988).	[53], observed on Google Earth [50]
Post-mining state (relevant for future impacts)				
(12) residue storage safety	suitability of new disposal site for safe storage:			
	unclarified	F3	NRR0: predicting long-term stability might be difficult. CRR1 & ERR2: new disposal site unknown.	[69]
(13) rehabilitation	level of detail of planning:			
	conceptual	F2	NRR0, CRR1, & ERR2: conceptual planning.	-

Table A19. Overall project rating with the UNFC-compliant categorisation matrix of the economic viability (E a).

Factor	Indicator	UNFC Rating	Justification	Source
Microeconomic aspects (relevant for project development)				
(14) economic viability	discounted cash flow over projected LOM:			
	positive (NPV >> 0€)	E3.1a	CRR1m & ERR2m: NPVs of EUR 73 mio. & EUR 172 mio., respectively, with mean price forecast.	-
	negative (NPV << 0€)	E3.3a	NRR0: costs of EUR 125 mio. incurred.	-
(15) economic uncertainty	uncertainty of cash flow in pessimistic scenario:			
	-	-	NRR0: no forecast performed.	-
	low (NPV in pessimistic scenario >> 0€)	E3.1a	ERR2p: NPV = EUR 73 mio.	-
	high (NPV in pessimistic scenario << 0€)	E3.3a	CRR1p: NPV = EUR −17 mio.	-

Table A19. Cont.

Factor	Indicator	UNFC Rating	Justification	Source
Financial aspects (relevant for project development)				
(16) investment conditions	country rank in the ease-of-doing-business Index.			
	-	-	NRR0: not applicable since company works on assignment basis.	-
	high (<75)	E3.1a	CRR1 & ERR2: country rank 22 (Germany). Good investment conditions assumed.	[121]
(17) financial support	probability of approval:			
	high	E3.1a	CRR1 & ERR2: research on TSF was funded publicly & positive results give rise to the assumption that follow-up project proposal REWIMET might be accepted.	-
	no financial support scheme available	E3.3a	NRR0: no financial support scheme known at the moment.	-

Table A20. Overall project rating with the UNFC-compliant categorisation matrix for the environmental viability (E b).

Factor	Indicator	UNFC Rating	Justification	Source
Environmental impacts during project execution				
(18) air emission	risk of dust emission:			
	unclarified	E3.3b	NRR0: unclarified if TSF needs to be drained prior to concrete injection, which could lead to wind erosion of the tailings.	-
	high (>80%)	E3.1b	CRR1 & ERR2: complete submersion of tailings in dredging operation.	-
(19) liquid effluent emission	risk of groundwater contamination:			
	low	E3.1b	NRR0, CRR1, & ERR2: status quo is expected to be retained.	-
(20) noise emission	expected degree of impact:			
	medium	E3.2b	NRR0, CRR1, & ERR2: constant noise emission from TSF in 2 working shifts from Mondays to Fridays. Noise is expected to be audible, especially in the surrounding mountain area & areas on the same plane. It is possible that the noise would not be audible in residential areas to topography.	based on observation on Google Earth [50]
Environmental impacts after projection execution			CRR1 & ERR2: the processing plant is to be soundproofed.	
(21) biodiversity	total number of protected species that are affected by mining activities & that will be resettled on post-mining land:			
	none (0%)	E3b	NRR0, CRR1, & ERR2: protected flora & fauna species were sighted during an on-site inspection. Capturing the exact types & number of species is required for planning a resettlement or other compensation measures.	[53]
(22) land use	freely available post-mining land:			
	some (50–80%)	E3.2b	NRR0: surface area of current wet cover is made available for reuse. CRR1 & ERR2: original topography is restored. NRR0, CRR1, & ERR2: it is expected that a solution for the collection & further treatment of the neutralisation sludge requires a permanent land use.	-
(23) material reactivity	reduction in reactive material's mass:			
	high (>80%)	E3.1b	CRR1: 84 wt% of sulphides leave the system boundaries as commodities. ERR2: all tailings are valorised.	-
	low (<50%)	E3.3b	NRR0: factually, reactive materials remain in place. Long-term stability difficult to predict.	[69]

Table A21. Overall project rating with the UNFC-compliant categorisation matrix for the social viability (E c).

Factor	Indicator	UNFC Rating	Justification	Source
Social impacts during project execution				
(24) local community	probability of approval through active commitment:			
	medium (50–80%)	E3.2c	CRR1 & ERR2: first indication of positive prospects by stakeholder assessment (local government, industry, university, & environmental NGOs). Local population's opinion unknown.	[52]
	unclarified	E3.3c	NRR0: no data available.	-
(25) health & safety	total number of complaints or prosecutions for non-compliance in planning phase:			
	none	E3.3c	NRR0, CRR1, & ERR2: plans have not been communicated publicly.	-
(26) human rights & business ethics	total number of complaints or prosecutions for non-compliance in planning phase:			
	none	E3.3c	NRR0, CRR1, & ERR2: plans have not been communicated publicly.	-
Social impacts due to project execution				
(27) wealth distribution	total number of complaints or prosecutions for non-compliance in planning phase:			
	none	E3.3c	NRR0, CRR1, & ERR2: plans have not been communicated publicly.	-
(28) investment in local human capital	percentage of employees sourced from local communities:			
	unclarified	E3.3c	NRR0: it can be expected that an external contractor must be hired due to the special character of the required services. Aftercare measures could be carried out by local workers. CRR1 & ERR2: unclarified how many local workers could be employed.	-
(29) degree of RM recovery	residue disposal:			
	complete residue valorisation	E1c	ERR2: no loss since all tailings are valorised.	-
	mixed disposal	E3.2c	CRR1: it is assumed that the site for the disposal of new residues has no option to store different residues separately.	-
	sterilisation	E3.3c	NRR0: access to RM potential for future generations with reasonable effort prevented.	-
(30) RM valorisation	total mass reduction as percentage of original tailings mass:			
	high (>80%)	E3.1c	ERR2: all tailings are valorised.	-
	low (<50%)	E3.3c	NRR0: no valorisation takes place. CRR1: 38 wt% of tailings are valorised.	-
Social impacts after project execution				
(31) aftercare	duration of aftercare measures:			
	short-term (up to 5 years)	E1c	CRR1 & ERR2: aftercare assumed to be complete after 1 year	-
	long-term (more than 30 years)	E3c	NRR0: long-term behaviour difficult to predict & long-term monitoring might be necessary.	[69]
(32) landscape	impact on the environment:			
	non-perceptible	E1c	CRR1 & ERR2: former topography is restored.	-
	partially perceptible	E2c	NRR0: is expected to be well integrated into landscape with an according surface design. Main dam remains perceptible.	-

Table A22. Overall project rating with the UNFC-compliant categorisation matrix for the legal viability (E d).

Factor	Indicator	UNFC Rating	Justification	Source
Legal situation (relevant for project development)				
(33) right of mining	state of development: application not begun or unclarified	E3.3d	NRR0, CRR1, & ERR2: no concrete activities initiated.	-
(34) environmental protection	state of development: application not begun or unclarified	E3.3d	NRR0, CRR1, & ERR2: no concrete activities initiated.	-
(35) water protection	state of development: application not begun or unclarified	E3.3d	NRR0, CRR1, & ERR2: no concrete activities initiated.	-

Table A23. Rating of individual RMs with the UNFC-compliant categorisation matrix for the degree of confidence in the geological estimates (G).

Factor	Indicator	UNFC Rating	Justification	Source
Geological conditions (relevant for project development)				
(36) quantity	degree of geological certainty: medium	G2	CRR1 & ERR2: knowledge on $BaSO_4$, Cu, FeS_2, Pb, Zn, & inert material (Wissenbach shales, ankerit) with medium confidence in both ponds.	[53,54]
	low	G3	CRR1 & ERR2: knowledge on Co, Ga, & In with medium confidence in lower pond. Co, Ga, & In quantity in upper pond inferred.	[53]
(37) quality	degree of geological certainty: medium	G2	CRR1 & ERR2: knowledge on $BaSO_4$, Cu, FeS_2, Pb, Zn, & inert material (Wissenbach shales, ankerit) with medium confidence in both ponds.	[53,54]
	low	G3	CRR1 & ERR2: knowledge on Co, Ga, & In with medium confidence in lower pond. Co, Ga, & In quantity in upper pond inferred.	[53]
(38) homogeneity	degree of geological certainty: medium	G2	CRR1 & ERR2: knowledge on the distribution of $BaSO_4$, Cu, FeS_2, Pb, Zn, & inert material (Wissenbach shales, ankerit) with medium confidence.	[53,54]
	low	G3	CRR1 & ERR2: knowledge on the distribution of Co, Ga, & In with medium confidence in lower pond. Knowledge on Co, Ga, & In in upper pond inferred.	[53]

Table A24. Rating of individual RMs with the UNFC-compliant categorisation matrix for the technical feasibility (F).

Factor	Indicator	UNFC Rating	Justification	Source
Mine planning considerations (relevant for project execution)				
(39) recoverability	percentage of RM which is extracted from the tailings: high (>80%)	F1	CRR1 & ERR2: FeS_2 (87 wt% recovered in mixed sulphide concentrate), inert material (Wissenbach shales, ankerit) (93 wt% are recovered with the new residues).	[60]
	medium (50–80%)	F2	CRR1 & ERR2: $BaSO_4$ (74 wt%), Cu (74 wt%), Pb (68 wt%), Zn (70 wt%).	[60]
	low (>50%)	F3	CRR1, ERR2: Co (12 wt%), Ga (2 wt%), In (26 wt%).	[60]

Table A25. Rating of individual RMs with the UNFC-compliant categorisation matrix for the economic viability (E a).

Factor	Indicator	UNFC Rating	Justification	Source
Microeconomic aspects (relevant for project development)				
(40) demand	favourable conditions for RM extraction:			
	yes	E3.1a	CRR1 & ERR2: there is a demand for $BaSO_4$, Cu, Pb, Zn, Co, Ga, & In	[122]
	conditionally	E3.2a	CRR1 & ERR2: Fe & H_2SO_4 could theoretically be produced from $CuFeS_2$ & FeS_2.	[123]
	no	E3.3a	CRR1 & ERR2: residues theoretically usable in construction materials, but experiments are necessary. Currently, there is per se not a demand for residues so that a potential application of the inert fraction (Wissenbach shales, ankerit) of the new residues needs to be clarified.	-
(41) RM criticality	allocation to EC's criticality assessment:			
	CRM	E1a	CRR1 & ERR2: $BaSO_4$, Co, Ga, & In.	[112]
	high economic importance or supply risk	E2a	CRR1 & ERR2: Cu, Pb, S (from $CuFeS_2$ & FeS_2), & Zn.	[112]
	no criticality	E3a	CRR1 & ERR2: inert material (Wissenbach shales, ankerit).	
(42) price development	forecasted mean price development over the project's duration:			
	-	-	CRR1 & ERR2: FeS_2 is recovered as a non-paid co-product, & no price forecast was performed for the inert material (Wissenbach shales, ankerit).	-
	positive trend	E3.1a	CRR1 & ERR2: $BaSO_4$, Co, In.	Figures S3, S4 and S7
	stagnant trend	E3.2a	CRR1 & ERR2: Pb, Zn.	Figures S8 and S9
	negative trend	E3.3a	CRR1 & ERR2: Cu, Ga.	Figures S5 and S6

Table A26. Rating of individual RMs with the UNFC-compliant categorisation matrix for the environmental viability (E b).

Factor	Indicator	UNFC Rating	Justification	Source
Impacts after project execution				
(43) solid matter	concentration of RM solid matter in new residues to qualify for class DK 0 (inert waste) according to German Landfill Regulation DepV [61]:			
	-	-	NRR0: not applicable since no new residues are produced. ERR2: not applicable since no new residues are disposed of.	-
	non-hazardous material	E1b	CRR1 & ERR2: inert material (Wissenbach shales, ankerit).	-
	threshold value not exceeded	E3.1b	CRR1: Cu, Zn.	[60]
	threshold value exceeded	E3.2b	CRR1: Pb.	[60]
(44) eluate	concentration of RM in eluate from new residues to qualify for class DK 0 (inert waste) according to German Landfill Regulation DepV [61]:			
	-	-	NRR0: not applicable since no new residues are produced. ERR2: not applicable since no new residues are disposed of.	-
	non-hazardous material	E1b	CRR1 & ERR2: inert material (Wissenbach shales, ankerit).	-
	threshold value not exceeded	E3.1b	CRR1: Ba, Cu, Zn.	[60]
	threshold value exceeded	E3.2b	CRR1: Pb.	[60]

References

1. Henckens, M.L.C.M.; Driessen, P.P.J.; Worrell, E. Molybdenum resources: Their depletion and safeguarding for future generations. *Resour. Conserv. Recycl.* **2018**, *134*, 61–69. [CrossRef]
2. Kleijn, R.; van der Voet, E.; Kramer, G.J.; van Oers, L.; van der Giesen, C. Metal requirements of low-carbon power generation. *Energy* **2011**, *36*, 5640–5648. [CrossRef]
3. Maung, K.N.; Hashimoto, S.; Mizukami, M.; Morozumi, M.; Lwin, C.M. Assessment of the Secondary Copper Reserves of Nations. *Environ. Sci. Technol.* **2017**, *51*, 3824–3832. [CrossRef]
4. Watari, T.; McLellan, B.C.; Giurco, D.; Dominish, E.; Yamasue, E.; Nansai, K. Total material requirement for the global energy transition to 2050: A focus on transport and electricity. *Resour. Conserv. Recycl.* **2019**, *148*, 91–103. [CrossRef]
5. Elshkaki, A.; Graedel, T.E.; Ciacci, L.; Reck, B.K. Resource Demand Scenarios for the Major Metals. *Environ. Sci. Technol.* **2018**, *52*, 2491–2497. [CrossRef]
6. Valenta, R.K.; Kemp, D.; Owen, J.R.; Corder, G.D.; Lèbre, É. Re-thinking complex orebodies: Consequences for the future world supply of copper. *J. Clean. Prod.* **2019**, *220*, 816–826. [CrossRef]
7. Fellner, J.; Lederer, J.; Scharff, C.; Laner, D. Present Potentials and Limitations of a Circular Economy with Respect to Primary Raw Material Demand. *J. Ind. Ecol.* **2017**, *21*, 494–496. [CrossRef]
8. Revuelta, M.B. *Mineral Resources*; Springer International Publishing: Cham, Germany, 2018; 653p.
9. Schoenberger, E. Environmentally sustainable mining: The case of tailings storage facilities. *Resour. Policy* **2016**, *49*, 119–128. [CrossRef]
10. Wei, Z.; Yin, G.; Wang, J.G.; Wan, L.; Li, G. Design, construction and management of tailings storage facilities for surface disposal in China: Case studies of failures. *Waste Manag. Res.* **2013**, *31*, 106–112. [CrossRef]
11. Giurco, D.; Cooper, C. Mining and sustainability: Asking the right questions. *Miner. Eng.* **2012**, *29*, 3–12. [CrossRef]
12. Franks, D.M.; Boger, D.V.; Côte, C.M.; Mulligan, D.R. Sustainable development principles for the disposal of mining and mineral processing wastes. *Resour. Policy* **2011**, *36*, 114–122. [CrossRef]
13. Lottermoser, B. *Mine Wastes*, 3rd ed.; Springer: Berlin/Heidelberg, Germany, 2010; 400p.
14. Worrall, R.; Neil, D.; Brereton, D.; Mulligan, D. Towards a sustainability criteria and indicators framework for legacy mine land. *J. Clean. Prod.* **2009**, *17*, 1426–1434. [CrossRef]
15. Laurence, D. Establishing a sustainable mining operation: An overview. *J. Clean. Prod.* **2011**, *19*, 278–284. [CrossRef]
16. Roche, C.; Thygesen, K.; Baker, E. *Mine Tailings Storage: Safety Is No Accident: A UNEP Rapid Response Assessment.* United Nations Environment Programme and GRID-Arendal, Nairobi and Arendal; 2017; ISBN 978-827-701-170-7. Available online: https://www.grida.no/publications/383 (accessed on 10 January 2020).
17. Anawar, H.M. Sustainable rehabilitation of mining waste and acid mine drainage using geochemistry, mine type, mineralogy, texture, ore extraction and climate knowledge. *J. Environ. Manag.* **2015**, *158*, 111–121. [CrossRef]
18. Luptakova, A.; Ubaldini, S.; Macingova, E.; Fornari, P.; Giuliano, V. Application of physical–chemical and biological–chemical methods for heavy metals removal from acid mine drainage. *Process Biochem.* **2012**, *47*, 1633–1639. [CrossRef]
19. Silva Rotta, L.H.; Alcântara, E.; Park, E.; Negri, R.G.; Lin, Y.N.; Bernardo, N.; Mendes, T.S.G.; Souza Filho, C.R. The 2019 Brumadinho tailings dam collapse: Possible cause and impacts of the worst human and environmental disaster in Brazil. *Int. J. Appl. Earth. Obs. Geoinf.* **2020**, *90*, 102119. [CrossRef]
20. Lyu, Z.; Chai, J.; Xu, Z.; Qin, Y.; Cao, J. A Comprehensive Review on Reasons for Tailings Dam Failures Based on Case History. *Adv. Civ. Eng.* **2019**, *2019*, 1–18. [CrossRef]
21. World Information System on Energy Uranium Project (WISE). Chronology of Major Tailings Dam Failures. 2021. Available online: http://www.wise-uranium.org/mdaf.html (accessed on 2 February 2021).
22. Lèbre, É.; Stringer, M.; Svobodova, K.; Owen, J.R.; Kemp, D.; Côte, C.; Arratia-Solar, A.; Valenta, R.K. The social and environmental complexities of extracting energy transition metals. *Nat. Commun.* **2020**, *11*, 4823. [CrossRef] [PubMed]
23. Owen, J.R.; Kemp, D.; Lèbre, É.; Svobodova, K.; Pérez Murillo, G. Catastrophic tailings dam failures and disaster risk disclosure. *Int. J. Disaster Risk Reduct.* **2020**, *42*, 101361. [CrossRef]
24. Žibret, G.; Lemiere, B.; Mendez, A.-M.; Cormio, C.; Sinnett, D.; Cleall, P.; Szabó, K.; Carvalho, M.T. National Mineral Waste Databases as an Information Source for Assessing Material Recovery Potential from Mine Waste, Tailings and Metallurgical Waste. *Minerals* **2020**, *10*, 446. [CrossRef]
25. European Commission (EC). Towards a Circular Economy: A Zero Waste Programme for Europe: COM(2014) 398 Final. 2014. Available online: https://ec.europa.eu/environment/circular-economy/pdf/circular-economy-communication.pdf (accessed on 4 August 2019).
26. European Commission (EC). A New Circular Economy Action Plan for a Cleaner and More Competitive Europe. COM(2020) 98 Final. 2020. Available online: https://eur-lex.europa.eu/resource.html?uri=cellar:9903b325-6388-11ea-b735-01aa75ed71a1.0017.02/DOC_1&format=PDF (accessed on 14 January 2021).
27. Nuss, P.; Blengini, G.A. Towards better monitoring of technology critical elements in Europe: Coupling of natural and anthropogenic cycles. *Sci. Total Environ.* **2018**, *613*, 569–578. [CrossRef] [PubMed]
28. Falagán, C.; Grail, B.M.; Johnson, D.B. New approaches for extracting and recovering metals from mine tailings. *Miner. Eng.* **2017**, *106*, 71–78. [CrossRef]

29. Kuhn, K.; Meima, J.A. Characterization and Economic Potential of Historic Tailings from Gravity Separation: Implications from a Mine Waste Dump (Pb-Ag) in the Harz Mountains Mining District, Germany. *Minerals* **2019**, *9*, 303. [CrossRef]
30. López, F.; García-Díaz, I.; Rodríguez Largo, O.; Polonio, F.; Llorens, T. Recovery and Purification of Tin from Tailings from the Penouta Sn–Ta–Nb Deposit. *Minerals* **2018**, *8*, 20. [CrossRef]
31. Niu, H.; Abdulkareem, M.; Sreenivasan, H.; Kantola, A.M.; Havukainen, J.; Horttanainen, M.; Telkki, V.-V.; Kinnunen, P.; Illikainen, M. Recycling mica and carbonate-rich mine tailings in alkali-activated composites: A synergy with metakaolin. *Miner. Eng.* **2020**, *157*. [CrossRef]
32. Pashkevich, M.A.; Alekseenko, A.V. Reutilization Prospects of Diamond Clay Tailings at the Lomonosov Mine, Northwestern Russia. *Minerals* **2020**, *10*, 517. [CrossRef]
33. Tang, C.; Li, K.; Ni, W.; Fan, D. Recovering Iron from Iron Ore Tailings and Preparing Concrete Composite Admixtures. *Minerals* **2019**, *9*, 232. [CrossRef]
34. Alfonso, P.; Tomasa, O.; Garcia-Valles, M.; Tarragó, M.; Martínez, S.; Esteves, H. Potential of tungsten tailings as glass raw materials. *Mater. Lett.* **2020**, *228*, 456–458. [CrossRef]
35. Okereafor, U.; Makhatha, M.; Mekuto, L.; Mavumengwana, V. Gold Mine Tailings: A Potential Source of Silica Sand for Glass Making. *Minerals* **2020**, *10*, 448. [CrossRef]
36. Zheng, W.; Cao, H.; Zhong, J.; Qian, S.; Peng, Z.; Shen, C. CaO–MgO–Al2O3–SiO2 glass-ceramics from lithium porcelain clay tailings for new building materials. *J. Non-Cryst. Solids* **2015**, *409*, 27–33. [CrossRef]
37. Attila Resources. Attila to Acquire the Century zinc Mine. 2017. Available online: https://www.newcenturyresources.com/wp-content/uploads/2018/01/170301-AYA-Acquisition-of-Century-ASX-Ann.pdf (accessed on 22 May 2021).
38. Campbell, M.D.; Absolon, V.; King, J.; David, C.M. *Precious Metal Resources of the Hellyer Mine Tailings*; 2015; Available online: http://www.i2massociates.com/downloads/I2MHellyerTailingsResourcesMar9-2015Rev.pdf (accessed on 22 May 2021).
39. Cronwright, M.; Gasela, I.; Derbyshire, J. *Kamativi Lithium Tailings Project*; 2018; Available online: http://sectornewswire.com/NI43-101TechnicalReport-Kamativi-Li-Nov-2018.pdf (accessed on 22 May 2021).
40. Johansson, N.; Krook, J.; Eklund, M.; Berglund, B. An integrated review of concepts and initiatives for mining the technosphere: Towards a new taxonomy. *J. Clean. Prod.* **2013**, *55*, 35–44. [CrossRef]
41. Corder, G. Mining and sustainable development. In *Mining in the Asia-Pacific*; O'Callaghan, T., Graetz, G., Eds.; Springer International Publishing: Cham, Germany, 2017; pp. 253–269.
42. United Nations General Assembly. Transforming Our World: The 2030 Agenda for Sustainable Development (A/RES/70/1). 2015. Available online: https://www.un.org/ga/search/view_doc.asp?symbol=A/RES/70/1&Lang=E (accessed on 16 May 2021).
43. Suppes, R.; Heuss-Aßbichler, S. How to Identify Potentials and Barriers of Raw Materials Recovery from Tailings? Part I: A UNFC-Compliant Screening Approach for Site Selection. *Resources* **2021**, *10*, 26. [CrossRef]
44. United Nations Economic Commission for Europe (UNECE). United Nations Framework Classification for Resources—Update 2019. 2020, p. 20. Available online: https://www.unece.org/fileadmin/DAM/energy/se/pdfs/UNFC/publ/UNFC_ES61_Update_2019.pdf (accessed on 13 November 2020).
45. Committee for Mineral Reserves International Reporting Standards (CRIRSCO). International Reporting Template for the Public Reporting of Exploration Results, Mineral Resources and Mineral Reserves. 2019. Available online: http://www.crirsco.com/templates/CRIRSCO_International_Reporting_Template_November_2019.pdf (accessed on 9 June 2020).
46. Winterstetter, A.; Heuss-Assbichler, S.; Stegemann, J.; Kral, U.; Wäger, P.; Osmani, M.; Rechberger, H. The role of anthropogenic resource classification in supporting the transition to a circular economy. *J. Clean. Prod.* **2021**, *297*, 126753. [CrossRef]
47. Heuss-Aßbichler, S.; Kral, U.; Løvik, A.; Mueller, S.; Simoni, M.; Stegemann, J.; Wäger, P.; Horváth, Z.; Winterstetter, A. Strategic Roadmap on Sustainable Management of Anthropogenic Resources. 2020. Available online: https://zenodo.org/record/3739269#.X6WBG1Bo3b1 (accessed on 6 November 2020).
48. Lederer, J.; Kleemann, F.; Ossberger, M.; Rechberger, H.; Fellner, J. Prospecting and Exploring Anthropogenic Resource Deposits: The Case Study of Vienna's Subway Network. *J. Ind. Ecol.* **2016**, *20*, 1320–1333. [CrossRef]
49. Suppes, R.; Heuss-Aßbichler, S. Resource potential of mine wastes: A conventional and sustainable perspective on a case study tailings mining project. *J. Clean. Prod.* **2021**, 126446. [CrossRef]
50. Google Earth. Available online: https://www.google.com/earth/ (accessed on 6 July 2021).
51. Stegmann, R.; Heyer, K.-U.; Hupe, K. Landfill Aftercare—Options for Action, Duration, Costs and Quantitative Criteria for the Discharge from Aftercare; Hamburg (Germany). 2006. Available online: http://www.ifas-hamburg.de/PDF/UFOPLAN_IFAS.pdf (accessed on 29 July 2020). (In German).
52. Bleicher, A.; David, M.; Rutjes, H. When environmental legacy becomes a resource: On the making of secondary resources. *Geoforum* **2019**, *101*, 18–27. [CrossRef]
53. Goldmann, D.; Zeller, T.; Niewisch, T.; Klesse, L.; Kammer, U.; Poggendorf, C.; Stöbich, J. *Recycling of Mine Processing Wastes for the Extraction of Economically Strategic Metals Using the Example of Tailings at the Bollrich in Goslar (REWITA): Final Report*; TU Clausthal: Clausthal-Zellerfeld, Germany, 2019; Available online: https://www.tib.eu/de/suchen/id/TIBKAT:1688127496/ (accessed on 22 July 2020). (In German).
54. Woltemate, I. Assessment of the Geochemical and Sedimentpetrographic Significance of Drilling Samples from Flotation Tailings in Two Tailing Ponds of the Rammelsberg Ore Mine. Ph.D. Thesis, University of Hanover, Hanover, Germany, 5 November 1987. (In German).

55. Brunner, P.H.; Rechberger, H. *Practical Handbook of Material Flow Analysis*; Lewis Publishers: Boca Raton, FL, USA, 2004; 318p.
56. Zhou, X.; Lin, H. Sensitivity analysis. In *Encyclopedia of GIS*; Shekhar, S., Xiong, H., Zhou, X., Eds.; Springer International Publishing: Cham, Germany, 2017; pp. 1884–1887.
57. Eichhorn, P. *Ore Processing Rammelsberg—Origin, Operation, Comparison; Goslar (Germany)*; 2012; Available online: https://docplayer.org/16359673-Erzaufbereitung-rammelsberg.html (accessed on 30 August 2020). (In German)
58. European Commission. Communication on the 2017 list of Critical Raw Materials for the EU. COM (2017) 490 final. 2017. Available online: https://eur-lex.europa.eu/legal-content/EN/TXT/PDF/?uri=CELEX:52017DC0490&from=EN (accessed on 14 August 2019).
59. European Commission (EC). Critical Raw Materials Resilience: Charting a Path towards Greater Security and Sustainability. COM (2020) 474 final. 2020. Available online: https://eur-lex.europa.eu/legal-content/EN/TXT/PDF/?uri=CELEX:52020DC0474&from=EN (accessed on 21 December 2020).
60. Roemer, F. Investigations into the Processing of Deposited Flotation Residues at the Bollrich Tailings Pond with Special Regard to the Extraction of Raw Materials of Strategic Economic Importance. Ph.D. Thesis, Technical University of Clausthal, Clausthal-Zellerfeld, Germany, 4 February 2020. (In German)
61. German Federal Ministry of Justice and Consumer Protection. Ordinance on Landfills and Long-Term Storage Facilities (Landfill Ordinance—DepV)—Landfill ordinance of 27 April 2009 (BGBl. I p. 900), last amended by Article 2 of the ordinance of 30 June 2020 (BGBl. I p. 1533). 2009. Available online: https://www.gesetze-im-internet.de/depv_2009/DepV.pdf (accessed on 11 April 2021). (In German)
62. Wellmer, F.-W.; Dalheimer, M.; Wagner, M. *Economic Evaluations in Exploration*, 2nd ed.; Springer: Berlin, Germany, 2008.
63. Federal Office of Justice. Federal Soil Protection and Contaminated Sites Ordinance (BBodSchV). 1999. Available online: https://www.gesetze-im-internet.de/bbodschv/anhang_2.html (accessed on 4 April 2021). (In German)
64. District of Goslar | Environmental Service. Map of contaminated Ground. 2020. Available online: https://www.landkreis-goslar.de/index.phtml?mNavID=1749.35&sNavID=1749.35&La=1 (accessed on 30 September 2020). (In German)
65. Ackers, W.; Pechmann, S. Integrated Urban Development Concept Goslar 2025. Goslar (Germany). 2011. Available online: https://www.goslar.de/stadt-buerger/stadtentwicklung/isek-2025 (accessed on 28 July 2020). (In German)
66. Gesellschaft für Grundbau und Umwelttechnik mbH (GGU). *Gelmke Dam Safety Report*; Braunschweig, Germany, Unpublished material; 2003. (In German)
67. Poggendorf, C.; Rüpke, A.; Gock, E.; Saheli, H.; Kuhn, K.; Martin, T. Utilisation of the Raw Material Potential of Mining and Metallurgical Dumps Using the Example of the Western Harz Region. 2015, p. 22. Available online: https://www.researchgate.net/profile/Tina_Martin5/publication/303941732_Nutzung_des_Rohstoffpotentials_von_Bergbau_und_Huttenhalden_am_Beispiel_des_Westharzes/links/575fbf8d08aed884621bbfa3/Nutzung-des-Rohstoffpotentials-von-Bergbau-und-Huettenhalden-am-Beispiel-des-Westharzes.pdf (accessed on 13 November 2020). (In German)
68. Expert Group on Resource Management (EGRM). United Nations Framework Classification for Resources—Draft Update Version 2019 EGRM-10/2019/INF.2. 2019. Available online: https://www.unece.org/fileadmin/DAM/energy/se/pdfs/egrm/egrm10_apr2019/UNFC_Update_2019_2.1_clean_rev.pdf (accessed on 1 June 2020).
69. CL:AIRE Technology and Research Group. *Contaminated Land Remediation*; 2010; Available online: https://www.claire.co.uk/ (accessed on 14 August 2020).
70. European Commission (EC). *Critical Raw Materials for Strategic Technologies and Sectors in the EU—A Foresight Study*; European Union: Luxembourg, 2020; 100p.
71. Krzemień, A.; Riesgo Fernández, P.; Suárez Sánchez, A.; Diego Álvarez, I. Beyond the pan-european standard for reporting of exploration results, mineral resources and reserves. *Resour. Policy* **2016**, *49*, 81–91. [CrossRef]
72. Norgate, T.; Haque, N. Energy and greenhouse gas impacts of mining and mineral processing operations. *J. Clean. Prod.* **2010**, *18*, 266–274. [CrossRef]
73. Bouchard, J.; Sbarbaro, D.; Desbiens, A. Plant automation for energy-efficient mineral processing. In *Energy Efficiency in the Minerals Industry*; Awuah-Offei, K., Ed.; Springer International Publishing: Cham, Germany, 2018; pp. 233–250.
74. Soofastaei, A.; Karimpour, E.; Knights, P.; Kizil, M. Energy-efficient loading and hauling operations. In *Energy Efficiency in the Minerals Industry*; Awuah-Offei, K., Ed.; Springer International Publishing: Cham, Germany, 2018; pp. 121–146.
75. Sözen, S.; Orhon, D.; Dinçer, H.; Ateşok, G.; Baştürkçü, H.; Yalçın, T.; Öznesil, H.; Karaca, C.; Allı, B.; Dulkadiroğlu, H.; et al. Resource recovery as a sustainable perspective for the remediation of mining wastes: Rehabilitation of the CMC mining waste site in Northern Cyprus. *Bull. Eng. Geol. Environ.* **2017**, *76*, 1535–1547. [CrossRef]
76. Esteves, A.M. Mining and social development: Refocusing community investment using multi-criteria decision analysis. *Resour. Policy* **2008**, *33*, 39–47. [CrossRef]
77. Moomen, A.-W.; Lacroix, P.; Bertolotto, M.; Jensen, D. The Drive towards Consensual Perspectives for Enhancing Sustainable Mining. *Resources* **2020**, *9*, 147. [CrossRef]
78. Huber, F.; Fellner, J. Integration of life cycle assessment with monetary valuation for resource classification: The case of municipal solid waste incineration fly ash. *Resour. Conserv. Recycl.* **2018**, *139*, 17–26. [CrossRef]
79. Pell, R.; Tijsseling, L.; Palmer, L.W.; Glass, H.J.; Yan, X.; Wall, F.; Zeng, X.; Li, J. Environmental optimisation of mine scheduling through life cycle assessment integration. *Resour. Conserv. Recycl.* **2019**, *142*, 267–276. [CrossRef]

80. Reid, C.; Bécaert, V.; Aubertin, M.; Rosenbaum, R.K.; Deschênes, L. Life cycle assessment of mine tailings management in Canada. *J. Clean. Prod.* **2009**, *17*, 471–479. [CrossRef]
81. Durucan, S.; Korre, A.; Munoz-Melendez, G. Mining life cycle modelling: A cradle-to-gate approach to environmental management in the minerals industry. *J. Clean. Prod.* **2006**, *14*, 1057–1070. [CrossRef]
82. Figueiredo, J.; Vila, M.C.; Góis, J.; Biju, B.P.; Futuro, A.; Martins, D.; Dinis, M.L.; Fiúza, A. Bi-level depth assessment of an abandoned tailings dam aiming its reprocessing for recovery of valuable metals. *Miner. Eng.* **2019**, *133*, 1–9. [CrossRef]
83. Wang, A.; Liu, H.; Hao, X.; Wang, Y.; Liu, X.; Li, Z. Geopolymer Synthesis Using Garnet Tailings from Molybdenum Mines. *Minerals* **2019**, *9*, 12. [CrossRef]
84. Ahmari, S.; Zhang, L. Production of eco-friendly bricks from copper mine tailings through geopolymerization. *Constr. Build. Mater.* **2012**, *29*, 323–331. [CrossRef]
85. Statista. Electricity Prices for Commercial and Industrial Customers in Germany from 2010 to 2020. 2020. Available online: https://de.statista.com/statistik/daten/studie/154902/umfrage/strompreise-fuer-industrie-und-gewerbe-seit-2006/ (accessed on 20 December 2020).
86. Statista. Average Price of Diesel Fuel in Germany from 1950 to 2020. 2020. Available online: https://de.statista.com/statistik/daten/studie/779/umfrage/durchschnittspreis-fuer-dieselkraftstoff-seit-dem-jahr-1950/ (accessed on 20 December 2020).
87. U.S. Geological Survey (USGS). Mineral. Commodity Summaries 2020. 2020. Available online: https://pubs.usgs.gov/periodicals/mcs2020/mcs2020.pdf (accessed on 6 March 2021).
88. U.S. Geological Survey (USGS). Mineral. Commodity Summaries 2019. 2019. Available online: https://prd-wret.s3-us-west-2.amazonaws.com/assets/palladium/production/atoms/files/mcs2019_all.pdf (accessed on 6 March 2021).
89. U.S. Geological Survey (USGS). Mineral. Commodity Summaries 2016. 2016. Available online: https://s3-us-west-2.amazonaws.com/prd-wret/assets/palladium/production/mineral-pubs/mcs/mcs2016.pdf (accessed on 6 March 2021).
90. U.S. Geological Survey (USGS). Mineral. Commodity Summaries 2021. 2021. Available online: https://pubs.usgs.gov/periodicals/mcs2021/mcs2021.pdf (accessed on 6 March 2021).
91. U.S. Geological Survey (USGS). Mineral. Commodity Summaries 2012. 2012. Available online: https://s3-us-west-2.amazonaws.com/prd-wret/assets/palladium/production/mineral-pubs/mcs/mcs2012.pdf (accessed on 6 March 2021).
92. U.S. Geological Survey (USGS). Mineral. Commodity Summaries 2008. 2008. Available online: https://s3-us-west-2.amazonaws.com/prd-wret/assets/palladium/production/mineral-pubs/mcs/mcs2008.pdf (accessed on 6 March 2021).
93. U.S. Geological Survey (USGS). Mineral. Commodity Summaries 2004. 2004. Available online: https://s3-us-west-2.amazonaws.com/prd-wret/assets/palladium/production/mineral-pubs/mcs/mcs2004.pdf (accessed on 6 March 2021).
94. IndexMundi. Copper, Grade A Cathode Monthly Price. 2021. Available online: https://www.indexmundi.com/commodities/?commodity=copper (accessed on 31 March 2021).
95. IndexMundi. Lead Monthly Prices. 2021. Available online: https://www.indexmundi.com/commodities/?commodity=lead (accessed on 31 March 2021).
96. IndexMundi. Zinc Monthly Price. 2021. Available online: https://www.indexmundi.com/commodities/?commodity=zinc (accessed on 31 March 2021).
97. Azapagic, A. Developing a framework for sustainable development indicators for the mining and minerals industry. *J. Clean. Prod.* **2004**, *12*, 639–662. [CrossRef]
98. Garbarino, E.; Orveillon, G.; Saveyn, H.G.M.; Barthe, P.; Eder, P. *Best Available Techniques (BAT) Reference Document for the Management of Waste from Extractive Industries, in Accordance with Directive 2006/21/EC*; Publications Office of the European Union: Luxembourg, 2018; Available online: https://op.europa.eu/en/publication-detail/-/publication/74b27c3c-0289-11e9-adde-01aa75ed71a1/language-en (accessed on 15 April 2021).
99. Govindan, K. Application of multi-criteria decision making/operations research techniques for sustainable management in mining and minerals. *Resour. Policy* **2015**, *46*, 1–5. [CrossRef]
100. United Nations Economic Commission for Europe (UNECE). Safety Guidelines and Good Practices for Tailings Management Facilities; New York and Geneva. 2014. Available online: https://unece.org/environment-policy/publications/safety-guidelines-and-good-practices-tailings-management-facilities (accessed on 15 April 2021).
101. Hartman, H.L.; Mutmansky, J.M. *Introductory Mining Engineering*, 2nd ed.; Wiley: Hoboken, NJ, USA, 2002; 570p.
102. European Commission (EC). *A Guide to EU Funding*; Luxembourg, 2017; 20p, Available online: https://op.europa.eu/de/publication-detail/-/publication/7d72330a-7020-11e7-b2f2-01aa75ed71a1 (accessed on 20 May 2021).
103. Park, J.K.; Clark, T.; Krueger, N.; Mahoney, J. A Review of Urban Mining in the Past, Present and Future. *Adv. Recycling Waste Manag.* **2017**, *2*, 4.
104. Prno, J.; Slocombe, D.S. Exploring the origins of 'social license to operate' in the mining sector: Perspectives from governance and sustainability theories. *Resour. Policy* **2012**, *37*, 346–357. [CrossRef]
105. Bächtold, H.G.; Schmid, W.A. *Contaminated Sites and Spatial Planning—A European Challenge*; vdf-Hochschulverl. AG an der ETH: Zurich, Switzerland, 1995. (In German)
106. Lèbre, É.; Corder, G. Integrating Industrial Ecology Thinking into the Management of Mining Waste. *Resources* **2015**, *4*, 765–786. [CrossRef]
107. Liessmann, W. *Historical Mining in the Harz Mountains*, 3rd ed.; Springer: Berlin/Heidelberg, Germany, 2010; 470p. (In German)

108. Mohr, K. *Geology and Mineral Deposits of the Harz Mountains: With 37 Tables in Text and on 5 Folded Inserts and 2 Overview Tables on the Inside Pages of the Cover*, 2nd ed.; Schweizerbart: Stuttgart, Germany, 1993; 496p. (In German)
109. Climate-Data.org. Climate Goslar (Germany), n. d. Available online: https://de.climate-data.org/europa/deutschland/niedersachsen/goslar-22981/ (accessed on 23 August 2020).
110. State Office for Mining Energy and Geology (LBEG). NIBIS® map server: Climate. Available online: https://nibis.lbeg.de/cardomap3/# (accessed on 27 March 2021).
111. Large, D.; Walcher, E. The Rammelsberg massive sulphide Cu-Zn-Pb-Ba-Deposit, Germany: An example of sediment-hosted, massive sulphide mineralisation. *Miner. Deposita* **1999**, *34*, 522–538. [CrossRef]
112. European Commission (EC). CRM list 2020. 2021. Available online: https://rmis.jrc.ec.europa.eu/?page=crm-list-2020-e294f6 (accessed on 22 April 2021).
113. DIN. DIN 1054:2010-12—subsoil: Verification of the safety of earthworks and foundations. 2010. Available online: https://www.beuth.de/de/norm/din-1054/135236978 (accessed on 19 April 2021).
114. Federal State Working Group on Waste (LAGA). Requirements for the recycling of mineral waste—Part II: Technical Rules for Recycling, 1.2 Soil Material (LAGA TR Boden II). 2004. Available online: https://mluk.brandenburg.de/sixcms/media.php/land_bb_test_02.a.189.de/tr_laga2.pdf (accessed on 17 April 2021). (In German)
115. European Central Bank. Pound sterling (GBP). 2020. Available online: https://www.ecb.europa.eu/stats/policy_and_exchange_rates/euro_reference_exchange_rates/html/eurofxref-graph-gbp.en.html (accessed on 14 August 2020).
116. European Central Bank. US dollar (USD). 2020. Available online: https://www.ecb.europa.eu/stats/policy_and_exchange_rates/euro_reference_exchange_rates/html/eurofxref-graph-usd.en.html (accessed on 4 August 2020).
117. InfoMine USA Inc. *Mine and Mill Equipment Costs: An Estimator's Guide*; CostMine: Spokane Valley, WA, USA, 2016.
118. Bray, R.N. *A Guide to Cost Standards for Dredging Equipment*, 2nd ed.; Construction Industry Research & Information Ass: London, UK, 2009.
119. Figueiredo, J.; Vila, M.C.; Fiúza, A.; Góis, J.; Futuro, A.; Dinis, M.L.; Martins, D. A Holistic Approach in Re-Mining Old Tailings Deposits for the Supply of Critical-Metals: A Portuguese Case Study. *Minerals* **2019**, *9*, 638. [CrossRef]
120. Kieckhäfer, K.; Breitenstein, A.; Spengler, T.S. Material flow-based economic assessment of landfill mining processes. *Waste Manage.* **2017**, *60*, 748–764. [CrossRef]
121. World Bank. Doing Business 2020: Comparing Business Regulation in 190 Economies. Washington, DC, USA. 2020. Available online: http://documents1.worldbank.org/curated/en/688761571934946384/pdf/Doing-Business-2020-Comparing-Business-Regulation-in-190-Economies.pdf (accessed on 13 November 2020).
122. Bastian, D.; Brandenburg, T.; Buchholz, P.; Huy, D.; Liedtke, M.; Schmidt, M.; Sievers, H. *DERA List of Raw Materials*; German Mineral Resources Agency (DERA) in the Federal Institute for Geosciences and Natural Resources (BGR): Berlin, Germany, 2019; 116p, ISBN 978-3-943566-61-1. (In German) Available online: https://www.deutsche-rohstoffagentur.de/DE/Gemeinsames/Produkte/Downloads/DERA_Rohstoffinformationen/rohstoffinformationen-40.pdf?__blob=publicationFile (accessed on 9 June 2021)
123. Yang, C.; Chen, Y.; Peng, P.; Li, C.; Chang, X.; Wu, Y. Trace element transformations and partitioning during the roasting of pyrite ores in the sulfuric acid industry. *J. Hazard. Mater.* **2009**, *167*, 835–845. [CrossRef]

Article

Life Cycle Sustainability Assessment of a Novel Bio-Based Multilayer Panel for Construction Applications

Aitor Barrio [1], Fernando Burgoa Francisco [2], Andrea Leoncini [3], Lars Wietschel [4,*] and Andrea Thorenz [4]

1. TECNALIA, Basque Research and Technology Alliance (BRTA), Area Anardi 5, 20730 Azpeitia, Spain; aitor.barrio@tecnalia.com
2. CARTIF Technology Centre, Parque Tecnológico de Boecillo, 205 Boecillo, 47151 Valladolid, Spain; ferbur@cartif.es
3. RINA Consulting S.p.A., Via A.Cecchi 6, 16129 Genoa, Italy; andrea.leoncini@rina.org
4. Resource Lab, Institute for Materials Resource Management, University of Augsburg, Universitätsstraße 16, 86399 Augsburg, Germany; andrea.thorenz@mrm.uni-augsburg.de
* Correspondence: lars.wietschel@wiwi.uni-augsburg.de; Tel.: +49-821-598-3950

Citation: Barrio, A.; Francisco, F.B.; Leoncini, A.; Wietschel, L.; Thorenz, A. Life Cycle Sustainability Assessment of a Novel Bio-Based Multilayer Panel for Construction Applications. *Resources* **2021**, *10*, 98. https://doi.org/10.3390/resources10100098

Academic Editor: Carlo Ingrao

Received: 26 July 2021
Accepted: 15 September 2021
Published: 29 September 2021

Publisher's Note: MDPI stays neutral with regard to jurisdictional claims in published maps and institutional affiliations.

Copyright: © 2021 by the authors. Licensee MDPI, Basel, Switzerland. This article is an open access article distributed under the terms and conditions of the Creative Commons Attribution (CC BY) license (https://creativecommons.org/licenses/by/4.0/).

Abstract: The bioeconomy can be integral to transforming the current economic system into one with reduced environmental and social impacts of material consumption. This work describes a bio-based multi-layer panel that is based on residual coniferous bark. To ensure that the presented bio-based panel positively contributes to environmental protection while remaining competitive with conventional products and meeting high social standards, the development of the panel is accompanied by a life cycle sustainability assessment. This study performs a comparative LCA and LCC of the developed panel to conventional benchmark panels, as well as a qualitative social life cycle assessment. While the panel performs only economically marginally weaker than the benchmarks, the results are more heterogeneous for the environmental dimension with benefits of the bio-based panel in categories such as climate change, acidification, and ozone formation and detriments in categories including eutrophication. The S-LCA analysis shows that all of the involved companies apply social principles in direct proximity; however, social responsibility along the supply chain could be further promoted. All results need to be viewed with the caveat that the manufacturing processes for the new panel have been implemented, to date, on a pilot scale and further improvements need to be achieved in terms of upscaling and optimisation cycles.

Keywords: bark-based biorefinery; bio-based material; construction material; insulation; polyurethane; biophenolic resin

1. Introduction

The building sector accounts for about 38% of the annual greenhouse gas emissions and around 40% of the global energy demand [1,2]. About 70 to 90% of the energy consumption corresponds to the operation of buildings, while the embodied energy accounts for 10 to 30% [3]. New policies and strategies such as net-zero buildings try to reduce the energy consumption of the service life of buildings, although they exclude the embodied carbon of the construction materials, which can account for up to 11% of the global GHG emissions [4].

In addition to the energy needs of the building sector, it consumes large amounts of resources. In 2018, resource use dedicated to housing and infrastructure accounted for 40% of the total global material use [5]. The construction industry is among the primary water users [6], while construction and demolition waste is the most significant waste stream in the EU [7]. With a rising world population, from 7.7 billion in 2019 to a projected 8.5 billion in 2030 [8], housing needs will further stress an already strained resource situation. In summary, the construction sector faces risks related to climate change, such as rising carbon taxes, and has great potential to contribute to decarbonising the economy, reducing resource use, and reducing construction and demolition waste [9].

Bio-based materials can contribute to CO_2 reduction during the service-life of buildings through better thermal insulation, while materials themselves can act as a temporary carbon sink and the manufacturing process of the materials can be even more energy-efficient than conventional building materials [10,11]. Since first-generation bio-based materials can strongly contribute to direct and indirect land-use changes and thus to high environmental impacts, second-generation bio-based materials such as lignocellulose residues, in particular, are seen as having great potential to replace conventional products with significantly lower environmental impacts [12–14]. The challenges in the material and energy use of lignocellulose relate to the resistance in breaking down into individual components including lignin, cellulose, hemicellulose, and tannin; the large variability of the different structures and chemical compositions due to genetic and other environmental influences; and the large number of different sugars released from the breakdown of cellulose and hemicellulose [15]. While much research in recent years has focused on the use of agricultural residues [16,17], less attention has been paid to the use of bark in high-value applications. Bark could also be a valuable raw material for producing high-quality products such as adhesives, resins, plastics, or bioethanol [18,19]. Lacoste et al. [20], for example, showed that spruce bark-based insulation foam from condensed tannins has excellent mechanical resistance at low thermal conductivity and low density. Arias et al. [21] compared the environmental impacts of different wood panel adhesives and found the lowest impacts for a residual bark-based solution. Santos et al. [22] performed a life cycle analysis on a cross-insulated timber panel with a conventional PU foam as an inner insulation layer. Sinka et al. [23] presented different magnesium-hemp multi-layer construction panel types and compared their GHG emissions to conventional references. While many approaches in the field of bio-based construction materials adequately address the environmental and economic dimension [11,14,24], Ingrao and colleagues [25] observed a shortcoming in the assessment of the social dimension, mainly due to a lack of data for modelling the social dimension. To avoid a shift of burden, the transition to a post-fossil society, which uses bio-based residues to substitute conventional materials, must be evaluated from the perspective of the three dimensions of sustainability [26].

This paper's main goal is to describe the development of a novel and bio-based multi-layer construction panel and to present both a comparative life cycle sustainability assessment and a mechanical property assessment. The forestry residue coniferous bark serves as the main feedstock and delivers all chemically essential components for producing the plywood board based on bio-resins, the insulation foam based on bioesterpolyol, and the bioesterpolyol-based adhesives for the final panel bonding. The production of all components is based on processing methods that were upscaled by different companies and research facilities throughout the REHAP project from 2016 to 2021 [27]. This work contributes through a description of the whole production process of a bio-based multi-layer panel, starting with the resource provision, the value-adding steps for the production of the functional components, and up to the production of the final product.

Since bark is the focal feedstock, this work also provides insight into how the full utilisation of the individual bark components can take place in a bark-based biorefinery. In order to compare the technical, economic, and environmental characteristics of the developed multi-layer panel with conventional benchmark panels, a life cycle sustainability assessment (LCSA) and a measurement of the technical characteristics are carried out. Regarding the social perspective, no benchmarking is carried out due to the difficulty of adequately assessing the benchmark products' social aspects. The LCSA is performed by standard methodologies to highlight potential benefits and burdens of the novel product compared to two selected conventional benchmarks. This work is a comprehensive and multidimensional assessment of a future-oriented construction material based on the raw material bark, and attempts to answer the following research question:

RQ: How does a novel bio-based multi-layer panel perform technically, economically, and environmentally compared to conventional, fossil resource-based benchmark panels?

The introduction is followed by an explanation of our methodology that describes the study's goal and scope, the system boundaries and the considered value chain steps, the life cycle sustainability assessment, and the measurement of the technical properties. The results follows this section. The article ends with a discussion and conclusion that situates the findings within the existing literature and places them in a larger scope.

2. Materials and Methods

The research question is answered by a description of the production process, a comparative Life Cycle Sustainability Assessment (LCSA) with selected benchmark systems, and an analysis of the technical properties. The LCSA is carried out by a Life Cycle Assessment (LCA), a Life Cycle Costing (LCC), and a Social Life Cycle Assessment (S-LCA), which are methodologies that follow the Life Cycle Assessment framework defined in the ISO 14040 and ISO 14044 standards. LCA frameworks consist of four steps: Goal and Scope Definition, Life Cycle Inventory, Impact Assessment, and Interpretation of the Results. LCA is a standard method for studying the environmental impacts of construction products, with the first studies dating back to the 1980s [28]. LCC studies for the construction sector have also been conducted, with the first definitions of this methodology dating back to the 1960s [29]. Social impact evaluations are a more incipient methodology and there are still discussions on the most suitable methodology [30]. However, several studies already exist in which different methodologies or frameworks have been developed to evaluate social impacts in the construction sector [31,32]. More recently, the integrated LCSA has been used in the construction sector. In [33], for example, the authors applied a holistic LCSA to the production and recycling of concrete and timber production. The methodology has also been used to assess bio-based products across different sectors [34–36].

2.1. Goal and Scope

The goal of this study is the evaluation of the sustainability performance of a bio-based, multi-layer building panel compared to benchmarks primarily based on fossil resources. The system boundaries of the presented layer and the benchmark systems follow a cradle-to-gate approach and include both the raw materials extraction and the production process. The environmental and economic features of the developed multi-layer panel are compared to two different benchmarks with similar characteristics by LCA, and LCC Benchmark 1 is a conventional sandwich panel (i.e., SIP panel) composed of plywood (density of 450 kg/m^3), XPS foam, and a fossil-based PU adhesive. Benchmark 2 is a multi-layer panel similar to the bio-based one but is based on fossil-based components. Indeed, in the latter case, PU components (foam and adhesive), including fossil-based esterpolyols and plywood boards with phenol-formaldehyde resins as binders, are considered. In particular, a different formulation for fossil-based esterpolyols is taken into account, depending on their target application, i.e., foams or adhesives. While fossil-based esterpolyols for foams include diethylene glycol and phthalic anhydride as monomers, benchmark esterpolyols for adhesives include fossil-based 1,4-BDO and adipic acid. For the social LCA, the assessment of the social impacts linked to the bio-based panel was performed. The results of the social assessment are only given qualitatively as comparable social assessments were not available for the conventional benchmark panels.

An appropriate functional unit is the key to providing a consistent comparison between different product systems. In the case of the presented multi-layer insulation panel, it is vital to consider the properties and functionality of the panel. Consequently, the functional unit selected for this study is the production of 1 m^2 of multi-layer panels with technical properties that exceed the Compressive and Tensile Strength UNE EN Standards. The selection of this functional unit is in line with similar LCA studies on composite structural panels [23,37] and is the common functional unit in the EPD program for construction products.

Primary data was provided first-hand by the industrial companies involved in the different stages of the production chain. For the LCA, background data was mainly sourced from LCI databases such as the Ecoinvent databases version3.6 [38] or GaBi databases, scientific papers, and industry-average and government statistics sources (e.g., Eurostat). The geographical scope of the project is the EU but was not limited to EU countries. For places where no geographical information was available, relevant European databases were used (e.g., Eurostat). Regarding the temporal representativeness, primary data used in this study represents current production technologies and standard operations for the biorefining sector, whereas secondary data retrieved from the GaBi database is representative for the years from 2010 to the present. The different scenarios were modelled and evaluated using the software GaBi, version 9 [39]. The production of co-products was handled using subdivision or system expansion when possible (ILCD). When these alternatives were not available, economic allocation was used to allocate the impact to the different co-products in the process. For the LCC, the use of economic allocation was convenient due to the intrinsic economic nature of the assessment [40]. Finally, for the SLCA, allocation was deemed irrelevant due to the nature and scope of social data [41]. The target audience includes researchers from academia, industry representatives of the biorefining and construction sector, environmental and socio-economic analysts, and other potential stakeholders interested in both the circular economy and the sustainability of new materials.

2.2. System Boundaries and Description of the Production System

Figure 1 shows the value chain of the multi-layer panel production with focus on the foreground processes, namely the refining of bark. The primary refining process extracts tannins, lignin, and sugars from spruce bark [42]. After a preliminary shredding step, the ground sawmill spruce bark is processed through hot water extraction. A water-bark mixture is heated up to 90 °C, thus separating tannin-rich fractions and a bark residue containing mainly lignin, residual tannins, and sugars. Tannins-rich fractions are clarified by centrifugation and tannins are then concentrated through a falling film evaporator to obtain a purified tannins extract.

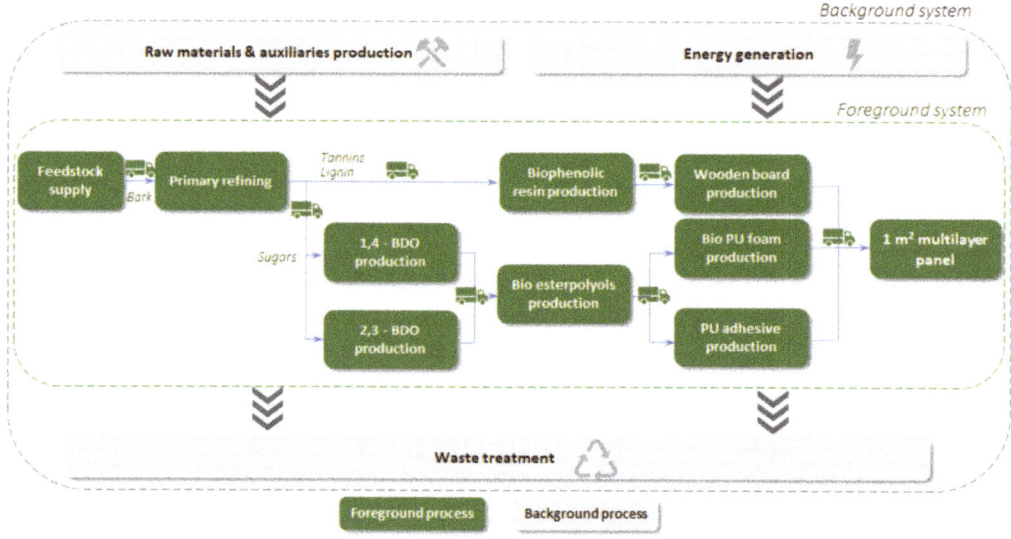

Figure 1. Simplified description of the bio-based multi-layer panel production system.

The bark residue undergoes a soda cooking step, where it is fed together with NaOH solution (50% w/w) and water, and the mixture is heated up to 160 °C by direct steam injection (6–7 bar). The resulting soda slurry is fed into a decanter to separate the lignin-rich black liquor from cellulose fibres. While the black liquor is sent for lignin separation steps, the cellulose fibres are firstly washed and then sent to a saccharification process with a specific enzyme cocktail to obtain a sugars-rich solution from which sugars are extracted and concentrated (sugar stream: 78.2% w/w dry matter; 321.7 g/L glucose; 49.65 g/L xylose). The resulting carbohydrates fraction is then converted through fermentation processes into the diols 1,4-BDO and 2,3-BDO to be used as bio-based intermediates and bio-based dicarboxylic acids for the production of bioesterpolyols for BioPU foams and bioPU adhesives, respectively. The phenolic fraction derived from bark (Lignin I) is used in the production process of biophenolic resins, during which this bio-based component is mixed and reacts with phenol and formaldehyde; in particular, a 50% substitution of the phenol mass with biophenolic fraction is applied, compared to a benchmark product, i.e., phenol-formaldehyde resin.

The biophenolic resins are used as glue in the production process of plywood boards. In the latter process, pine logs are debarked and chipped to obtain wooden chips with the required dimensions to undergo the subsequent steps, i.e., chip digestion with water and defibration. After these steps, biophenolic resins are blended with the fibres before the pressing phases, during which preliminary plywood boards are obtained. The final plywood boards are the result of the following stabilisation, sanding, and cutting stages. A dedicated boiler is used to burn several wood-based residual streams and thus provide steam, hot thermal oil, and hot gases to the main processes.

Figure 2 shows the bio-based multi-layer panel prototype with a detailed description of the functional components. The multi-layer is composed of three bark-based components, namely the plywood board based on lignin-derived biophenolic resin, the bioPU foam based on bioesterpolyols, and the bioPU adhesive based on bioesterpolyols, which bond the plywood board and foam. For the final production of the multi-layer panel, the whole system undergoes a hydraulic press for 4 h at a pressure of 3 kg/cm^2. The process is similar to a SIP panel production based on plywood (with a different density compared to the one used in the bio-based multi-layer panel) and an XPS foam.

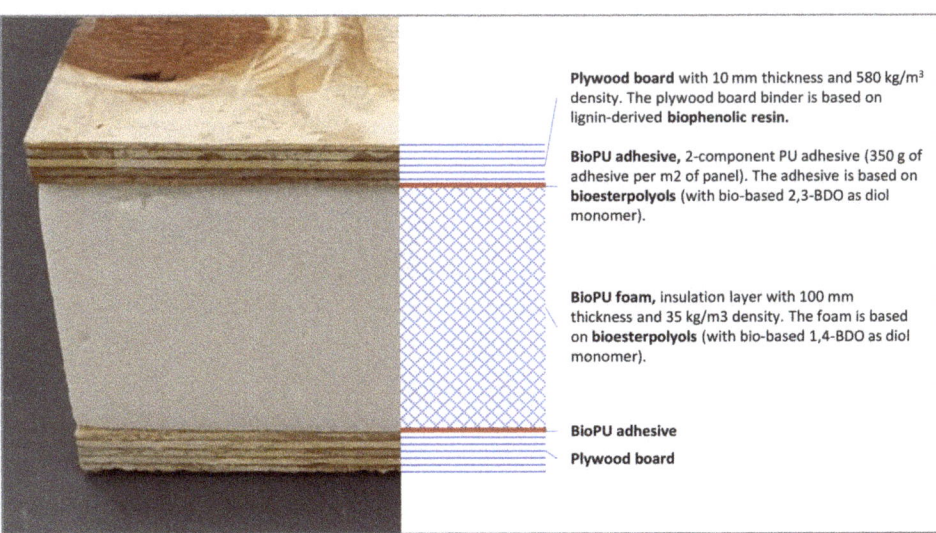

Figure 2. Picture of the multi-layer panel with detailed descriptions of the structure and functional components.

The system boundaries include the phases of the product's life cycle from the raw material extraction to the factory gate ("cradle-to-gate" approach). The system is divided into two subsystems: a foreground system and a background system, as schematised in Figure 1. The foreground system consists of processes under the control of the decision-maker for which the study is carried out on, i.e., bioPU adhesive, plywood board, bioPU foam value chains, and panel manufacturing. The background system represents all up and downstream processes connected to the foreground system, namely the raw material and energy production, transportation, and waste treatment and disposal.

2.3. Feedstock Availability

Coniferous bark is the focal feedstock for producing the functional components of the multi-layer panel, the plywood board, the bioPU foam, and the bioPU adhesive. Coniferous bark typically consists of about 25% cellulose, around 10% hemicellulose, about 30% lignin, and around 10% tannin. Bark as feedstock for biorefineries has several advantages. These include the mainly low-value use as combustion for energy recovery or surface mulching, its low price, and its high availability and accumulation in large volumes at discrete locations (sawmills and both pulp and paper mills), which facilitates the feedstock collection. Since spruce and pine, in particular, are used in large quantities among softwoods, the residual bark stream can furthermore be described as relatively homogeneous, which is beneficial for its valorisation in biorefineries [43]. The average annual bioeconomic potential of coniferous bark in the EU28 from 2015 to 2018 is supposed to be around 14.6 Mt dry matter, of which 7.6 Mt is supposed to be spruce bark and 7 Mt is supposed to be pine bark (calculation based on [43,44]). In the bioeconomic bark potential, all wood industry residues (excluding firewood) are considered, hence bark residues of sawmills, pulp and paper mills, and other wood industries. Figure 3 shows the regional distribution in the EU28. The main spruce bark potentials can be found in Sweden (2.2 Mt), Germany (1.5 Mt), and Finland (1.3 Mt), and the main pine bark potentials are in Poland (1.4 Mt), Sweden (1.3 Mt), and Finland (1.2 Mt). The tannin and lignin for producing the biophenolic resins, and the sugar for the production of 1,4 and 2,3-BDO of the underlying research work are obtained in a primary refining step from Scandinavian sawmill spruce bark. Pinewood is used for the production of the plywood board.

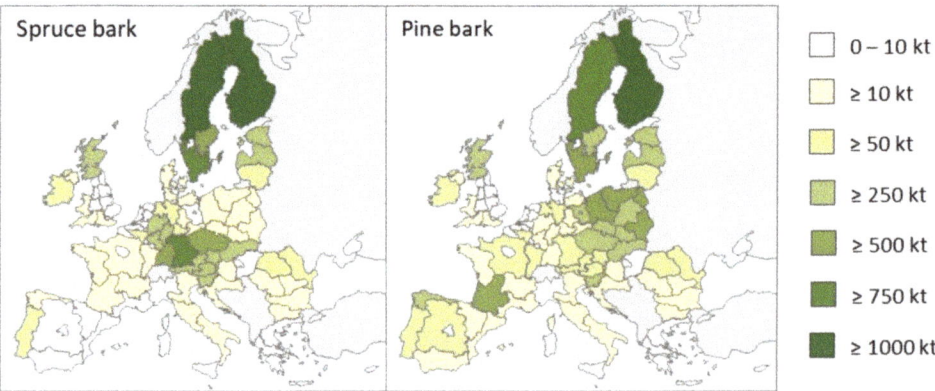

Figure 3. Average annual bioeconomic potential of spruce and pine bark on the NUTS-1 level between 2015 and 2018 (calculation basesd on [43,44]).

2.4. Technical/Physical Properties

For a meaningful comparison of the bio-based panel with the two benchmarks, comparable technical and physical properties with regard to the functional unit are necessary. Therefore, this study includes the evaluation of the technical properties of the multi-layer panel. In this study, the performance tests focused on the two properties tensile strength and compression strength. The sandwich panel samples consist of plywood covers on both sides and the bio-based polyurethane core bonded with the bio-based adhesive. To determine the tensile strength, the test performed was the tensile test perpendicular to the panel faces according to the test procedure described in UNE-EN 14509:2014 (amended version, 2016). The self-aligning plates were bonded to the surfaces of each specimen using two-component epoxy adhesive. The tensile test was carried out at room temperature and the tensile strength perpendicular to the panel sides evaluated the tensile modulus. Breaking modes of the sample can be cohesive (core breaking or higher adhesive line bonding) or adhesive (detachment between different components).

To determine the compressive strength, the compression resistance and modulus test were performed according to the test procedure described in UNE-EN 14509:2014 (amended version, 2016). The compression test was carried out, calculating the compression strength and compression modulus at 10% of deformation. When breaking occured before that 10%, the values given were the ones that correspond to that point. For both tests, ambient conditions were kept constant at a temperature of 23 °C and 50% relative to the humidity. The mechanical testing was performed on six different samples which varied in terms of the adhesive quantity, the bonding pressure, and the formulation of the BioPU foam (PU 1 with $\rho = 30.53 \pm 0.53$ kg/m^3 and PU 2 with $\rho = 26.35 \pm 0.98$ kg/m^3). The selection of the different test samples was based on achieving the best configuration of materials (foam/adhesive/plywood board) and the optimum conditions of application, curing, and pressing, considering the physical and chemical properties of the different elements. Sample preparation started with cutting the PU foams and plywood boards to the desired size to manufacture the specimens according to the size required by the standards. This step was followed by sanding the plywood surfaces to improve adhesion in the next stage, i.e., applying the adhesive. Finally, the multi-layer panel was assembled using a pressing process. Table 1 shows the different test samples and the different pressures that were tested to determine the optimum application pressure.

Table 1. Test samples for the performance of the mechanical/physical testing. The size of each test sample is 100 mm × 100 mm and the pressing time is four hours.

Test Sample	Cover	Core	Adhesive Quantity (g/m^2)	Bonding Pressure (kg/m^2)	Curing Time
#1	Poplar plywood	PU 2	350	1	3–4 days
#2	Poplar plywood	PU 2	250	1	3–4 days
#3	Poplar plywood	PU 1	150	2	12–24 h
#4	Poplar plywood	PU 1	350	2	3–4 days
#5	Poplar plywood	PU 1	250	0.4	3–4 days
#6	Poplar plywood	PU 2	250	0.4	3–4 days

2.5. Life Cycle Sustainability Assessment

In the following sections, the three pillars, namely the Life Cycle Assessment, Life Cycle Costing, and Social Life Cycle Assessment of the integrated assessment methodology, are described and reported separately. Each section includes a Life Cycle Inventory (LCI), a section on the assumptions and limitations, and a description of the applied Life Cycle Impact Assessment (LCIA) method. The central part of the data associated with this work was collected throughout the REHAP project.

2.5.1. LCA

Life Cycle Assessment (LCA) is a structured, comprehensive, and internationally recognised technique for assessing the environmental aspects of a process or a product (i.e., good or service) and the potential environmental impacts throughout the product's life cycle. The goal of this study's LCA was to evaluate the environmental performance of a bio-based, multi-layer building panel compared to specific benchmarks from a cradle-to-gate perspective. Primary data was collected through questionnaires, while secondary data was gathered through LCI databases (GaBi database (ts) and Ecoinvent database v3.6), as well as through literature research. Table 2 contains all the data used for the assessment of both the bio-based and benchmark solutions, but not referring to the functional units and net of recycled streams and recovered energy.

Table 2. Life Cycle Inventory of the production of the bio-based multi-layer panel and benchmark solutions.

	Category	Flow	Amount	Unit	Dataset	Source
Bio-based panel	Raw material	BioPU foam	3.5	kg	Based on own modelling	Primary data
	Raw material	BioPU adhesive	0.35	kg	Based on own modelling	Primary data
	Raw material	Plywood board with biophenolic resin	11.6	kg	Based on own modelling	Primary data
	Energy	Electricity	96	kWh	ES: electricity grid mix (ts)	Secondary data
Benchmark 1	Raw material	XPS foam	2.5	kg	EU-28: extruded polystyrene (XPS) (EN15804 A1-A3) (ts)	Primary data
	Raw material	PU adhesive	0.35	kg	DE: thermoplastic polyurethane (TPU and TPE-U) adhesive (ts)	Primary data
	Raw material	Plywood	4.5	kg	EU-28: plywood board (EN15804 A1-A3) (ts)	Primary data
	Energy	Electricity	96	kWh	ES: electricity grid mix (ts)	Secondary data
Benchmark 2	Raw material	PU foam with fossil-based esterpolyols	3.5	kg	Based on own modelling	Primary data
	Raw material	PU adhesive with fossil-based esterpolyols	0.35	kg	Based on own modelling	Primary data
	Raw material	Plywood board with phenol-formaldehyde resin	11.6	kg	Based on own modelling	Primary data
	Energy	Electricity	96	kWh	ES: electricity grid mix (ts)	Secondary data
	Product	Panel	1	m^2	-	-

In the framework of environmental assessment through LCA, several assumptions were introduced at different stages of the assessment. In particular, bark was considered as a secondary material with no environmental impacts/benefits (e.g., due to carbon storage) associated with it ("zero burden approach") [45]. Moreover, the calculation of the biogenic carbon uptake and release linked to the raw materials production followed the Product Environmental Footprint (PEF) approach, as described in [46] (i.e., only biogenic methane emissions are modelled). Considering the need to properly integrate several processing steps implemented at different scales along the value chains, some assumptions regarding the scale-up and optimisation of the targeted processes were introduced, in particular regarding the treatment and recirculation of liquid residual streams, as well as the incineration of solid residual streams left after wastewater treatments. Moreover, solid residual streams originated from within the primary refining (e.g., after solid–liquid separation steps) were assumed to be sent to an "internal" boiler for energy recovery (efficiency 0.9) and their calorific values were estimated according to [47]. Bio-based adipic acid (data retrieved from [48]) was considered as a monomer for bioesterpolyols production instead of as bio-based azelaic acid, for which no data or datasets were available. Regarding multifunctional processes (e.g., primary refining and 1,4-BDO production), economic allocation among the main and co or by-products was applied.

Among the impact categories recommended by the PEF Guide 2013 [46], the eight impact categories acidification, climate change, aquatic freshwater eutrophication, aquatic marine eutrophication, terrestrial eutrophication, ozone depletion, photochemical ozone formation, and resource-use energy carriers were selected for this study based on their relevance for the assessed processes. These are related to resource use and emissions of environmentally damaging substances (e.g., greenhouse gases and toxic chemicals), which may affect human health. Impact assessment methods use models for quantifying the causal relationships between the material/energy inputs and the emissions associated with the product life cycle for each impact category considered. Each category hence referred to a particular stand-alone impact assessment model. The two optional steps of the Impact Assessment phase, namely normalisation and weighting, were not considered in this analysis.

2.5.2. LCC

Similar to LCA, Life Cycle Costing (LCC) is a method that summarises all the costs associated with the life cycle of a product (or service) that are directly covered by one or more of the actors involved in the product life cycle (e.g., supplier, producer, user/consumer, and end-of-life actor). Life Cycle Costing can be used as a stand-alone tool or can be used in the broader context of the sustainable development of a product, together with environmental LCA and social LCA. As for LCA, primary data for LCC were gathered from REHAP partners through questionnaires, while secondary data were mainly retrieved through the literature research. Table 3 contains all of the data and information that were used for the assessment of both the bio-based and benchmark solutions, addressing the raw material and energy costs, as well as the personnel and equipment costs.

Table 3. Life Cycle Costing Inventory of the production of the bio-based multi-layer panel production and the benchmark solutions.

	Category	Flow	Amount	Unit	Comments	Source
Bio-based panel	Raw material	BioPU foam	3.94	€/kg	-	Calculations based on primary data
	Raw material	BioPU adhesive	3.651	€/kg	-	Calculations based on primary data
	Raw material	Plywood board with biophenolic resin	0.579	€/kg	-	Calculations based on primary data
	Energy	Electricity	0.1076	€/kWh	Reference country: Spain	Secondary data: Eurostat
Benchmark 1	Raw material	XPS foam	90	€/m^3	Density: 25 kg/m^3	Primary data
	Raw material	PU adhesive	3.5	€/kg	-	Primary data
	Raw material	Plywood	523	€/m^3	Density: 450 kg/m^3	Primary data
	Energy	Electricity	0.1076	€/kWh	Reference country: Spain	Secondary data: Eurostat
Benchmark 2	Raw material	PU foam with fossil-based esterpolyols	2.7	€/kg	-	Calculations based on secondary data
	Raw material	PU adhesive with fossil-based esterpolyols	2.46	€/kg	-	Calculations based on secondary data
	Raw material	Plywood board with phenol-formaldehyde resins	0.6	€/kg	-	Calculations based on secondary data
	Energy	Electricity	0.1076	€/kWh	Reference country: Spain	Secondary data: Eurostat
Generic	Personnel	Unskilled worker	25	€/h	2 workers; 1760 h/y	Calculations based on primary data
	Maintenance	-	-	-	1% of CAPEX cost	Assumption
	Equipment	Press	40,000	€	Depreciation time: 10 years	Estimation

The assumptions and estimations introduced for LCA were also valid for the LCC. Specific hypotheses have also been considered in the framework of the economic assessment through LCC, particularly related to the annual maintenance costs. Those were assumed to be 1% of the overall CAPEX for bio phenolic resin production and multi-layer panel assembly (value based on primary data provided by REHAP project partners [27]), and 3% for all other processes upstream from the value chain (literature-based average value for different biorefinery concepts [49,50]).

2.5.3. S-LCA

The Social Life Cycle Assessment (S-LCA) is a social impact assessment technique that aims to assess the social and socio-economic aspects of products and their potential positive and negative impacts along their life cycle. The S-LCA of this study is based on UNEP/SETAC guidelines. The guidelines provide impact indicators that are differentiated in stakeholder categories, which are supposed to be the main groups possibly impacted by a product's life cycle [51]. The relevant stakeholder categories were identified according to the current study's research objectives, system boundaries, and data availability. Due to selecting a "cradle-to-gate" approach, the stakeholder category "Consumers" was not included in the system boundary and therefore not considered.

Table 4 reports the considered stakeholders, the subcategories, and the inventory indicators that have been applied in this study. For each of the identified social indicators, tailored questions were prepared for the industrial partners involved in the bio-based multi-layer panel value chain (related, for example, to the presence of specific policies or procedures within the company, focused on some social aspects such as local employment or supplier relationships). Based on the selected S-LCA subcategories, a questionnaire was developed and answered by the industrial partners involved in panel manufacturing. Due to the accessibility of questionnaire recipients, the S-LCA of the multi-layer panel was limited to the manufacturing process of the functional components and the end product.

Table 4. Selected subcategories for the S-LCA study.

SH	Subcategory	Indicators
Local Community	Delocalisation and migration	• Strength of organisational policies related to resettlement • Strength of organisational procedures for integration of migrant workers into the community
	Community engagement	• Presence of policies regarding community engagement at the company level • Organisational support for community initiatives
	Local employment	• Presence of policies on local hiring preferences • Percentage of workforce hired locally • Percentage of spending on locally based suppliers
	Access to im-material resourses	• Presence of community education initiatives and community service programmes
	Access to material resources	• Existence of projects to improve community infrastructure • Strength of potential material resource conflicts
	Safe and healthy living conditions	• Presence of certified environmental management systems • Management effort to improve the environmental and safety performance of the organisation
Value Chain Actors	Fair competition	• Presence of anti-competitive behaviour or violation of anti-trust and monopoly legislation can be linked to the organisation • Presence of policies to prevent anti-competitive behaviour • Presence of actions to increase employee awareness related to fair competition
	Respect of intellectual property rights	• Organisation's policy and practice regarding the respect of intellectual property rights
	Supplier relationships	• Interaction of the company with suppliers (payment on time, sufficient lead time, appropriate communication, and collaboration regarding quality issues)
	Promoting social responsibility	• Membership in an initiative that promotes social responsibility along the supply chain • Integration of ethical, social, environmental, and gender equality criteria in purchasing policy, distribution policy, and contract signatures

Table 4. Cont.

SH	Subcategory	Indicators
Workers	Freedom of association and collective bargaining	• Evidence of restriction to freedom of association and collective bargaining in the company • Presence of unions within the organisation and adequately support for them • Employee/union representatives are invited to contribute to the planning of larger changes in the company, which will affect the working conditions
	Fair salary	• Wage of the lowest paid worker compared to the minimum wage • Regular and documented worker payments
	Hours of work	• Number of hours effectively worked by employees • Number of holidays effectively used by employees • Flexibility • Respect of contractual agreements concerning overtime
	Equal opportunities/discrimination	• Presence of formal policies on equal opportunities • Occurrence of discrimination in the company • Share of women and minorities within the personnel • Ratio of basic salary of men to women by employee category
	Health and safety	• Number of injuries or fatal accidents in the organisation • Presence of a formal policy regarding health and safety • Preventive measures and emergency protocols regarding safety aspects (accidents and injuries, and chemical exposure) • Appropriate PPE required in all applicable situations • Training, counselling, prevention, and risk control programmes in place to assist workforce members
	Social benefit	• Social benefits provided to workers (e.g., health insurance, pension funding, and childcare)
Society	Public commitments regarding sustainability	• Presence of publicly available commitments, agreements, and codes of conduct regarding sustainable issues (and complaints to the non-fulfilment of these commitments)
	Contribution to economic development	• Contribution of the product/sector/company to economic development
	Corruption	• Presence of an anti-corruption program in the company • Presence of cooperation with internal and external controls to prevent corruption • Evidence of active involvement of the organisation in corruption and bribery
	Technology development	• Involvement of the company in technology transfer programmes or projects • Presence of partnerships regarding R&D programmes or projects • Investments in technology development

3. Results

The following chapter shows the absolute results for the LCA, LCC, and S-LCA, and for the technical/physical properties of both the bio-based panel and the two benchmark panels, as well as visualizes the relative comparison of the different systems. Due to confidentiality, S-LCA results are presented in aggregate and qualitative form.

3.1. LCA

Table 5 and Figure 4 report the environmental impacts of the production of the bio-based multi-layer panel compared with the benchmark solutions, where conventional components (including fossil-based ones) were used. For this comparison, impacts related to the transportation routes in the bio-based case were not considered.

Table 5. Life Cycle Impact Assessment results for the selected impact categories per functional unit (1 m^2).

Impact Category	Bio-Based Panel	BENCHMARK 1	BENCHMARK 2	Unit
Acidification terrestrial and freshwater	4.17×10^{-2}	6.45×10^{-2}	8.43×10^{-2}	Mole of H+ eq.
Climate change	$2.66 \times 10^{+1}$	$2.43 \times 10^{+1}$	$2.98 \times 10^{+1}$	kg CO$_2$ eq.
Eutrophication freshwater	1.62×10^{-3}	6.55×10^{-5}	4.05×10^{-4}	kg P eq.
Eutrophication marine	2.26×10^{-2}	1.63×10^{-2}	2.16×10^{-2}	kg N eq.
Eutrophication terrestrial	2.30×10^{-1}	1.72×10^{-1}	2.16×10^{-1}	Mole of N eq.
Ozone depletion	1.91×10^{-5}	3.08×10^{-3}	1.99×10^{-5}	kg CFC-11 eq.
Photochemical ozone formation, regarding human health	5.10×10^{-2}	6.13×10^{-2}	6.59×10^{-2}	kg NMVOC eq.
Resource use and energy carriers	$5.96 \times 10^{+2}$	$5.40 \times 10^{+2}$	$6.21 \times 10^{+2}$	MJ

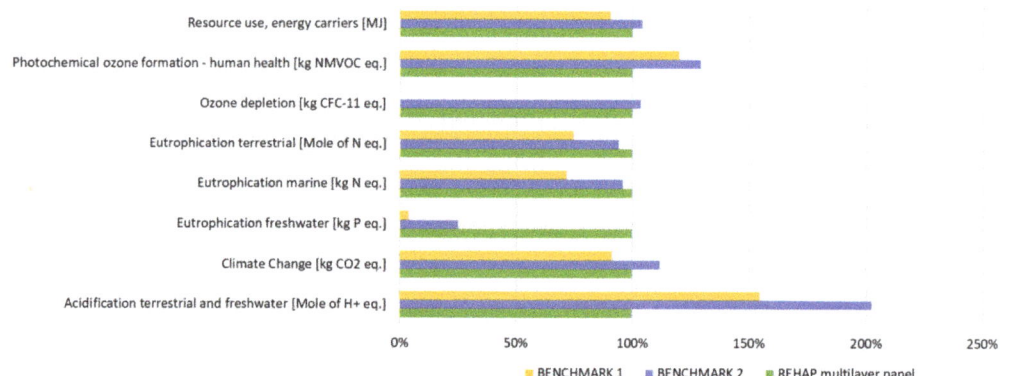

Figure 4. Relative LCA results of the bio-based multi-layer panel (REHAP multi-layer panel) compared to the benchmarks.

The bio-based multi-layer panel, which consists of the assembly of the bio-based products (i.e., BioPU foam, plywood board, and bioPU adhesive), presented quite similar results compared with the "benchmark 2" panel, i.e., a panel including the same components as the bio-based case but with a "fossil-based" origin (PU foam and PU adhesive with fossil-based esterpolyols, and plywood board with phenol-formaldehyde resin). Differences between 4% and 6% are shown for the following impact categories: resource use, ozone depletion, eutrophication terrestrial, and eutrophication marine. Concerning the other impact categories, the bio-based panel entailed lower impacts in terms of climate change (−12%), photochemical ozone formation (−29%), and acidification (−102%, mainly associated with the primary refining processes within bio-based value chains), and presented a considerable increase for the eutrophication freshwater category. This was mainly due to the PU-based components, i.e., the PU foam and PU adhesive, and the bio-based monomers (1,4-BDO and 2,3-BDO) used in bioesterpolyols formulation. While 2,3-BDO impacts on eutrophication freshwater are to be associated with the activated carbon and solvents (e.g., ethyl acetate) employed in the process, impacts on such categories linked to 1,4-BDO production can be mainly related to the glucose used, as additional sugars' sources along with second-generation sugars were derived from bark.

The "benchmark 1" panel (SIP with plywood and XPS foam) entailed slightly better environmental performances in almost all impact categories if compared with the other solutions (especially for the categories ozone depletion and eutrophication freshwater).

However, the bio-based panel showed lower impacts in photochemical ozone formation and acidification potential.

In the framework of the bio-based multi-layer panel production, most of the impacts were related to the electricity used in the process, followed by the BioPU foam production. In the latter, the MDI used in BioPU foam formulation entailed relevant environmental impacts in almost all impact categories, except for eutrophication freshwater, for which the glucose used in 1,4-BDO production entailed a high share of the impacts.

Figure 5 shows that most of the impacts for "benchmark 1" were attributed to the electric energy demand, followed by XPS foam and plywood manufacturing. For "benchmark 2", most of the impacts were allocated to PU foam production (due to its MDI content), followed by electricity and plywood board. In both benchmark solutions, impacts associated with the PU adhesive were very low compared with the other components.

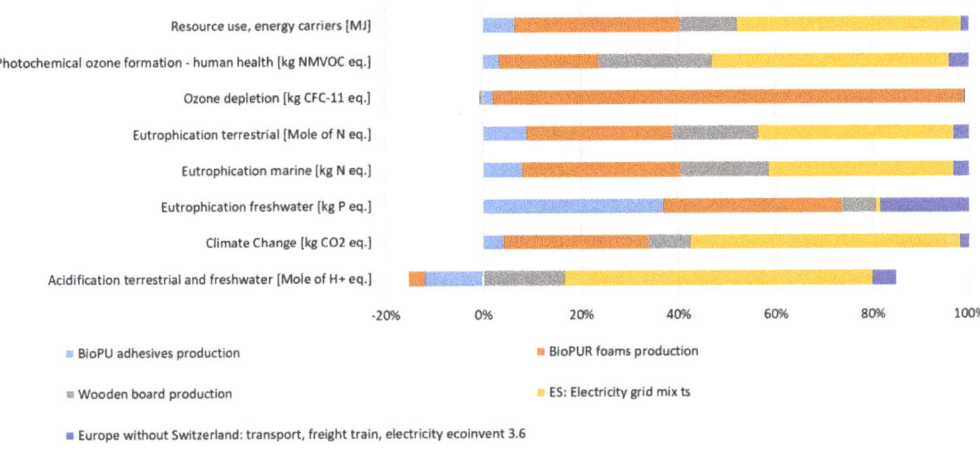

Figure 5. Contributions of the different components to the LCA result of the bio-based panel.

3.2. LCC

The resulting production cost of the bio-based multi-layer panel was equal to 58.59 €/m² when the assessment outcomes of the pilot-scale processes within bio-based value chains were included. Indeed, the results obtained from the LCC of the bio-based components (i.e., BioPU foam, bioPU adhesive, and plywood board) showed that the highest costs were related to the processes for which pilot-scale implementation is carried out: primary refining and 2,3-BDO production. Indeed, further improvements and optimisation actions are still needed to lower the production costs of sugars, lignin, and 2,3-BDO in order to make such intermediates competitive on the market.

For this reason, a competitive price of 1 €/kg for sugars and 0.5 €/kg for lignin, along with a market price for 2,3-BDO (to be used in the formulation of bioesterpolyols for PU adhesives) was considered to estimate the prices of bio-based components for the multi-layer panel production (value based on primary data provided by REHAP project partners [27]). In the latter case, the results became more promising, as shown in Table 6, which shows the LCC results of the bio-based multi-layer panel compared with the selected benchmark solutions, i.e., the SIP panel (benchmark 1) and panel with fossil-based components (benchmark 2). For this comparison, impacts related to the transportation routes in the bio-based case were not considered.

Table 6. LCC results for the bio-based multi-layer panel and the benchmarks per functional unit (1 m² panel).

	BIO BASED	BENCHMARK 1	BENCHMARK 2	Unit
Multilayer panel	51.97	50.87	47.46	$€/m^2$

Even though the differences among the assessed cases were not relevant, the best results in terms of economic impacts were associated with the "benchmark 2" case, while the bio-based panel presented higher economic impacts, with a slight increase if compared with the SIP panel production costs. Focusing on the bio-based case, the cost breakdown of the multi-layer panel production is shown in Figure 6. Considering also the transportation cost, the total economic impacts increased to 53.61 $€/m^2$. The highest contribution was associated with the personnel (46%), followed by the raw materials (41%) and energy (8%). Other contributions were covered by equipment and transportation, which accounted for 2% and 3% respectively. BioPU foams entailed 63% of the economic impacts among the raw materials, while plywood board and bioPU adhesives accounted for 31% and 6%, respectively. If compared with the benchmark cases, the economic impact of BioPU foam is higher than both XPS foam (benchmark 1) and fossil-based PU foam (benchmark 2). This depends on the higher production cost of bioesterpolyols compared with the fossil-based counterparts, which in turn is linked to the higher price of bio-based monomers. Additionally, the plywood board entailing biophenolic resins showed economic benefits compared with the plywood used in SIP panel (benchmark 1) and with the fossil-based plywood board (benchmark 2), while the adhesive contribution was similar in all of the cases.

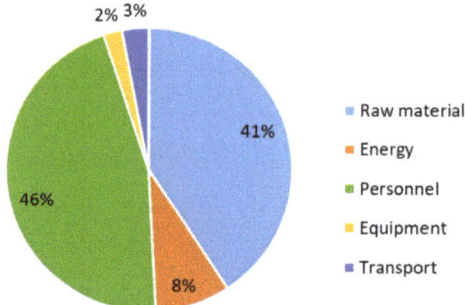

Figure 6. LCC cost breakdown for bio-based panel production.

3.3. S-LCA

The results of the social assessment of the bio-based multi-layer panel were based on the associated companies' responses to the questionnaire. The answers were evaluated qualitatively and suggestions for improvement were made to the companies based on their specific S-LCA results. According to the outcomes of the S-LCA of each bio-based component, good results in terms of social impacts were shown for all the processes related to the production of intermediates and bio-based components.

In particular, some companies achieved good impacts for the stakeholder category "Local Community" mainly because of their strong relationship of collaboration and open dialogue with the local community. Some companies were indeed committed to the sustainable growth of the areas near their factories and considered the people living there, for example, by favouring the hiring of local people and local suppliers. In particular, the smaller companies assessed within the REHAP project would need to increase their engagement in initiatives for the community (e.g., education initiatives) and commit clearly to local employment. Some companies paid great attention to technology development and sustainability issues by publishing environmental reports on the implementation of the company's environmental policy and on the consistency between the goals proposed. Specific documents aimed to enhance ethical principles such as transparency, fairness, equality, social responsibility, and sustainability were also encountered, thus driving and supervising the relationships with suppliers, employees, and customers. Examples of such documents include the Ethical Code, Code of Conduct, Supplier Code of Conduct, Code of Ethics, and Annual Sustainability Report. Concerning the stakeholder group "Society", the development of public documents regarding commitments, agreements, or codes of conduct on sustainability issues and anti-corruption measures may increase the positive effects on the society.

3.4. Technical/Physical Properties

The results for the tensile and compression strength tests for the different samples evaluated in the study are shown in Table 7 (the detailed list of results for each test sample is included in Appendix A). Initial estimates considered 120–150 kPa for the stress limit of the PU core of the panel, based on the UNE-EN 14509:2014 standard. The compressive tests for the core (test samples 5 and 6) showed that the stress limit ranged between 150 and 225 kPa. This proves that the initial estimate of the properties was too conservative and shows that the properties of the material exceeded the original expectations. The tensile strength tests performed over the different samples resulted in stress limits between 205 to 209 kPa for samples with sufficient adhesives. The initial estimates considered 100–120 kPa as the limit of the PU core of the panel (based on the UNE-EN 14509:2014 standard), which also showed that the material's tensile strength exceeded the initial expectations. The compressive and tensile strength of the plywood was far higher than expected, which shows that the plywood is not the limiting material in the performed tests. Based on the bio-based panel's technical/physical properties testing, the panel is technically competitive with the two benchmark systems. This result emphasizes that the comparative LCC and LCA study between the developed and benchmark panels with the selected functional unit is valid. The best sample is the plywood with 250 g/m2 of bioPU adhesive based on the results obtained. A pressure of 0.4 bar is sufficient to achieve good adhesion between the faces and the core without causing breaking.

Table 7. Results of the tensile and compression tests. The detailed results of the tensile test can be found in Tables A1–A6 and the results of the compression tests can be found in Tables A7 and A8, both in Appendix A.

Test Sample	Tensile		Compression	
	Resistance (MPa)	Modulus (MPa)	Resistance at 10% (MPa)	Modulus (MPa)
# 1	0.127	11.4	-	-
# 2	0.073	8.7	-	-
# 3	0.110	12.1	-	-
# 4	0.124	13.3	-	-
# 5	0.205	18.3	0.225	12.2
# 6	0.209	12.6	0.150	9.23

4. Discussion

The LCA results show that the bio-based and "benchmark 2" panel have quite similar environmental impacts, even though the bio-based solution entailed better environmental performances in terms of climate change (−12%), acidification potential (−102%), and photochemical ozone formation (−29%), but higher impacts in the eutrophication freshwater potential (+75%). The acidification benefits of the bio-based panel are to be attributed to the primary refining step within the value chains towards bio-based components (in particular, due to the incineration of residual solid streams after wastewater treatment), while the lower impacts in the other impact categories are mainly due to the substitution of fossil-based raw materials (i.e., phenol in resins used in the plywood board, fossil-based esterpolyols for the PU foam and adhesive) with bio-based materials. The higher impacts for the category eutrophication freshwater can be allocated to the raw materials employed in producing bio-based monomers for bio-based esterpolyols, i.e., 2,3-BDO and 1,4-BDO. The SIP panel showed slightly better environmental impacts than the bio-based and "benchmark 2" panel, except for the photochemical ozone formation and acidification categories. The results from LCC have shown that the bio-based solution entails slightly higher costs compared with both benchmark cases. This discrepancy mainly depends on the BioPU foam, which entails an economic impact significantly higher than for the XPS foam (benchmark 1) and fossil-based PU foam (benchmark 2). The higher cost for BioPU foam is mainly due to the bioesterpolyols whose increased production cost, compared with fossil-based counterparts, is mainly related to the higher price of bio-based monomers.

While the presented bio-based panel performs only economically marginally weaker than the conventional panels, the result is more heterogeneous for the environmental dimension. The findings that bio-based products often outperform conventional fossil-based products in categories such as climate change and tend to underperform in categories such as eutrophication are consistent with other research findings [14]. Regarding the economic and environmental results, it is essential to underline that the processes used for the production of the novel panels have so far been implemented on a pilot scale and not yet on a large industrial scale. Furthermore, the processes are novel and have not yet been subjected to optimisation cycles. With an upscaling and further optimisation of the processes, an improvement of the LCA and LCC results can be expected. This is in line with the conclusions of similar existing works, highlighting that the differences in scale and technology readiness are a constraint in this type of study [21]. This limits the validity of the direct comparison with conventional products, which are based on processes that have, in some cases, been optimised for decades.

The environmental and economic competitiveness of the bio-based solution versus the fossil-based benchmarks are influenced by several factors along the entire value chain, from the primary refining of biomass to the downstream processes, which have the potential to be further optimised. Efficient recycling and energy recovery from residual and waste materials can increase both environmental and economic benefits. This is particularly interesting in primary refining, where many liquid and solid residues are generated. Regarding the production of bio-based intermediates, particularly 2,3-BDO production (intermediate in the bioPU adhesive value chain) urgently needs reduced raw material requirements (including solvents), while maintaining the same product output. This increase in efficiency should take place when scaling up from the pilot to the industrial scale. Although costs related to the bio-based panel are already at the level of the benchmarks and the cost calculations are based on the laboratory or pilot scale, further improvements should be targeted. In particular, the total raw materials costs could be reduced by investigating alternative processes or substances. For example, the price of bio-based azelaic acid used in the bioesterpolyols formulation mainly contributes to the overall production costs for bio-esterpolyols. In addition, it should be emphasised that the bio-based materials have other advantages which were not subject to the underlying research. The produced bio-polyesterpolyols showed better fire performance than the fossil benchmarks, which implies a reduction in the use of flame retardants for specific applications and would significantly reduce the overall cost of the final product (polyurethane foam in this case).

From the S-LCA analysis, it can be concluded that although all companies involved in the REHAP project respect and apply social principles, further improvements can be implemented in some cases to explicitly indicate the company's engagements. Further commitments could improve topics such as promoting social responsibility along the supply chain and the ensuring the involvement of local communities. In particular, the small companies involved in the production process have the potential to further develop their social benefits for different stakeholders. In particular, the adoption of explicit and written procedures and measures may significantly raise social benefits linked to the company's activities and could enhance the engagement of local communities and value chain actors including suppliers, workers, and society in bio-based operations, thus fostering a wider diffusion of bio-based value chains across Europe. A comparison of the social performance of the bio-based panel with conventional panels was not performed in this study but would have been desirable. The lack of recognised methods and standards for evaluating the social dimension do not yet allow for such a comparison. It was also found that the participating companies only provided data on social aspects under the assurance of confidentiality. The fear of bad social responsibility ratings seems to be very high among the companies, which considerably limits transparency. Even with more recognised and standardised S-LCA models, the accessibility of sensitive social data will be a limiting factor. Ingrao et al. (2021) drew the same conclusion from the results of a recent Special Issue [25]. In order to make small steps towards more transparency, the S-LCA results are presented qualitatively. Furthermore, the conclusion of the S-LCA was internally fed back directly to the participating companies in the form of recommendations for improvement. In order to increase the transparency of social aspects, the disclosure of specific data on social aspects could be a condition for companies in publicly financed projects.

In conclusion, this work presents an already marketable and economically competitive product with a technical performance comparable to the benchmark panels and is within both the stress limit and tensile strength standards. From a technical point of view, the bio-based panel has the properties to substitute conventional insulation panels and thereby can reduce the demand for fossil resources. The 14.6 Mt coniferous bark (spruce plus pine bark) would serve enough feedstock to produce 0.926 Mil t PU foam or 2.327 Mil t PU adhesives (both products are based on esterpolyol derived from coniferous bark-based carbohydrates) and additionally 145.504 Mil t biophenolic resin (based on lignin and tannin). The coniferous bark-derived carbohydrates needed for PU foam and PU adhesive production are the limiting feedstock. The EU's bioeconomic coniferous bark potentials would allow for a total multi-layer production volume of about 250 Mil m^2. Although the panel presented in this study was produced on a laboratory and pilot scale, it exceeds conventional benchmarks in many environmental categories and has a competitive price. With additional research and development to implement the proposed process improvements, the bio-based panel can further increase its sustainability significantly.

Author Contributions: Conceptualisation, A.B. and A.T.; methodology, A.L. and F.B.F.; software, A.L.; validation, A.B., L.W. and A.T.; data curation, A.L.; writing—original draft preparation, A.L., F.B.F. and L.W.; writing—review and editing, A.B. and L.W.; visualisation, A.L., F.B.F. and L.W.; supervision A.B. and A.T.; project administration, A.B.; funding acquisition, A.B. All authors have read and agreed to the published version of the manuscript.

Funding: The research leading to these results has received funding from the European Union's Horizon 2020 research and innovation program under grant agreement number 723670, with the title "Systemic approach to reduce energy demand and CO2 emissions of processes that transform agroforestry waste into high added value products (REHAP)".

Data Availability Statement: Data are derived from Ecoinvent [38] and GaBi Life Cycle Inventory Databases [39], and industry data are derived from project partners.

Conflicts of Interest: The authors declare no conflict of interest.

Appendix A

Detailed results of the mechanical tests performed on the multi-layer panel are shown in Tables A1–A6 for the tensile tests and in Tables A7 and A8 for the compression test.

Table A1. Results of the tensile test for test sample 1.

Test Sample 1				
Test Specimen	Maximum Force (N)	Strength (MPa)	Modulus (MPa)	Breakage Mode
1	1216.17	0.122	11.512	95% cohesive (core) + 5% adhesive between core and board
2	1496.07	0.150	11.703	100% cohesive (core)
3	1084.38	0.110	11.095	90% cohesive (core) + 10% adhesive between core and board
Average	-	0.127	11.4	-
Uncertainty	-	0.027	1.11	-

Table A2. Results of the tensile test for test sample 2.

Test Sample 2				
Test Specimen	Maximum Force (N)	Strength (MPa)	Modulus (MPa)	Breakage Mode
1	1075.32	0.108	8.689	95% cohesive (core) + 5% adhesive between core and board
2	708.76	0.071	9.303	100% cohesive (core)
3	384.18	0.038	8.208	100% cohesive (core)
Average	-	0.073	8.73	-
Uncertainty	-	0.041	1.12	-

Table A3. Results of the tensile test for test sample 3.

Test Sample 3				
Test Specimen	Maximum Force (N)	Strength (MPa)	Modulus (MPa)	Breakage Mode
1	1345.2	0.136	14.039	100% cohesive (core)
2	987.61	0.100	11.905	100% cohesive (core)
3	942.7	0.095	10.325	90% cohesive (core) + 10% adhesive between core and board
Average	-	0.11	12.1	-
Uncertainty	-	0.029	2.4	-

Table A4. Results of the tensile test for test sample 4.

Test Sample 4				
Test Specimen	Maximum Force (N)	Strength (MPa)	Modulus (MPa)	Breakage Mode
1	974.72	0.099	12.021	100% cohesive (core)
2	915.86	0.092	10.244	100% cohesive (core)
3	1595.38	0.162	12.575	100% cohesive (core)
4	1398.65	0.144	18.546	85% cohesive (core) + 15% adhesive between core and board
Average	-	0.124	13.3	-
Uncertainty	-	0.038	3.9	-

Table A5. Results of the tensile test for test sample 5.

Test Sample 5				
Test Specimen	Maximum Force (N)	Strength (MPa)	Modulus (MPa)	Breakage Mode
1	1879.72	0.194	18.829	95% cohesive (core) + 5% adhesive between core and board
2	2259.07	0.220	18.106	95% cohesive (core) + 5% adhesive between core and board
3	1154.31	0.115	16.334	95% cohesive (core) + 5% adhesive between core and board
4	2257.34	0.227	18.667	100% cohesive (core)
5	2724.02	0.267	19.328	100% cohesive (core)
Average	-	0.205	18.3	-
Uncertainty	-	0.055	2.0	-

Table A6. Results of the tensile test for test sample 6.

Test Sample 6				
Test Specimen	Maximum Force (N)	Strength (MPa)	Modulus (MPa)	Breakage Mode
1	1663.96	0.174	11.766	100% cohesive (core)
2	2509.34	0.250	13.082	100% cohesive (core)
3	1898.66	0.192	10.494	100% cohesive (core)
4	1610.09	0.164	13.311	100% cohesive (core)
5	2717.74	0.268	14.514	100% cohesive (core)
Average	-	0.209	12.6	-
Uncertainty	-	0.047	1.8	-

Table A7. Results of the compression test for test sample 5.

Test Sample 5						
Test Specimen	Breaking Force (N)	Breaking Resistance (MPa)	Deformation at Breaking (%)	Force at 10% (N)	Resistance at 10% (MPa)	Modulus (MPa)
1	2552.49	0.249	3.000	2440.92	0.238	13.124
2	2342.45	0.235	2.840	2240.49	0.224	12.259
3	2384.78	0.233	2.056	2309.98	0.225	12.412
4	2233.31	0.216	2.324	2175.87	0.210	11.054
Average	-	0.233	2.55	-	0.225	12.2
Uncertainty	-	0.014	1.16	-	0.012	1.3

Table A8. Results of the compression test for test sample 6.

Test Sample 6						
Test Specimen	Breaking Force (N)	Breaking Resistance (MPa)	Deformation at Breaking (%)	Force at 10% (N)	Resistance at 10% (MPa)	Modulus (MPa)
1	1536.11	0.151	1.974	1588.40	0.1560	9.2454
2	1480.33	0.145	1.911	1450.03	0.1423	9.2744
3	1624.18	0.160	1.867	1673.16	0.1649	10.444
4	1415.07	0.138	2.217	1391.95	0.1357	7.964
Average	-	0.149	1.99	-	0.15	9.23
Uncertainty	-	0.010	1.68	-	0.013	1.23

References

1. UNEP. Global Status Report for Buildings and Construction: Towards a Zero-Emission, Efficient and Resilient Buildings and Construction Sector. Nairobi. 2020. Available online: https://wedocs.unep.org/bitstream/handle/20.500.11822/34572/GSR_ES.pdf?sequence=3&isAllowed=y (accessed on 20 July 2021).
2. Berardi, U. A cross-country comparison of the building energy consumptions and their trends. *Resour. Conserv. Recycl.* **2017**, *123*, 230–241. [CrossRef]
3. Ingrao, C.; Arcidiacono, C.; Bezama, A.; Ioppolo, G.; Winans, K.S.; Koutinas, A.; Schmid, A.G. Sustainability issues of by-product and waste management systems, to produce building material commodities: A comprehensive review of findings from a virtual special issue. *Resour. Conserv. Recycl.* **2019**, *146*, 358–365. [CrossRef]
4. WGBC. Advancing Net Zero Status Report 2020. London. 2020. Available online: https://www.worldgbc.org/advancing-net-zero-status-report-2020 (accessed on 19 July 2021).
5. De Wit, M.; Ramkumar, J.H.S.; Douma, H.F.A. The Circularity Gap Report. An analysis of the circular state of the global economy. *Circ. Econ.* **2018**, 1–36. Available online: https://www.circle-economy.com/resources/the-circularity-gap-report-our-world-is-only-9-circular (accessed on 2 July 2021).
6. Arosio, V.; Arrigoni, A.; Dotelli, G. Reducing water footprint of building sector: Concrete with seawater and marine aggregates. *IOP Conf. Ser. Earth Environ. Sci.* **2019**, *323*. [CrossRef]
7. EEA. Construction and Demolition Waste: Challenges and Opportunities in a Circular Economy. 2020. Available online: https://www.eea.europa.eu/publications/construction-and-demolition-waste-challenges/construction-and-demolition-waste-challenges (accessed on 20 July 2021).
8. UN. World Population Prospects 2019: Highlights. 2019. Available online: https://www.un.org/development/desa/publications/world-population-prospects-2019-highlights.html#:~{}:text= (accessed on 22 June 2021).
9. Müller, M.; Krick, T.; Blohmke, J. Putting the construction sector at the core of the climate change debate I Deloitte Central Europe. *Deloitte* **2020**, 2–3. Available online: https://www2.deloitte.com/ce/en/pages/real-estate/articles/putting-the-construction-sector-at-the-core-of-the-climate-change-debate.html (accessed on 20 June 2021).
10. Pittau, F.; Lumia, G.; Heeren, N.; Iannaccone, G.; Habert, G. Retrofit as a carbon sink: The carbon storage potentials of the EU housing stock. *J. Clean. Prod.* **2019**, *214*, 365–376. [CrossRef]
11. Ingrao, C.; Scrucca, F.; Tricase, C.; Asdrubali, F. A comparative Life Cycle Assessment of external wall-compositions for cleaner construction solutions in buildings. *J. Clean. Prod.* **2016**, *124*, 283–298. [CrossRef]
12. Von Braun, J. Bioeconomy—The global trend and its implications for sustainability and food security. *Glob. Food Sec.* **2018**, *19*, 81–83. [CrossRef]
13. Lewandowski, I. Securing a sustainable biomass supply in a growing bioeconomy. *Glob. Food Sec.* **2015**, *6*, 34–42. [CrossRef]
14. Hjuler, S.V.; Hansen, S.B. LCA of Biofuels and Biomaterials. In *Life Cycle Assessment*; Springer International Publishing: Cham, Switzerland, 2018; pp. 755–782.
15. Balat, M. Production of bioethanol from lignocellulosic materials via the biochemical pathway: A review. *Energy Convers. Manag.* **2011**, *52*, 858–875. [CrossRef]
16. Uihlein, A.; Schebek, L. Environmental impacts of a lignocellulose feedstock biorefinery system: An assessment. *Biomass Bioenergy* **2009**, *33*, 793–802. [CrossRef]
17. Borrion, A.L.; McManus, M.C.; Hammond, G.P. Environmental life cycle assessment of lignocellulosic conversion to ethanol: A review. *Renew. Sustain. Energy Rev.* **2012**, *16*, 4638–4650. [CrossRef]
18. Kemppainen, K.; Siika-aho, M.; Pattathil, S.; Giovando, S.; Kruus, K. Spruce bark as an industrial source of condensed tannins and non-cellulosic sugars. *Ind. Crops Prod.* **2014**, *52*, 158–168. [CrossRef]
19. Kemppainen, K. Production of Sugars, Ethanol and Tannin from Spruce Bark and Recovered Fibres. Ph.D Thesis, Aalto University, Espoo, Finland, March 2015.
20. Lacoste, C.; Čop, M.; Kemppainen, K.; Giovando, S.; Pizzi, A.; Laborie, M.P.; Sernek, M.; Celzard, A. Biobased foams from condensed tannin extracts from Norway spruce (Picea abies) bark. *Ind. Crops Prod.* **2015**, *73*, 144–153. [CrossRef]
21. Arias, A.; González-García, S.; González-Rodríguez, S.; Feijoo, G.; Moreira, M.T. Cradle-to-gate Life Cycle Assessment of bio-adhesives for the wood panel industry. A comparison with petrochemical alternatives. *Sci. Total Environ.* **2020**, *738*, 140357. [CrossRef] [PubMed]
22. Santos, P.; Correia, J.R.; Godinho, L.; Dias, A.M.P.G.; Dias, A. Life cycle analysis of cross-insulated timber panels. *Structures* **2021**, *31*, 1311–1324. [CrossRef]
23. Sinka, M.; Korjakins, A.; Bajare, D.; Zimele, Z.; Sahmenko, G. Bio-based construction panels for low carbon development. *Energy Procedia* **2018**, *147*, 220–226. [CrossRef]
24. Ingrao, C.; Giudice, A.L.; Tricase, C.; Rana, R.; Mbohwa, C.; Siracusa, V. Recycled-PET fibre based panels for building thermal insulation: Environmental impact and improvement potential assessment for a greener production. *Sci. Total Environ.* **2014**, *493*, 914–929. [CrossRef]
25. Ingrao, C.; Arcidiacono, C.; Siracusa, V.; Niero, M.; Traverso, M. Life cycle sustainability analysis of resource recovery from waste management systems in a circular economy perspective Key Findings from This Special Issue. *Resources* **2021**, *10*, 32. [CrossRef]

26. Ingrao, C.; Bacenetti, J.; Bezama, A.; Blok, V.; Goglio, P.; Koukios, E.G.; Lindner, M.; Nemecek, T.; Siracusa, V.; Zabaniotou, A.; et al. The potential roles of bio-economy in the transition to equitable, sustainable, post fossil-carbon societies: Findings from this virtual special issue. *J. Clean. Prod.* **2018**, *204*, 471–488. [CrossRef]
27. Rehap. Moving towards a resource-efficient Europe. *SPIRE Project* 2021. Available online: https://www.rehap.eu.com/ (accessed on 20 July 2021).
28. Buyle, M.; Braet, J.; Audenaert, A. Life cycle assessment in the construction sector: A review. *Renew. Sustain. Energy Rev.* **2013**, *26*, 379–388. [CrossRef]
29. Asiedu, Y.; Gu, P. Product life cycle cost analysis: State of the art review. *Int. J. Prod. Econ.* **1998**, *36*, 883–908. [CrossRef]
30. Huarachi, D.A.R.; Piekarski, C.M.; Puglieri, F.N.; de Francisco, A.C. Past and future of Social Life Cycle Assessment: Historical evolution and research trends. *J. Clean. Prod.* **2020**, *264*, 121506. [CrossRef]
31. Liu, S.; Qian, S. Evaluation of social life-cycle performance of buildings: Theoretical framework and impact assessment approach. *J. Clean. Prod.* **2019**, *213*, 792–807. [CrossRef]
32. Fatourehchi, D.; Zarghami, E. Social sustainability assessment framework for managing sustainable construction in residential buildings. *J. Build. Eng.* **2020**, *32*, 101761. [CrossRef]
33. Visentin, C.; Trentin, A.W.d.S.; Braun, A.B.; Thomé, A. Life cycle sustainability assessment: A systematic literature review through the application perspective, indicators, and methodologies. *J. Clean. Prod.* **2020**, *270*. [CrossRef]
34. Martin, M.; Røyne, F.; Ekvall, T.; Moberg, Å. Life Cycle Sustainability Evaluations of Bio-based Value Chains: Reviewing the Indicators from A Swedish Perspective. *Sustainability* **2018**, *10*, 547. [CrossRef]
35. Broeren, M.L.M.; Molenveld, K.; van den Oever, M.J.A.; Patel, M.K.; Worrell, E.; Shen, L. Early-stage sustainability assessment to assist with material selection: A case study for biobased printer panels. *J. Clean. Prod.* **2016**, *135*, 30–41. [CrossRef]
36. Escobar, N.; Laibach, N. Sustainability check for bio-based technologies: A review of process-based and life cycle approaches. *Renew. Sustain. Energy Rev.* **2021**, *135*, 110213. [CrossRef]
37. Yılmaz, E.; Arslan, H.; Bideci, A. Environmental performance analysis of insulated composite facade panels using life cycle assessment (LCA). *Constr. Build. Mater.* **2019**, *202*, 806–813. [CrossRef]
38. Wernet, G.; Bauer, C.; Steubing, B.; Reinhard, J.; Moreno-Ruiz, E.; Weidema, B. The ecoinvent database version 3 (part I): Overview and methodology. *Int. J. Life Cycle Assess.* **2016**, *21*, 1218–1230. [CrossRef]
39. Sphera. GaBi Product Sustainability Software. Sphera. 2020. Available online: https://gabi.sphera.com/international/software/ (accessed on 20 July 2021).
40. Ciroth, A.; Huppes, G.; Klöpffer, W.; Rüdenauer, I.; Steen, B.; Swarr, T. *Environmental Life Cycle Costing*; Taylor & Francis: London, UK, 2008.
41. UNEP SETAC; Benoît, C.; Mazijn, B. (Eds.) *Guidelines for Social Life Cycle Assessment of Products*; UNEP/SETAC Life Cycle Initiative: Nairobi, Kenya, 2009.
42. Alkakurtti, S.; Grönqvist, S.; Niemelä, K.; Ruuskanen, M.; Tamminen, T. Extraction of Valuable Components from Bark. WO2020084196A1, 30 April 2020.
43. Thorenz, A.; Wietschel, L.; Stindt, D.; Tuma, A. Assessment of agroforestry residue potentials for the bioeconomy in the European Union. *J. Clean. Prod.* **2018**, *176*, 348–359. [CrossRef]
44. Eurostat. Roundwood Removals by Type of Wood and Assortment. *for_remov*. 2021. Available online: https://appsso.eurostat.ec.europa.eu/nui/submitViewTableAction.do (accessed on 28 June 2021).
45. Olofsson, J.; Börjesson, P. Residual biomass as resource—Life-cycle environmental impact of wastes in circular resource systems. *J. Clean. Prod.* **2018**, *196*, 997–1006. [CrossRef]
46. Fazio, S.; Castellani, V.; Sala, S.; Schau, E.; Secchi, M.; Zampori, L.; Diaconu, E. *Supporting Information to the Characterisation Factors of Recommended EF Life Cycle Impact Assessment Methods*; European Commission: Brussels, Belgium, 2018.
47. Phyllis2. Database for (Treated) Biomass, Algea, Feedstocks Fogios Production and Biochar. 2020. Available online: https://phyllis.nl/ (accessed on 28 June 2021).
48. Aryapratama, R.; Janssen, M. Prospective life cycle assessment of bio-based adipic acid production from forest residues. *J. Clean. Prod.* **2017**, *164*, 434–443. [CrossRef]
49. Anderson, J. Determining Manufacturing Costs. 2008. Available online: https://www.semanticscholar.org/paper/Determining-Manufacturing-Costs-Anderson/99a1a56dd9adc9f24f3fc1d7c25a1d2db3a1a784 (accessed on 13 July 2021).
50. German Federal Government. Biorefineries Roadmap. Federal Ministry of Food, Agriculture and Consumer Protection (BMELV). 2012. Available online: https://www.bmbf.de/SharedDocs/Publikationen/de/bmbf/pdf/biorefineries-roadmap.pdf?__blob=publicationFile&amp%3Bv=2 (accessed on 14 July 2021).
51. UNEP SETAC. The Methodological Sheets for Subcategories in Social Life Cicle Assessment (S-LCA). 2013. Available online: http://link.springer.com/10.1007/978-1-4419-8825-6 (accessed on 14 July 2021).

MDPI
St. Alban-Anlage 66
4052 Basel
Switzerland
Tel. +41 61 683 77 34
Fax +41 61 302 89 18
www.mdpi.com

Resources Editorial Office
E-mail: resources@mdpi.com
www.mdpi.com/journal/resources

www.ingramcontent.com/pod-product-compliance
Lightning Source LLC
LaVergne TN
LVHW070414100526
838202LV00014B/1456